Various Aspects of Silicon Polymer Chemistry: A Themed Issue in Honor of Professor Julian Chojnowski on the Occasion of His 85th Birthday

Various Aspects of Silicon Polymer Chemistry: A Themed Issue in Honor of Professor Julian Chojnowski on the Occasion of His 85th Birthday

Editors

Sławomir Rubinsztajn
Marek Cypryk
Wlodzimierz Stanczyk

MDPI • Basel • Beijing • Wuhan • Barcelona • Belgrade • Manchester • Tokyo • Cluj • Tianjin

Editors

Sławomir Rubinsztajn
Centre of Molecular and Macromolecular Studies of Polish Academy of Sciences
Poland

Marek Cypryk
Centre of Molecular and Macromolecular Studies of Polish Academy of Sciences
Poland

Wlodzimierz Stanczyk
Centre of Molecular and Macromolecular Studies of Polish Academy of Sciences
Poland

Editorial Office
MDPI
St. Alban-Anlage 66
4052 Basel, Switzerland

This is a reprint of articles from the Special Issue published online in the open access journal *Molecules* (ISSN 1420-3049) (available at: https://www.mdpi.com/journal/molecules/special_issues/Julian_Chojnowski).

For citation purposes, cite each article independently as indicated on the article page online and as indicated below:

LastName, A.A.; LastName, B.B.; LastName, C.C. Article Title. *Journal Name* **Year**, *Volume Number*, Page Range.

ISBN 978-3-0365-2253-1 (Hbk)
ISBN 978-3-0365-2254-8 (PDF)

Contents

About the Editors

Sławomir Rubinsztajn graduated from the Faculty of Chemistry at the Technical University of Łódź in 1979. In the same year, he began working in the group of Prof. Julian Chojnowski at the Centre for Molecular and Macromolecular Research at the Polish Academy of Sciences (CMMS PAS) in Łódź. He received his Ph.D. in Chemistry in 1986 for his work on the mechanism of the polycondensation of siloxane oligomers. From 1988–1992 he worked as a postdoctoral researcher in the group of Prof. Martel Zeldin at Indiana University–Purdue University in Indianapolis. In 1992, he started work at General Electric's Central Research Centre in Niskayuna, NY. In 1998, he was promoted to the principal chemist's position at GE Silicones. Dr. Rubinsztajn received the title of Dr. hab. in 2013 from the CMMS PAS for novel works in the field of polysiloxane chemistry. Dr. Rubinsztajn retired from GE Company in 2016. In the same year, he began work in CMMS PAS, where he directs a research team working in silicon chemistry. Dr. Rubinsztajn has co-authored 62 US patents and 63 publications.

Marek Cypryk was born in Łódź, Poland. He received his MSc degree in 1974 in Polymer Chemistry from the Technical University in Łódź and PhD in 1982 from the Centre of Molecular and Macromolecular Studies of Polish Academy of Sciences under the supervision of Professor Julian Chojnowski. He worked as a postdoctoral fellow at the Carnegie-Mellon University in Pittsburgh, PA, from 1990–1991 in the group of Professor Krzysztof Matyjaszewski. In 2002, he completed his DSc (habilitation) in Polymer Chemistry on the mechanisms of polymerization reactions in organosiloxane chemistry. He is currently a full professor and a head of the Research Group for Computer Modeling in the CMMS PAS. His main research interests are organosilicon polymer chemistry, reactions mechanisms in heterorganic chemistry, computational chemistry, and molecular modeling. He is an author or co-author of over 100 original papers and reviews.

Wlodzimierz Stanczyk is a leader of the Inorganic–Organic Composites Research Group at the Centre of Molecular and Macromolecular Studies of the Polish Academy of Sciences in Łódź. He obtained his MSc Eng., PhD, and DSc from the Chemistry Department of the Łódź University of Technology. He spent over two years working with Prof. Colin Eaborn at the School of Molecular Sciences of the University of Sussex. He has published over 100 papers on various aspects of organosilicon polymers and organometallic reaction mechanisms, liquid crystal polymers, and nanoconjugates. His current interests focus on silsesquioxanes as drug nanocarriers.

Preface to "Various Aspects of Silicon Polymer Chemistry: A Themed Issue in Honor of Professor Julian Chojnowski on the Occasion of His 85th Birthday"

This special edition is dedicated to Professor Julian Chojnowski on the occasion of his 85th birthday for his outstanding achievements in the field of organosilicon chemistry.

Professor Chojnowski was born on 17th of June, 1935, in Warsaw, Poland. He obtained his MSc degree in chemistry from Technical University of Łódź in 1957. He moved to the newly created Centre of Molecular and Macromolecular Studies of Polish Academy of Sciences in 1972, where he became the head of the Laboratory of Hetero-organic Polymers. In 1983, he obtained the academic title of Professor of Chemistry. He retired in 2005 but, to date, he continues research at the Centre of Molecular and Macromolecular Studies as a professor emeritus. Professor Chojnowski is the promoter of 15 doctoral dissertations. He published about 180 original papers, 16 review papers and chapters in books, as well as several Polish and US patents. He was also a co-editor of two books. In 2005, Professor Chojnowski was awarded with the Jan Zawidzki medal by the Polish Chemical Society for outstanding achievements in the field of physical chemistry. He serves as a member of the Advisory Board of The Journal of Inorganic and Organometallic Polymers and Materials, Silicon (Springer), and Main Group Chemistry.

Professor Chojnowski is an outstanding chemist who has significantly contributed to the development of many fields of organosilicon chemistry and organosilicon polymers chemistry. His fundamental research in the field of kinetic and thermodynamic aspects of polymerization processes and the substitution at silicon atom has led to understanding of the mechanisms of ring-opening polymerization reactions of cyclic siloxanes, as well as the discovery of the importance of inter- and intramolecular catalysis in siloxane polymerization and polycondensation processes. His expertise in the field of siloxane polymer chemistry has resulted in cooperation with the leading silicone materials manufacturers such as Dow Corning and General Electric Company. His collaboration with GE resulted in the discovery of several new reactions of hydride-functional silanes and siloxanes involving activation of Si-H bond by electron-deficient boranes. Recently, he extended his research interest into silicon-based materials. He developed the synthesis of a new class of polymers, poly(oxymultisilylene)s, and new processes leading to the synthesis of highly cross-linked polyorganosiloxane beads, which can be subsequently converted into ceramic materials.

This special edition of Molecules contains one review and eleven original articles written by prominent experts on various aspects of silicon polymer chemistry which reflect the tireless passion of Professor Chojnowski to the development of organosilicon chemistry.

Sławomir Rubinsztajn, Marek Cypryk, Wlodzimierz Stanczyk
Editors

molecules

MDPI

Review

Applications of Hybrid Polymers Generated from Living Anionic Ring Opening Polymerization

Jonathan Goff *, Santy Sulaiman and Barry Arkles

Gelest Inc., 11 Steel Road East, Morrisville, PA 19067, USA; ssulaiman@gelest.com (S.S.); executiveoffice@gelest.com (B.A.)
* Correspondence: jgoff@gelest.com

Abstract: Increasingly precise control of polymer architectures generated by "Living" Anionic Ring-Opening Polymerization (Living AROP) is leading to a broad range of commercial advanced material applications, particularly in the area of siloxane macromers. While academic reports on such materials remain sparse, a significant portion of the global population interacts with them on a daily basis—in applications including medical devices, microelectronics, food packaging, synthetic leather, release coatings, and pigment dispersions. The primary driver of this increased utilization of siloxane macromers is their ability to incorporate the properties of silicones into organic structures in a balanced manner. Compared to organic polymers, the differentiating properties of silicones—low T_g, hydrophobicity, low surface energy, and high free molal space—logically lend themselves to applications in which low modulus, release, permeability to oxygen and moisture, and tactile interaction are desired. However, their mechanical, structural and processing properties have until recently precluded practical applications. This review presents applications of "Living" AROP derived polymers from the perspective of historical technology development. Applications in which products are produced on a commercial scale—defined as not only offered for sale, but sold on a recurrent basis—are emphasized. Hybrid polymers with intriguing nanoscale morphology and potential applications in photoresist, microcontact printing, biomimetic soft materials, and liquid crystals are also discussed. Previously unreported work by the authors is provided in the context of this review.

Keywords: hybrid polymers; ring-opening polymerization; contact lenses; breathable films; membranes; high elongation elastomers; biomimetic polymers; photoresists

Citation: Goff, J.; Sulaiman, S.; Arkles, B. Applications of Hybrid Polymers Generated from Living Anionic Ring Opening Polymerization. *Molecules* **2021**, *26*, 2755. https://doi.org/10.3390/molecules26092755

Academic Editors: Sławomir Rubinsztajn, Marek Cypryk and Marek Cypryk

Received: 23 March 2021
Accepted: 22 April 2021
Published: 7 May 2021

1. Introduction

Functional siloxane polymers constitute a large class of reactive materials. Siloxanes with vinyl, silanol, and hydride substitution are the most widely utilized, serving as the basis for the majority of elastomeric silicone products. Functional siloxanes combined with organic monomers form "hybrid" polymers which, despite their demonstrated utility, have comparatively limited commercial applications. Most siloxane polymers are prepared by ring-opening polymerization with high degrees of polydispersion, thereby curtailing their ability to act as precise structural elements. On the other hand, the economics of both the basic building blocks and the polymerization process itself favor equilibrium-derived siloxane polymers. Figure 1 depicts the range of synthetic methods utilized to prepare siloxane polymers.

Due to the intrinsic process as well as the structural control that it provides, Living AROP-derived polymers provide the potential for a broad range of hybrid organic–inorganic materials. Briefly, AROP-derived materials provide a mechanism for translating macromolecular synthetic methods normally associated with organic polymers (and excluded from inorganic polymers) into hybrid polymer structures. Among these AROP-derived siloxane polymers, perhaps the most technologically significant are monofunctional and heterobifunctional silicone macromers. Siloxane macromers are defined as silicon-containing species with a single functional polymerizable group which, although used as monomers, possess sufficiently high molecular weight and enough internal monomer units to be considered polymeric. In another sense, they are siloxane building blocks derived from Living AROP. Siloxane macromers enable the use of technologies other than those associated with siloxane polymerization—e.g., techniques associated with the wider range of synthetic organic polymerization technologies—to incorporate siloxane-associated properties: e.g., techniques associated with the wider range of synthetic organic polymerization technologies. Notably, only controlled "living" AROP provides a path to siloxane polymers with sufficiently controlled structures and functionality to behave as macromers in polymerization with organic monomers. Siloxane macromers thus enable the introduction of selected siloxane properties into higher order structures via macromolecular engineering.

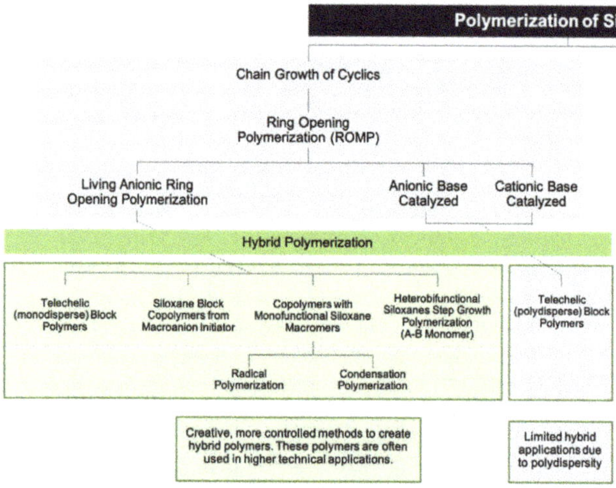

Figure 1. Synthetic methods for preparation of siloxane and siloxane-hybrid polymers.

Although the advent of living anionic ring-opening polymerization can be traced back approximately 50 years, commercial applications of this technology have only reached the marketplace within the last 20 and, by our estimate, production has exceeded 100 tons per annum only in the last 5. The availability and range of polymers with tailored molecular weight, polymer backbone structure and basic architecture, and both functional and non-functional alternatives have grown enormously, stimulated by symbiotic application and development efforts. This review relies heavily on patent literature; the global collection of patents, of which there are over 10,000,000 in the US alone, is one of the most comprehensive collections of technical information in the world but is often neglected in scholarly reviews. Nevertheless, the impact of patent technology on how materials are prepared and utilized may equal or exceed that of academic literature.

1.1. *Fundamentals/Building Blocks/Architectures*

The "living" anionic ring-opening polymerization of cyclosiloxanes is in fact more properly described as kinetically controlled ring-opening polymerization. The features that define the "living" aspects of the polymerization are: a quantitative initiation (as shown in Scheme 1), and the fact that the rate of polymerization propagation is significantly greater than that of the polymer chain randomization processes, particularly the reversion of the degree of polymerization driven by equilibration or "back-biting" processes (as shown in Scheme 2) [1].

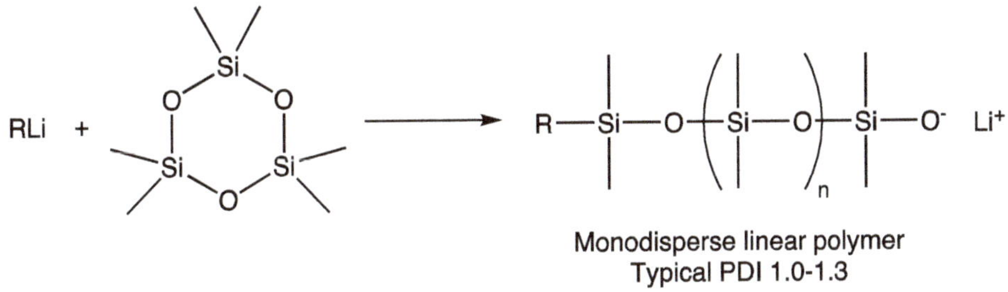

Scheme 1. Kinetically Controlled Polymerization.

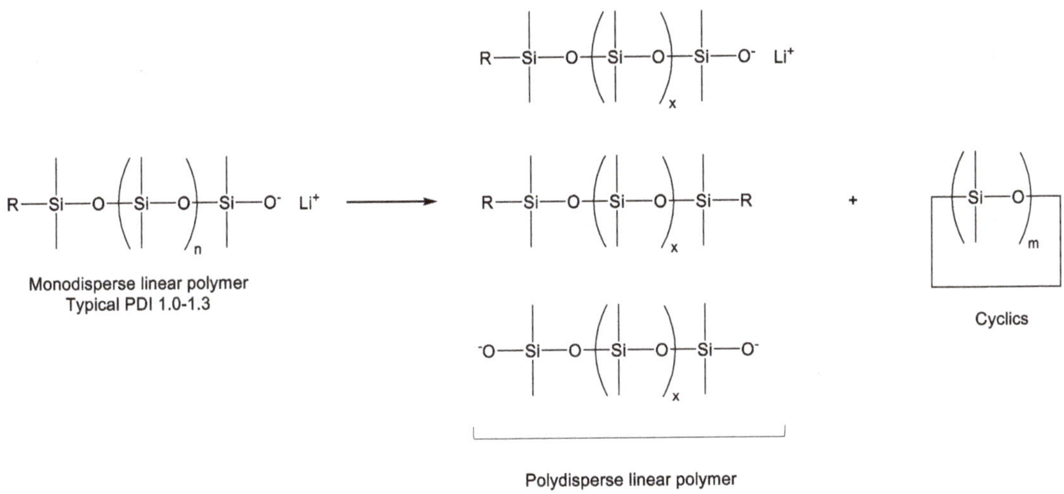

Scheme 2. Thermodynamic Equilibrium Driven Processes (i.e., Reversion, Backbiting, Chain-scrambling).

Both initiation and back-biting are driven by the catalytic opening of a siloxane bond. The evolution of Living AROP has depended on the recognition of classes of cyclic siloxane monomers that possess ring strain, as well as on "weak" catalysts which are able to rapidly cleave the Si-O-Si bonds of the strained monomers but are relatively slow and ineffective at cleaving Si-O-Si bonds in unstrained systems. The difference in reaction kinetics provides an opportunity to deactivate the catalyst before significant equilibration effects are observed, resulting in the scalable preparation of polymers with polydispersities approaching 1.

The differential polymerization of ring-strained cyclics, as opposed to unstrained cyclic siloxanes, was apparently observed in early industrial development, as is made particularly clear in the example of fluorinated silicones generated from 3,3,3-trifluoropropylmethylcyclic siloxanes [2]. Attempts to polymerize cyclic tetrasiloxanes were ineffective due to the fact that the reversion kinetics apparently matched those of polymerization for the unstrained cyclic tetramer. The polymerization of cyclic trisiloxanes, on the other hand, was effective due to the ability to deactivate the catalyst before significant reversion could occur. The potential for weak catalysts such as lithium phenoxide to produce polymers of ring-strained monomers was first recognized by McVannel [3], while the quantitative, selective formation of a lithium initiator generated from the reaction of organolithium reagents with cyclic trimers was studied by Frye [4], who, surprisingly, observed that, in a 1:1 molar stoichiometry, *n*-butyl lithium reacted with hexamethylcyclotrisiloxane (D_3) to form lithium n-butyldimethylsilanolate, leaving 2/3 of the D_3 unreacted. Lee and Frye also noted that polar "promoters" would then cause polymerization to proceed [5]. Finally,

Fessler reported the relative effectiveness of promoters in the living polymerization process and provided mechanistic insight into siloxane-silanolate reactions that could result in shifts between triad, Gaussian, and redistribution products, as shown in Scheme 3 [6].

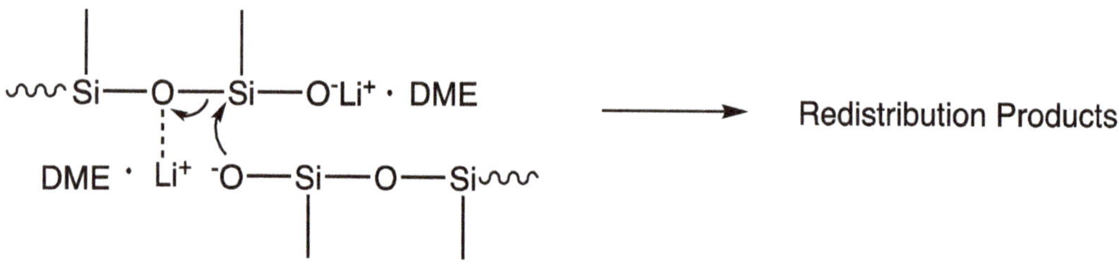

Scheme 3. Ion-pair Complexation by Dimethoxyethane (DME) Promoter Favors Chain-End Over Internal Scrambling [5].

A comparison of a GC (gas chromatograph) of siloxane macromers with equivalent MW and PDIs, showing triad and Gaussian distributions, is provided in Figure 2 (author's work). In contrast to anionic polymerization with K^+ and Na^+, in which there is little differentiation between chain scission points, the Li^+ redistribution mechanism favors chain termini: i.e., despite chain-end scrambling, narrow polydispersity is maintained.

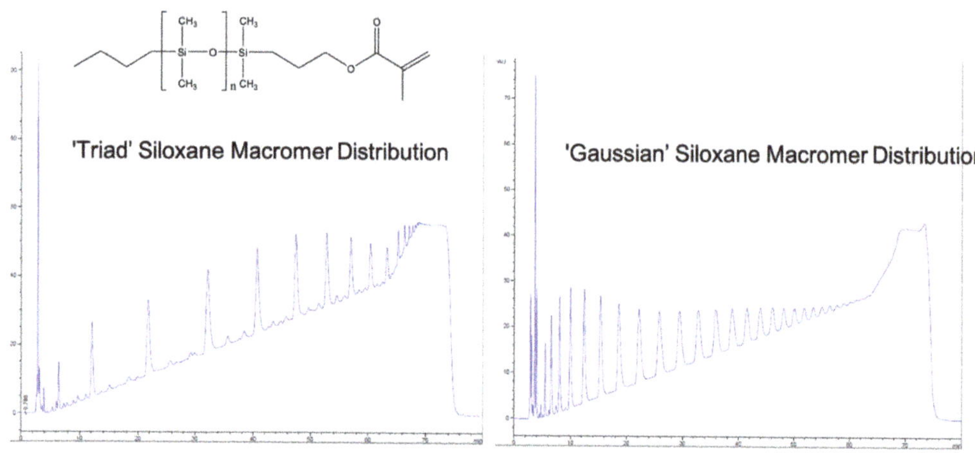

Figure 2. Comparison of silicone macromers demonstrating polymer termini scrambling (GC of methacrylate functional macromer (MCR-M11) with a nominal MW of 1000 showing shift from Triad to Gaussian distribution with a change in promoter. (Author's work)).

During the same period when Lee and Frye's report was published but separate from the discoveries relating to ring-strained siloxane monomers and lithium-based initiators, there was significant interest within the silicone industry in generating block copolymers, with lithium silanolate-based initiators being shown to lead to the sequential polymerization of cyclotrisiloxanes. The combination of the growing interest in forming block copolymers, the development of siloxanes with strained cyclic structures, and quantitative lithium catalyzed polymerization underlies the publication by Saam [7,8] that reviewers point to as establishing the potential of "living" AROP siloxanes—in which he clearly demonstrated initiation, promotion, narrow MW distribution, and the ability to form block copolymers. It was recognized at that time that these polymers had to be terminated before equilibrium processes dominated in order to maintain the target molecular weight, as visualized in Figure 3. Functionalized termination reagents were later used to create siloxane macromers [9,10], although a process for preparing these macromers was not reported until 10 years later [11].

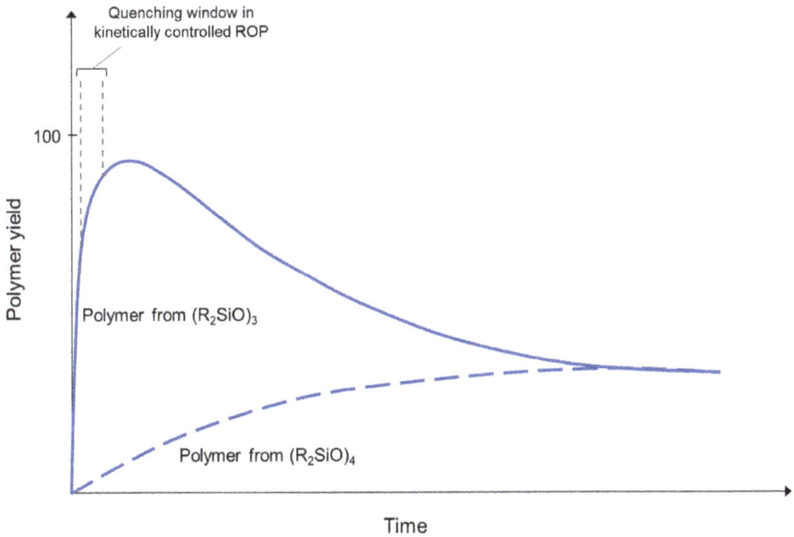

Figure 3. Time-dependent ROP polymer growth of strained versus unstrained cyclosiloxanes.

The term "macromer" is a contraction of the word macromonomer and refers to a relatively high molecular weight species with a single functional group which, although used as a monomer, has sufficient internal monomer units to be considered a polymer. The earliest commercial siloxane macromers contained methacrylate functionality and found commercial utility in the formation of organic–inorganic hybrid polymers (Scheme 4). Their termination, or "capping", functionality was derived from the use of methacryloxypropyldimethylchlorosilane. The general structure for a siloxane is depicted in Scheme 5. Variations of the basic structure are depicted in Schemes 6–8.

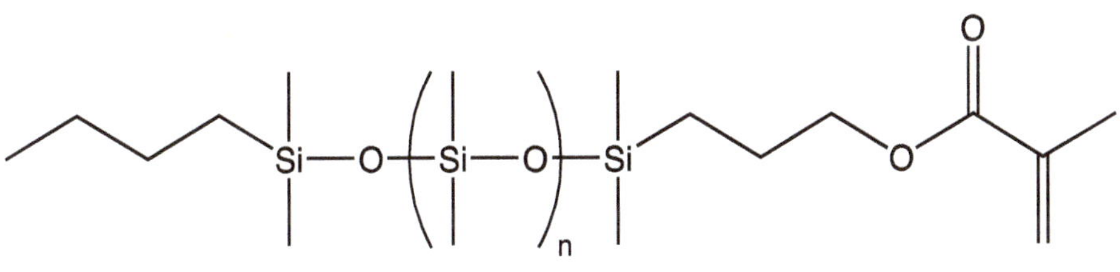

Scheme 4. Polydimethylsiloxane macromers with symmetric architectures, produced via the anionic ring-opening polymerization of a cyclic trisiloxane and subsequent coupling with a functional dihalogen-substituted silane reagent, have recently been described.

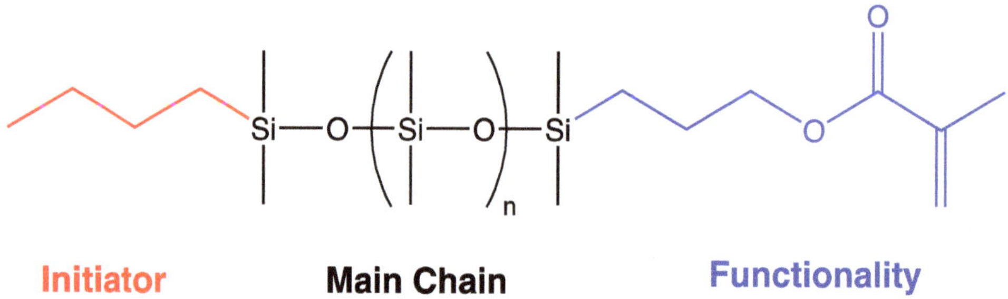

Scheme 5. Siloxane Macromer Structural Segments.

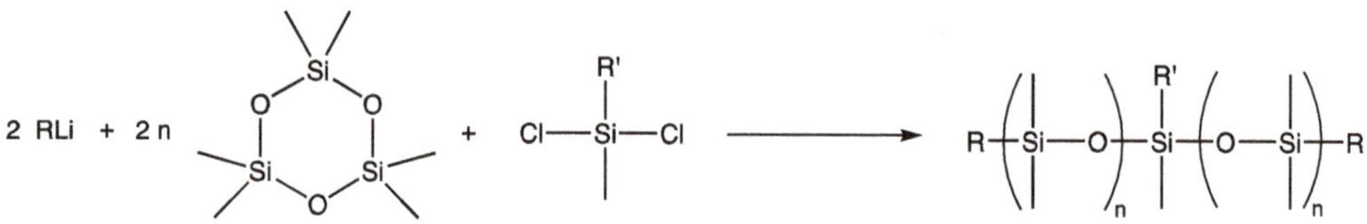

Scheme 6. Symmetric macromers can contain a pendant methacrylate group bisecting the polydimethylsiloxane backbone.

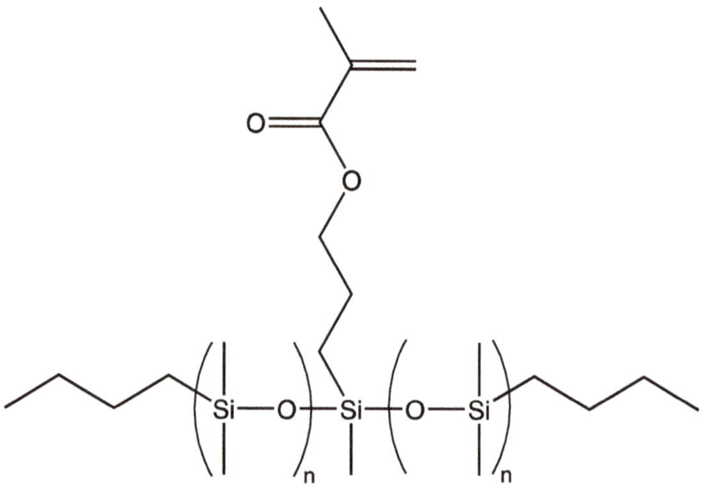

Scheme 7. Symmetric silicone macromer with methacrylate functionality bisection.

A second class of functionality can be introduced into macromers by using novel initiators, thereby yielding telechelic polymers in which the second functional class—e.g., hydroxyl—is located at the telechelic polymer termini, which are equidistant from the first functional class.

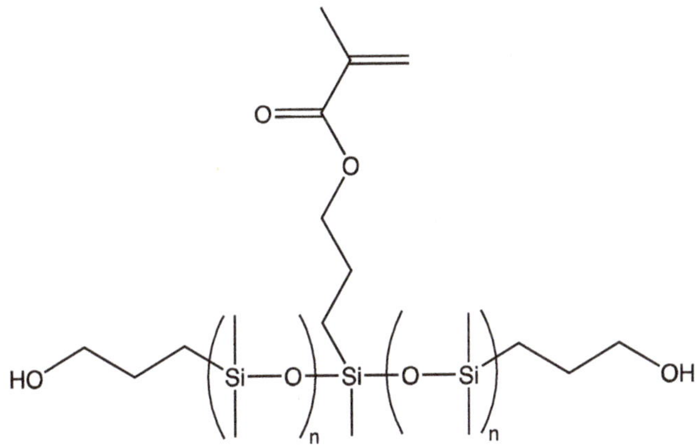

Scheme 8. Carbinol terminated telelechelic with macromeric methacrylate functionality.

More recently, living AROP has been combined with the concept of functional initiators to generate both monodisperse telechelic and heterobifunctional siloxanes.

This historic overview has only given a condensed description of the chemistry, structure, and function of siloxanes derived via living AROP. For those interested in more details regarding the chemical aspects of siloxanes derived from living AROP, the following references should be consulted [1,12,13]. While the bulk of the literature and commercial applications utilize a lithium anion as a weak base component of the initiator, an intriguing recent series of reports utilize substituted cyclic guanidines in combination with water or silanol that act as initiators for living AROP [14,15].

The following tables summarize monomeric building blocks (Table 1), the initiators (Table 2), functional terminations, and architectures that have shown practical utility in commercial applications.

1.2. Monomeric Building Blocks

The fact that there are only five monomeric building blocks for the living AROP-derived polymers that represent virtually all commercial applications as well as the vast majority of published reports is a consequence both of practical synthetic routes [16–18] and the sluggish rates of polymerization reported for cyclotrisiloxanes with greater organic substitution [19]. A preparation has been reported [20] for the simple and highly desirable monomer trimethylcyclotrisiloxane, but its practical isolation has not yet been described. This has led to interest in pentamethylcyclotrisiloxane [21] and hydridotetramethylsiloxanylethyl-substituted cyclotrisiloxanes [22], which can be used directly as monomers for polymerization or reacted with various olefins to form more elaborately substituted trisiloxanes. Similarly, vinylpentamethylcyclotrisiloxane has been prepared, and offers an advantage over trivinyltrimethylcyclotrisiloxane in cases where isolated vinyl substitution on the polymer chain is desired [23]. The primary cyclic monomers used in the production of macromers on both commercial and research levels are listed for convenience in Table 1. Other cyclic monomers have been reported in the literature, and include higher hexaalkylcyclotrisiloxanes [24], chloropropylmethylcyclotrisiloxanes [25], and substituted hexaarylcyclotrisiloxanes [26]. Specialty monomers with limited reference include D_4 [27], acrylate [28], cyclic ether [29], trimethylsiloxy-substituted [30] cyclotrisiloxanes, and similarly substituted strained carbosiloxanes [31].

Table 1. Cyclic Siloxane Macromer "Building Blocks".

Cyclic Siloxane Macromer Monomers					
Class	**Structure**	**MW**	**B.P.**	**CAS#**	**References**
		Primary			
Hexamethylcyclotrisiloxane	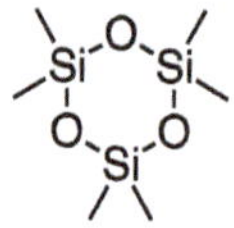	222.46	134°	541-05-9	Hyde [16]
Hexaphenylcyclotrisiloxane	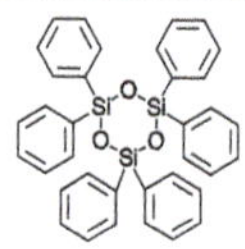	594.89	300°/1	512-63-0	Bostick [32]
Trivinyltrimethylcyclotrisiloxane	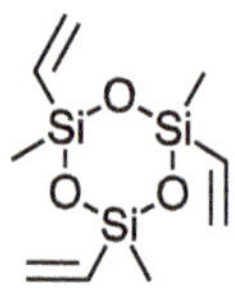	258.50	80°/20	3901-77-7	Pike [33]
Tris(3,3,3-trifluoropropyl)trimethylcyclotrisiloxane	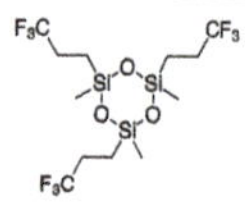	468.55	95°/3	2374-14-3	Pierce [18]
2,2,5,5-tetramethyl-2,5-disila-1-oxacyclopentane	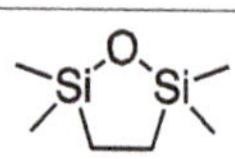	160.36	124°	7418-20-4	Piccolli [17]
		Minor			
Hexaethylcyclotrisiloxane	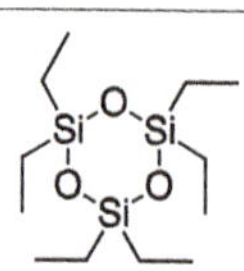	306.82	117°/10	2031-79-0	Dobay [34]
Vinylpentamethylcyclotrisiloxane	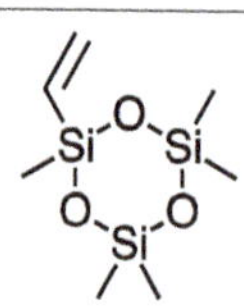	234.47	56°/23	18395-32-9	Rózga-Wijas [23]
Tris(tetramethylsiloxanylethyl) trimethylcyclotrisiloxane	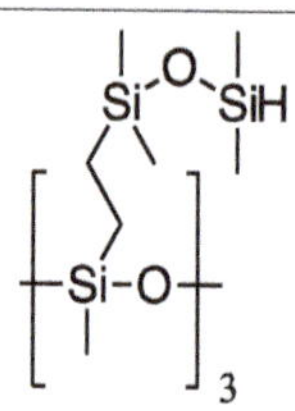	661.47	175°–185°/2	2378614-30-1	Goff [22]
1,3,5-[Tris(methoxyethoxyethoxypropyltetramethyldisiloxanyl)ethyl]-1,3,5-trimethylcyclotrisiloxane	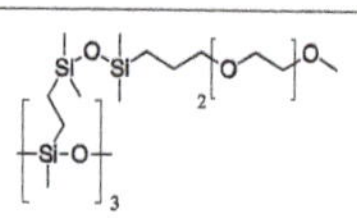	1142.10		2378623-20-0	Goff [35]

Table 2. Initiators.

Non-Functional Initiators			
Class	**Structure**	**Comments**	**References**
Organolithium			
n-Butyl lithium	$CH_3CH_2CH_2CH_2Li$	Most common initiator; stable and economic	Chojnowski [36] Cypryk [37] Maschke [38] Nakano [11] Razzano [39]
s-Butyl lithium	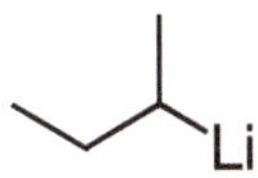	Less initial back-biting than n-butyl but less stable; used for substituted aryl-cyclic trisiloxane; used for higher alkylcyclo-trisiloxanes	Hawkridge [40] Bellas [41] Ninago [42] Lee [26] Molenberg [43–45]
t-Butyl lithium			Hempenius [46] Sieburth [47]
Phenyl lithium	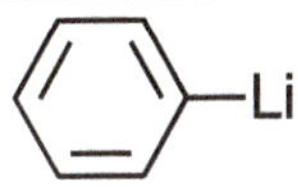	Initiator for PDMS MW standard	Arkles [48]
Trimethylsilylmethyl lithium	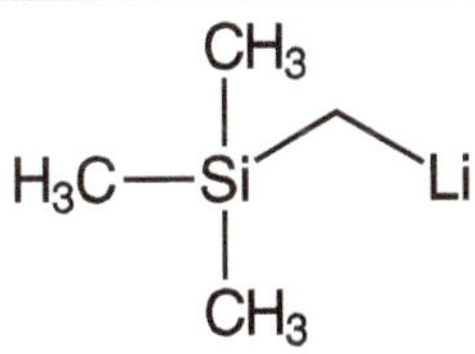	For diMeSiO-ViMeSiO block polymers	Boileau [49] Bauer [50]
Macroanions			
Polybutadienyl lithium			Molenberg [19]
Polystyryl lithium	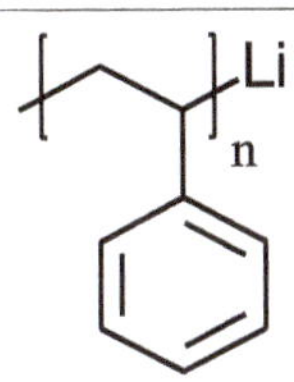	Formation of block and star polymers—varied block size for nanopatterning	Saam [7,51] Bellas [44] Georgopanos [52]
Poly a-methylstyryl lithium	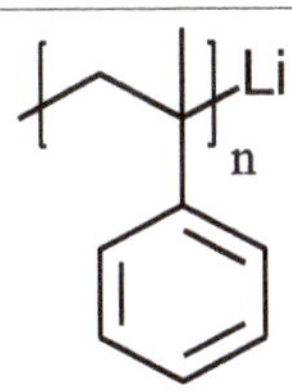		Saam [8] Noshay [53]

Table 2. *Cont.*

Non-Functional Initiators			
Class	**Structure**	**Comments**	**References**
Lithium Silanolate—Monoanion			
Trimethylsilanol lithium salt	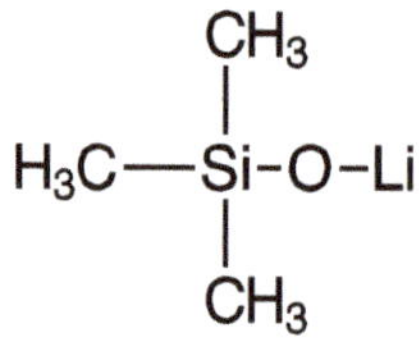	Formed from D_4 + Methyllithium; initiates tetramethyldisila-oxacyclopentane polymerization	Clemens [54] Kawakami [9] Awasthi [55]
n-Butyldimethylsilanol lithium salt	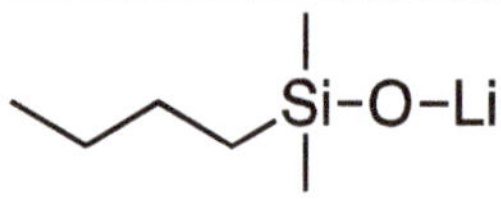	Formed from D_4 + n-Butyllithium	Fessler [6] Bostick [32] Cypryk [56] Leir [57]
Fluoroalkyldimethylsilanol lithium salt	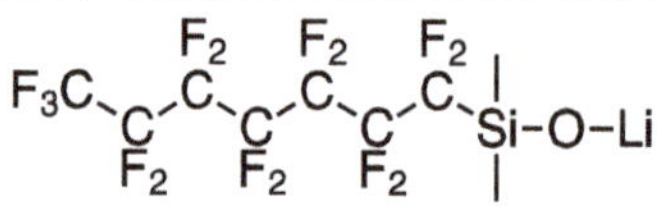	Formed in situ from silanol and n-Butyllithium	Saho [58]
Methoxypropyldimethylsilanol lithium salt	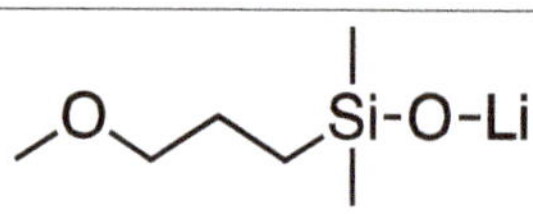		Arkles [59] Goff [60]
Methoxyethoxyethoxypropyld imethylsilanol lithium salt	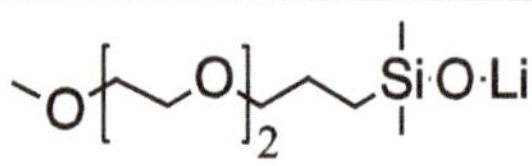	Hydrophilic silicone macromers for contact lenses	Goff [61] Arkles [62]
Methoxyethoxyethoxyetho xypropyldimethylsilanol lithium salt	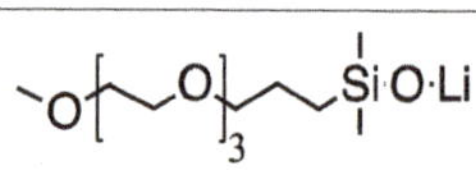	Hydrophilic additive for silicone elastomers	Goff [61] Arkles [62]
Tetrahydrofurfuryloxpropyl dimethylsilanol lithium salt	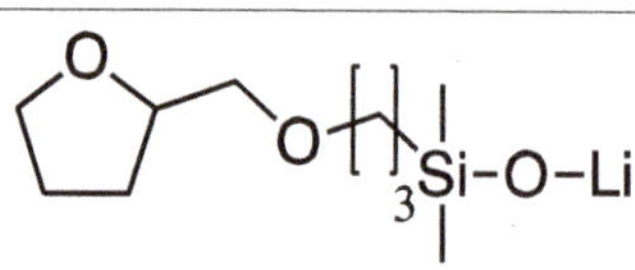	Hydrophilic additive for implantable silicones	Goff [61] Arkles [62]
Lithium Silanolate—Multiple Anions			
Diphenylsilanediol, dilithium salt	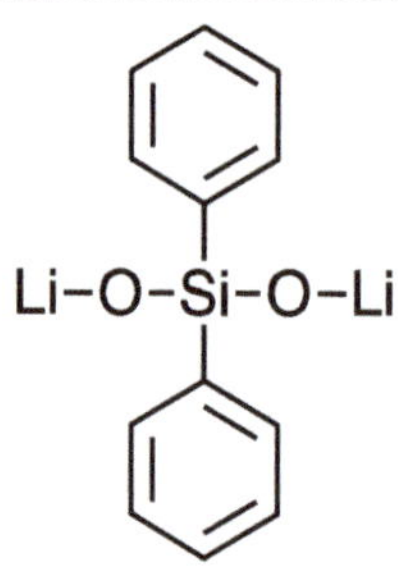	Initiator for pentamethylcyclo trisiloxane; initiator for trifluoropropyltrimethylcy-clotrisiloxane	Paulasaari [21] Cypryk [37]

Table 2. *Cont.*

Non-Functional Initiators			
Class	**Structure**	**Comments**	**References**
Lithium Silanolate—Multiple Anions			
Tetramethyldisiloxane-1,3 diol dilithium salt	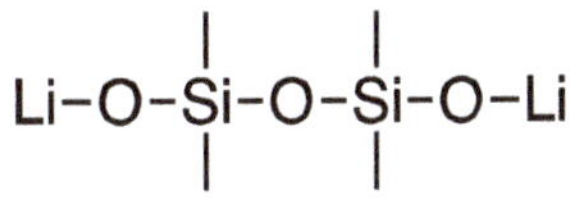		Gadda [63]
1,3-diphenyl-1,3-dimethylsiloxanediol, dilithium salt	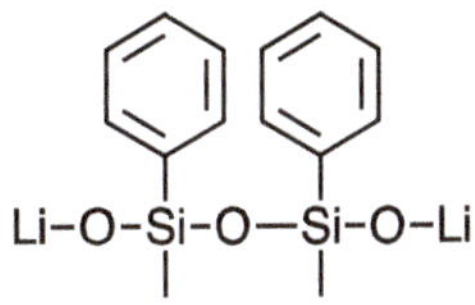		Bostick [32]
Bis(p-hydroxydimethylsilyl)phenyl ether, dilithium salt	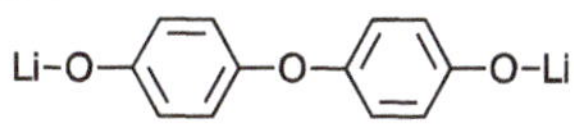	Used to form telechelic vinyl-terminated polymers	Kazama [64]
Branched siloxanes, multiple lithium salts		Branched vinyl-terminated siloxanes	Ogawa [65]
Branched carbosilanes, multiple lithium salts		Used to form multi-arm and dendrimer-like structures	Novozhilov [66] Vasilenko [67]
Other			
Dilithiobenzophenone	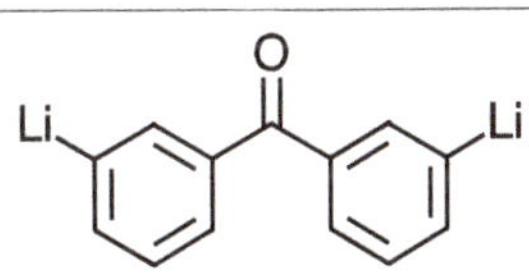	Siloxane methacrylate copolymers	Juliano [68]
Dilithiofluorenone	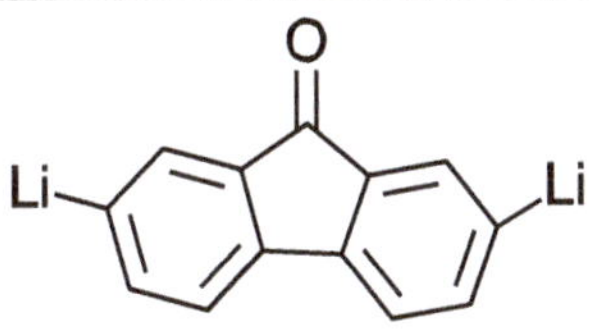	Color change when D_3 polymerization initiated	Juliano [68]
Functional Initiators			
Class	**Structure**	**Comments**	**References**
Organo Lithium			
Allyl lithium	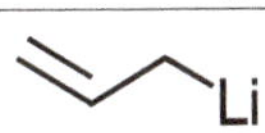	Functionalized polydiethylsiloxane	Out [69]
t-Butyldimethylsiloxypropyl lithium	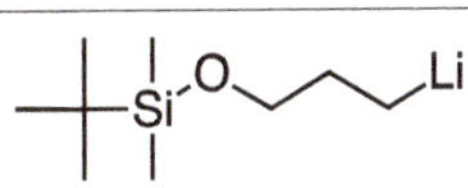	Deprotected to form hydroxyl termination	Arkles [70] Arkles [59]

Table 2. *Cont.*

Functional Initiators			
Class	**Structure**	**Comments**	**References**
N-Benzyl-N-methylaminopropyl lithium	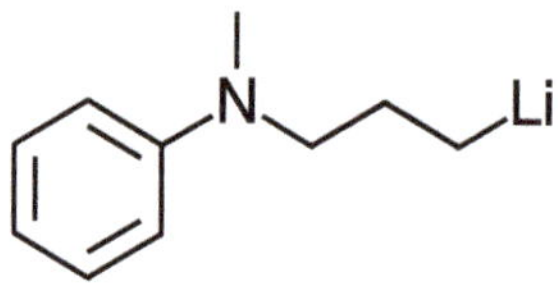	Can be deprotected to provide N-methylaminopropyl termination	Elkins [71]
Bis(trimethylsilyl)aminophenyl lithium	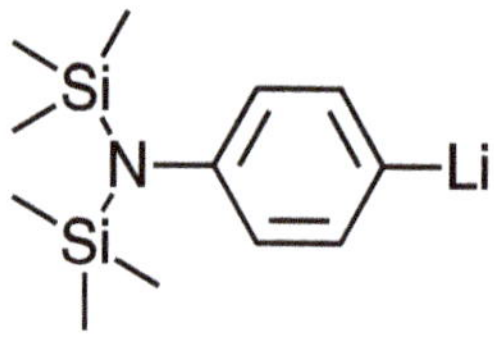	Deprotected to form aminophenyl termination	Babu [72]
Lithium Silanolate			
Vinyldimethylsilanol lithium salt	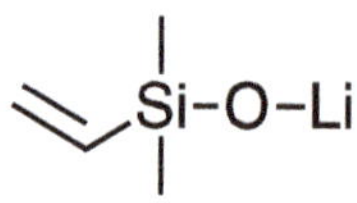		Goff [73] Okawa [73,74] Arkles [62]
Vinylbutylmethylsilanol lithium salt	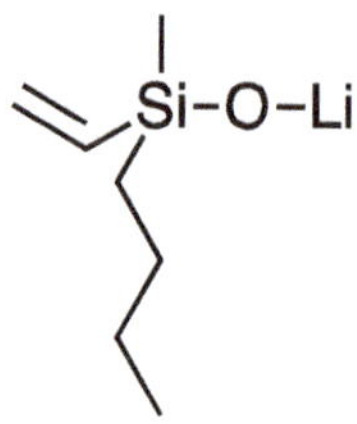		Arkles [62]
Hydridobutylmethylsilanol lithium salt	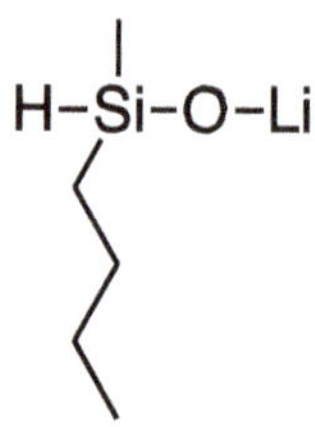		Arkles [62]
1-Hexenyl, 4-Hydroxyl-octamethyltetrasiloxane lithium salt	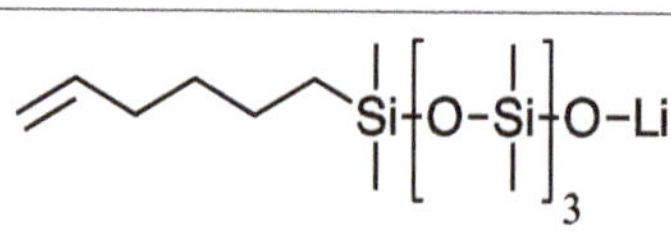		Okawa [74]

Table 2. *Cont.*

Functional Initiators			
Class	**Structure**	**Comments**	**References**
Other			
Lithium anion of Na pentanedionate	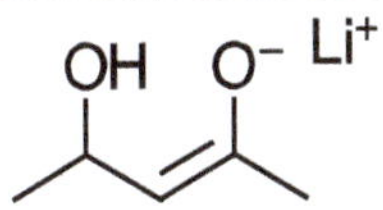	Forms Na AcAc-terminated polymer, but not stable after neutralization	Kumar [75]
1,5,7-Triazabicyclo [4.4.0] dec-5-ene	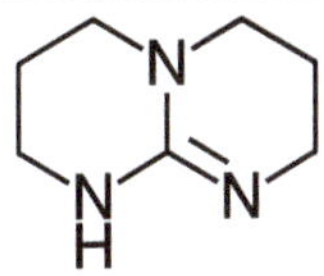	Used in the polymerization of monodisperse silanol-terminated siloxanes	Lohmeijer [14]
1,3-Trimethylene-n-propylguanidine	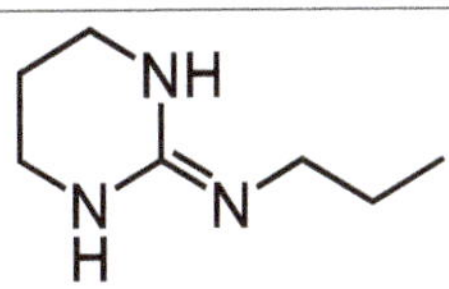	Used in combination with water and silanols	Fuchise [15,76]

1.3. Termination

Terminating molecules in living AROP can serve several functions. At a minimum, they serve to quench the polymerization before redistribution processes associated with reversion, interchain back-biting, and interchain scrambling; their potential functions can be further divided into the following categories:

Non-functional termination;
Non-functional coupling to effect non-functional doubling of MW;
Non-functional coupling with block insertion;
Functional termination to form asymmetric macromers;
Functional coupling to form symmetric macromers;
Functional coupling to form branched structures.

Terminators in AROP processes are usually chlorosilanes, although other halosilanes and alkoxysilanes have been reported. Apart from the terminators found in the literature associated with living AROP of siloxanes, terminators normally associated with the living anionic polymerization of olefins and other organic polymers can provide similar termination for siloxanes [77,78].

1.4. Macromers Commonly Reported in Literature

The structures and properties listed in Table 3 are an aggregate of the most common siloxane macromers reported in the scholarly and commercial literature. They should not be considered exact values, but rather nominal values for similar materials. Similarly, heterobifunctional macromers are reported in Table 4.

Table 3. Commonly Reported Macromers.

			Asymmetric Monofunctional Siloxanes			
Code *	**Description**	**Molecular Weight, Mn**	**Viscosity, cSt**	**Density**	**Refractive Index**	**Reference**
		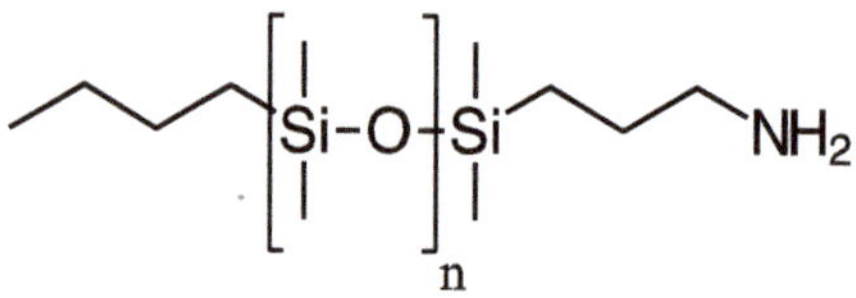				
MCR-A11	monoAMINOPROPYL TERMINATED POLYDIMETHYLSILOXANE	800–1000	8–12	0.82	1.411	[79–81]
MCR-A12	monoAMINOPROPYL TERMINATED POLYDIMETHYLSILOXANE	2000	18–25	0.97	1.411	[79,81]
		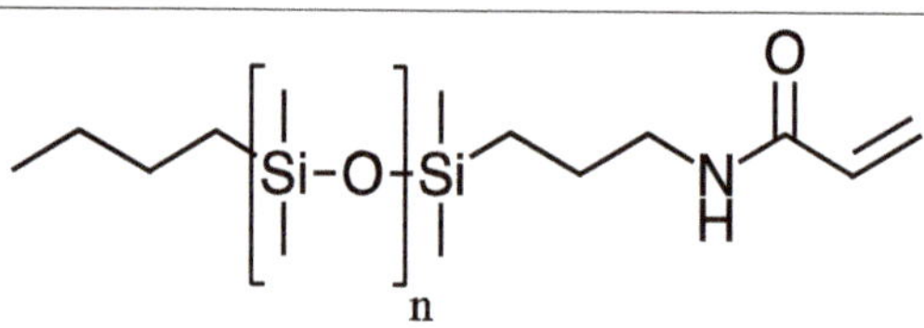				
MCR-W15	monoACRYLAMIDOPROPYL TERMINATED POLYDIMTHYLSILOXANE	1000–1500	50–75	0.96	1.418	[79]
		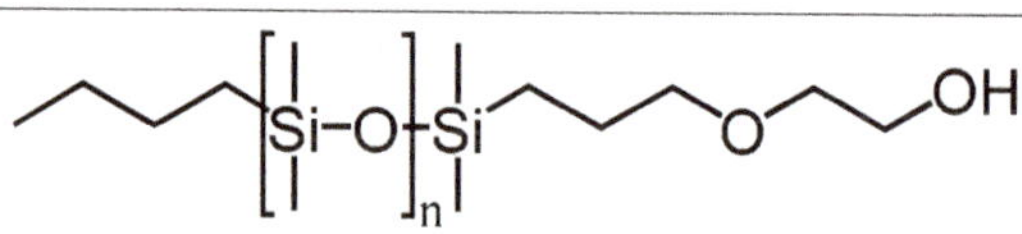				
MCR-C12	monoCARBINOL TERMINATED POLYDIMETHYLSILOXANE	1000	15–20	0.96	1.409	[79,82]
MCR-C18	monoCARBINOL TERMINATED POLYDIMETHYLSILOXANE	5000	80–90	0.97	1.405	[79,82–84]
MCR-C22	monoCARBINOL TERMINATED POLYDIMETHYLSILOXANE	10,000	250	0.98	1.404	[79,81,82,85]
		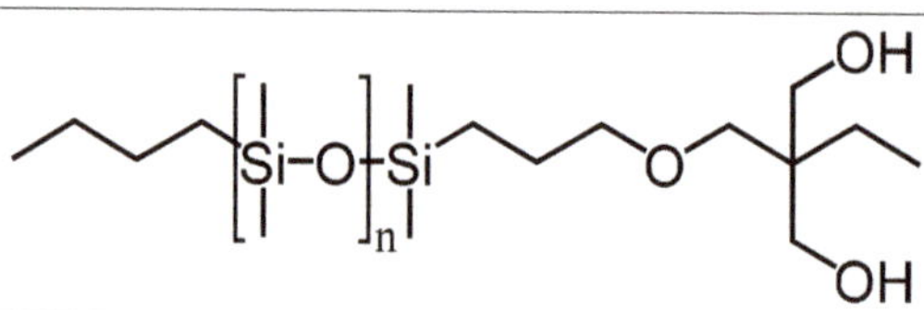				
MCR-C61	monoDICARBINOL TERMINATED POLYDIMETHYLSILOXANE	1000	50–60	0.97	1.417	[79,86]
MCR-C62	monoDICARBINOL TERMINATED POLYDIMETHYLSILOXANE	5000	100–125	0.97	1.409	[79,86,87]

Table 3. *Cont.*

Asymmetric Monofunctional Siloxanes						
Code *	Description	Molecular Weight, Mn	Viscosity, cSt	Density	Refractive Index	Reference
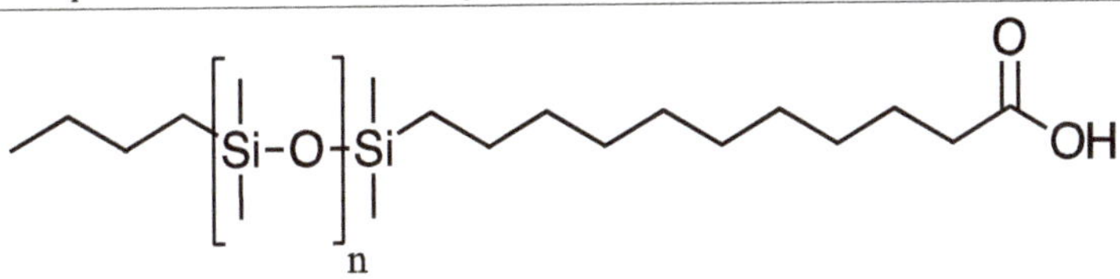						
MCR-B12	monoCARBOXYDECYL TERMINATED POLYDIMETHYLSILOXANE	1500	20	0.94	1.415	[79,88]
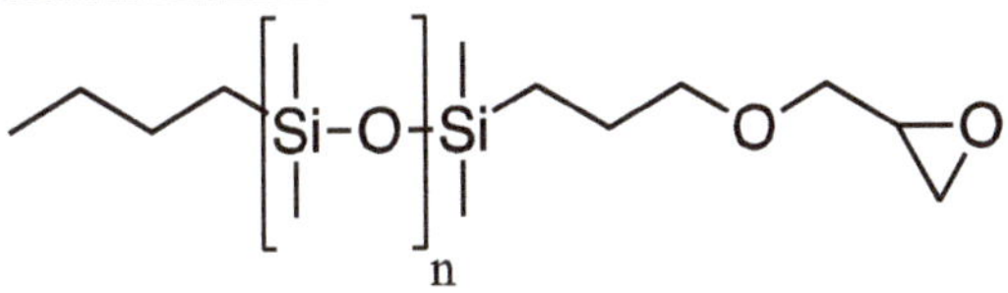						
MCR-E11	Mono(2,3-EPOXY)PROPYLETHER TERMINATED POLYDIMETHYLSILOXANE	1000	10–15	0.96	1.410	[79,89,90]
MCR-E21	Mono(2,3-EPOXY)PROPYLETHER TERMINATED POLYDIMETHYLSILOXANE	5000	120	0.97	1.408	[79]
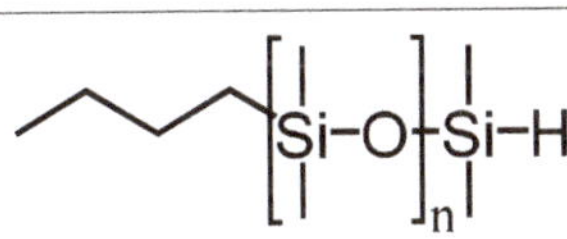						
MCR-H07	MonoHYDRIDE TERMINATED POLYDIMETHYLSILOXANE	800–900	5–8	0.96	1.404	[60,79]
MCR-H11	MonoHYDRIDE TERMINATED POLYDIMETHYLSILOXANE	900–1100	8–12	0.96	1.407	[60,79,91]
MCR-H21	MonoHYDRIDE TERMINATED POLYDIMETHYLSILOXANE	4500–5000	80–120	0.96	1.411	[79]
MCR-H22	MonoHYDRIDE TERMINATED POLYDIMETHYLSILOXANE	10,000	160–220	0.98	1.411	[79]
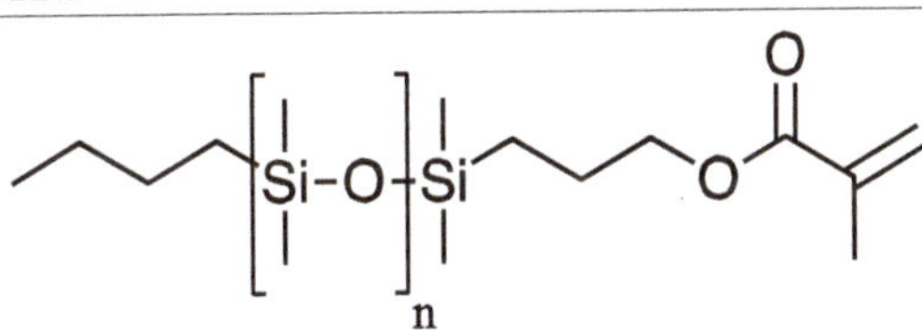						
MCR-M07	MonoMETHACRYLOXYPROPYL TERMINATED POLYDIMETHYLSILOXANE	600–800	6–9	0.96	1.416	[60,79,92–97]

Table 3. *Cont.*

Asymmetric Monofunctional Siloxanes						
Code *	Description	Molecular Weight, Mn	Viscosity, cSt	Density	Refractive Index	Reference
MCR-M11	MonoMETHACRYLOXYPROPYL TERMINATED POLYDIMETHYLSILOXANE	800–1000	10–11	0.96	1.411	[60,70,79,92,93,95–108]
MCR-M17	MonoMETHACRYLOXYPROPYL TERMINATED POLYDIMETHYLSILOXANE	5000	70–80	0.97	1.406	[54,70,79,89,90,93,95–97,108–110]
MCR-M22	MonoMETHACRYLOXYPROPYL TERMINATED POLYDIMETHYLSILOXANE	10,000	150–200	0.97	1.405	[54,79,93,94,96,97]
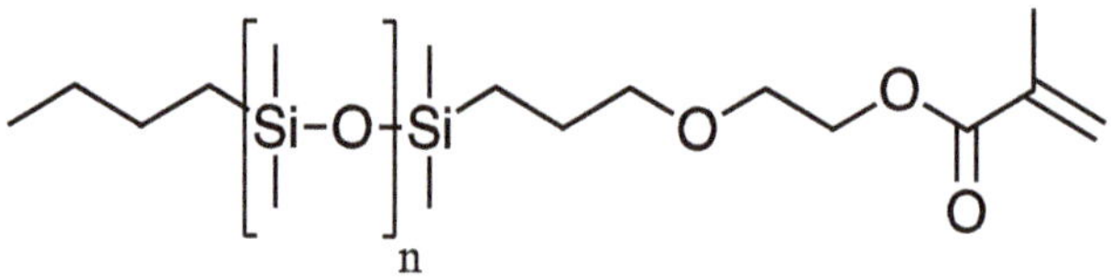						
MCR-MXe11	MonoMETHACRYLOXYETHOXY-PROPYL TERMINATED POLYDIMETHYLSILOXANE	900–1000	9–10	0.95	1.415	None?
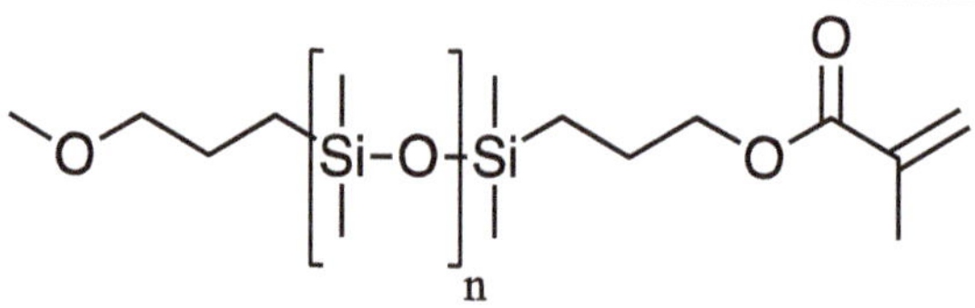						
MCR-MXt11	MonoMETHACRYLOXYPROPYL, MonoMETHOXYPROPYL TERMINATED POLYDIMETHYLSILOXANE	800–1000	8–12	0.96	1.430	[60,70]
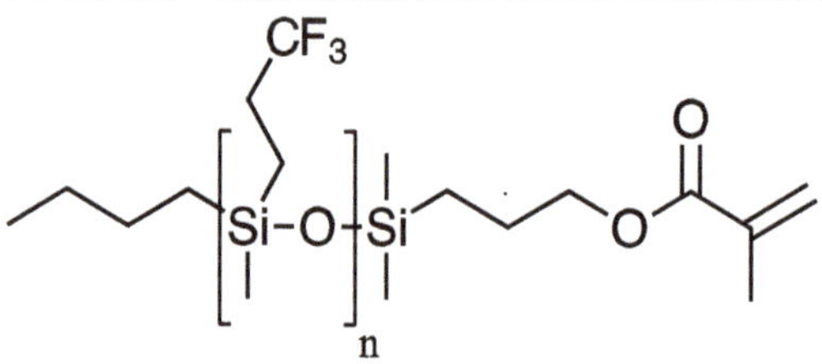						
MFR-M15	MonoMETHACRYLOXYPROPYL TERMINATED POLYTRIFLUOROPROPYL-METHYLSILOXANE	800–1000	50–70	1.09	1.398	[59,60,70,79,93,100]
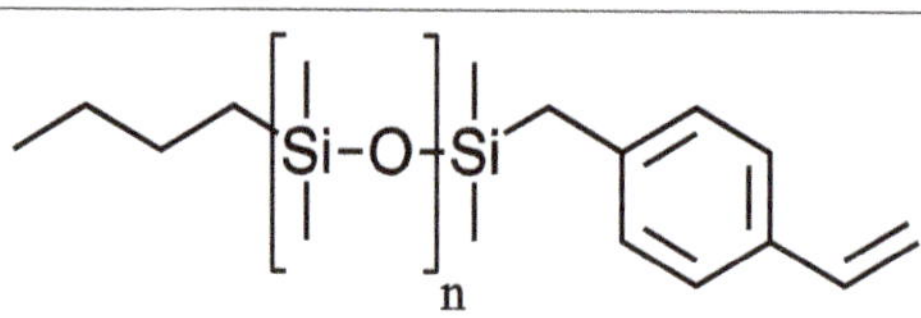						
MCR-ST11	MonoSTYRYL TERMINATED POLYDIMETHYLSILOXANE	800–1000	8–15	0.95	1.429	[99,102]

Table 3. *Cont.*

Code *	Description	Molecular Weight, Mn	Viscosity, cSt	Density	Refractive Index	Reference
Asymmetric Monofunctional Siloxanes						
		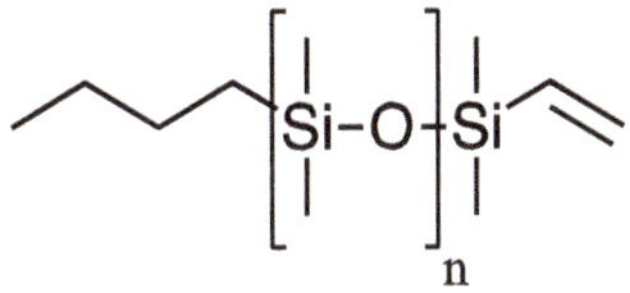				
MCR-V21	MonoVINYL TERMINATED POLYDIMETHYLSILOXANE	5500–6500	80–120	0.97	1.403	[79,91,111]
MCR-V25	MonoVINYL TERMINATED POLYDIMETHYLSILOXANE	15,000–20,000	400–600	0.97	1.403	[79]
MCR-V41	MonoVINYL TERMINATED POLYDIMETHYLSILOXANE	55,000–65,000	8000–12,000	0.98	1.404	[79]
		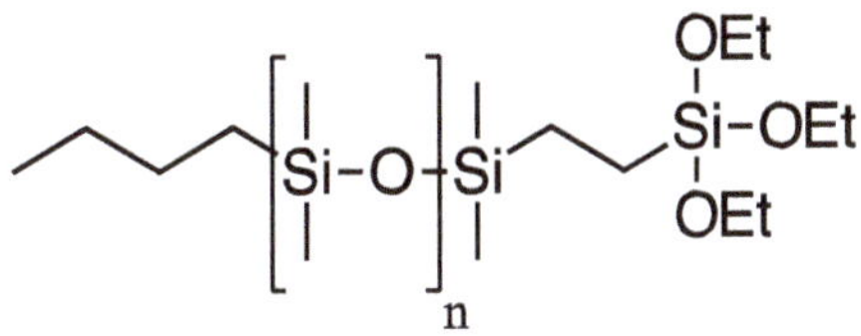				
MCR-XT11	MonoTRIETHOXYSILYLETHYL TERMINATED POLYDIMETHYLSILOXANE	500–1000	16–24	0.97	1.412	[79]
Symmetric Monofunctional Siloxanes						
		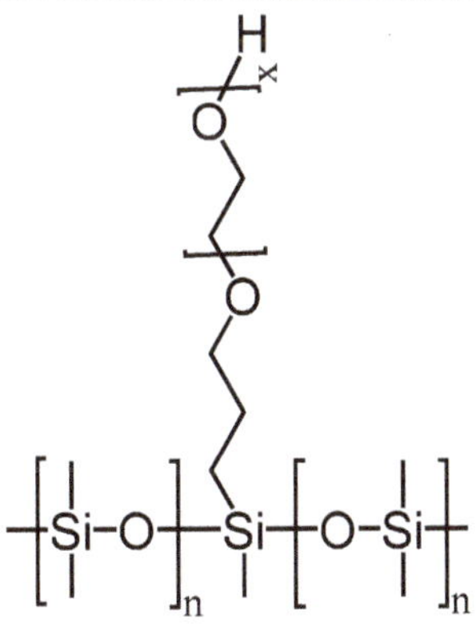				
MCS-C13	MonoCARBINOL FUNCTIONAL POLYDIMETHYLSILOXANE, symmetric	550–650	35–40	1.02	1.446	[79]

Table 3. *Cont.*

Code *	Description	Molecular Weight, Mn	Viscosity, cSt	Density	Refractive Index	Reference
Asymmetric Monofunctional Siloxanes						
		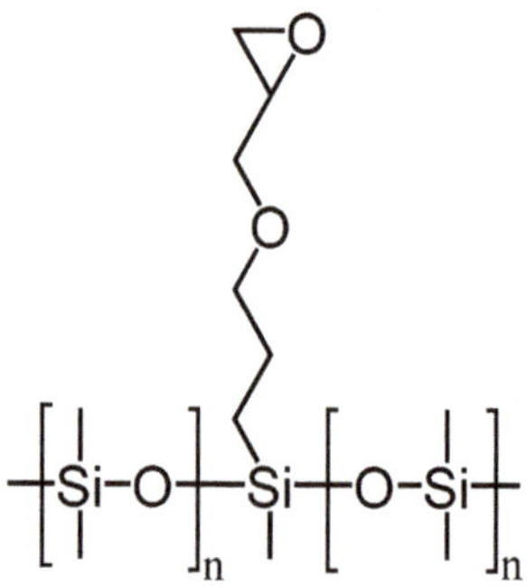				
MCS-E15	Mono(2,3-EPOXY)PROPYL ETHER FUNCTIONAL POLYDIMETHYLSILOXANE, symmetric	800–900	45–55	1.09	1.398	[59,70,79]
		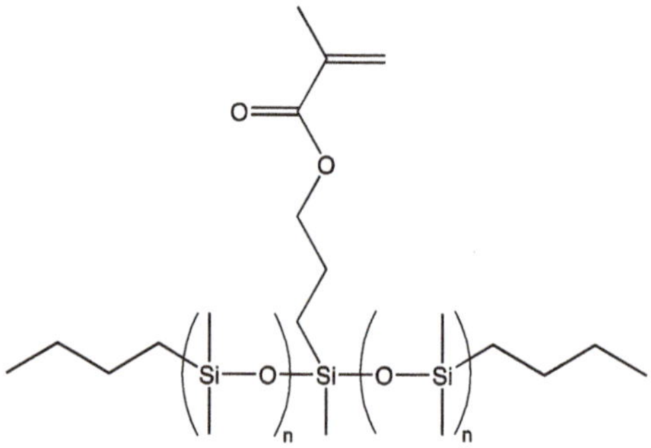				
MCS-M11	MonoMETHACRYLOXYPROPYL FUNCTIONAL POLYDIMETHYLSILOXANE, symmetric	800–1000	7–9	0.93	1.417	[59,60,70,79,93]
		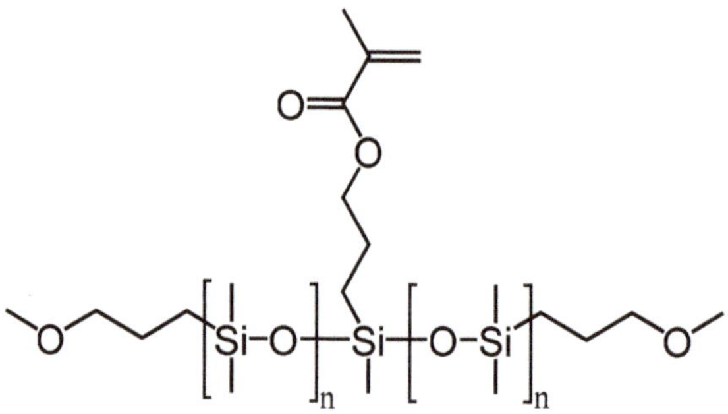				
MCS-MXt11	MonoMETHACRYLOXYPROPYL FUNCTIONAL POLYDIMETHYLSILOXANE, METHOXYPROPYL TERMINATED, symmetric	900–1100	8–12	0.96	1.43	[60,79]

Table 3. *Cont.*

Code *	Description	Molecular Weight, Mn	Viscosity, cSt	Density	Refractive Index	Reference
Asymmetric Monofunctional Siloxanes						
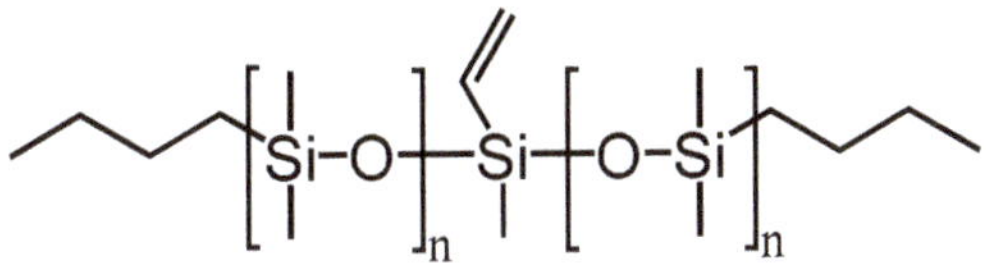						
MCS-V212	MonoVINYL TERMINATED POLYDIMETHYLSILOXANE, symmetric	1200–1400	16–24	0.97	1.419	[79]
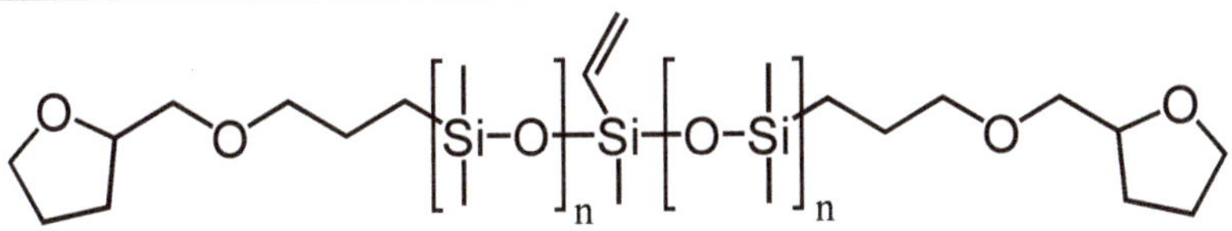						
MCS-VF14	monoVINYL FUNCTIONAL POLYDIMETHYLSILOXANE, TETRAHYDROFURFURY-LOXYPROPYL TERMINATED—symmetric	1000–1200	30–40	0.976	1.4141	[61,79,87,112]
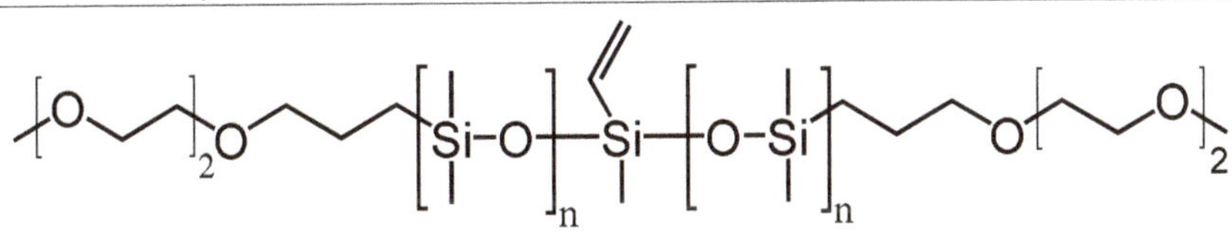						
MCS-VXt15	MonoVINYL FUNCTIONAL POLYDIMETHYLSILOXANE, METHOXY(DIETHYLENEOXIDE) PROPYL TERMINATED, symmetric	1000–1200	40–60	0.96	1.42	[61,79,87,112]

* In order to provide a concise description of the macromer systems discussed, the following designation system was adopted: the first two letters, MC, indicate macromer M structure with a C conventional dimethylsiloxane repeat unit; the third letter indicates whether it is asymmetric (R), symmetric (S), or T-structure (T). Macromer functionality and viscosity are indicated after a hyphen, as follows: (1) letters indicate functionality, where M is methacrylate, H is hydride, C is carbinol, V is vinyl, and E is methoxy; (2) numbers are the viscosity ranges of the macromer in powers of 10, followed by the first significant figure. For example, MCR-H25 indicates a macromer, an asymmetric architecture with hydride functionality, and a viscosity of $10^2 \times 5$ or 500 cSt. These designations provide a facile method of discussing comparative structures but are not intended to be detailed structural descriptions of the polymers. To remove ambiguity for ethyleneoxy designated X spacer groups, the lower-case letter t, for "tipped", or e, for "embedded", is used.

Table 4. Monodisperse Difunctional Macromers.

Code	Description	Molecular Weight, Mn	Viscosity, cSt	Density	Refractive Index	Reference
Heterobifunctional Siloxanes						
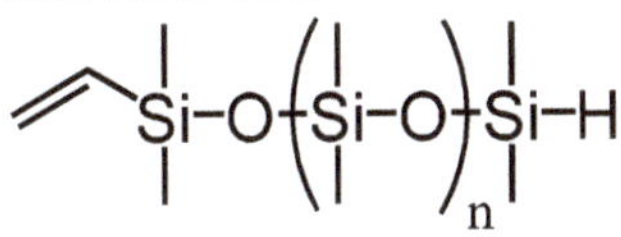						
DMS-HV15	monoVINYL-monoHYDRIDE terminated POLYDIMETHYLSILOXANE	2000–3000	40–60	0.96	1.404	[62,73,79,113,114]
DMS-HV22	monoVINYL-monoHYDRIDE terminated POLYDIMETHYLSILOXANE	10,000	150–250	0.97	1.403	[62,73,79,113–116]
DMS-HV31	monoVINYL-monoHYDRIDE terminated POLYDIMETHYLSILOXANE	25,000	600–1000	0.97	1.403	[62,73,79,114]

Table 4. *Cont.*

Heterobifunctional Siloxanes						
Code	**Description**	**Molecular Weight, Mn**	**Viscosity, cSt**	**Density**	**Refractive Index**	**Reference**
		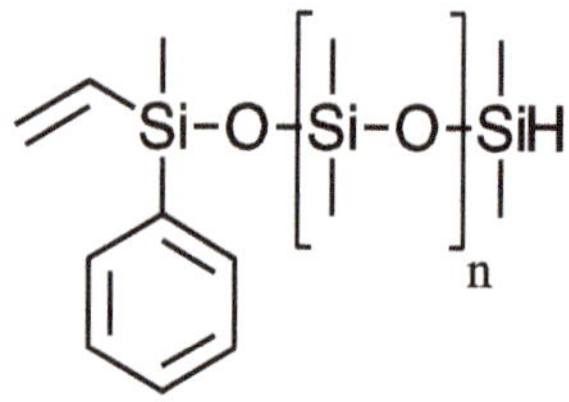				
PMM-HV12	α-monoVINYL,monoPHENYL-ω-monoHYDRIDE terminated POLYDIMETHYLSILOXANE	2000	20	0.97	1.414	[62,79]

2. Applications

2.1. Compatibility and Reactivity Introduction

The living polymerization of siloxanes provides the basis for synthesizing siloxane macromers capable of acting as precise structural elements, achieving both the requirements of copolymerization with organic monomers as well as the desired properties of copolymers. The incorporation of properties associated with siloxanes—including oxygen and moisture permeability, release, or thermal performance—would thereby improve the performance of a wide range of organic polymers. To date, however, the utilization of silicone macromers in combination with organic monomers has been primarily limited by economic considerations, as well as by structural and compositional challenges.

The most commonly used siloxane macromers are methacrylate-terminated, both in the literature and in commercial applications. Reported reactivity ratios [79] and previously unreported solubility in monomers are shown in Tables 5 and 6, respectively

Table 5. Reactivity Ratios of Methacrylate Functional Siloxane Macromers.

Siloxane Macromer	Comonomer	Reactivity Ratio (r1:r2) *
MCR-M11	Methylmethacrylate	nm **:1.60
MCR-M22		nm **:2.10
MCR-M11	Styrene	0.26:1.07
MCR-M11	Acrylonitrile	5.40:0.89

* Rate constants—M1M1/M1M2: M2M2/M2M1; ** no meaningful results.

Table 6. Solubility of Methacrylate Functional Silicone Macromers *.

Siloxane Macromer—Methacrylate Functional	Macromer Solubility in Hydrogel Monomers		Water Solubility in Macromer (ppm)
	Dimethylacrylamide (DMA)	Hydroxyethylmethacrylate (HEMA)	
MCR-M11 (asymmetric)	4%	1%	2000
MCS-M11 (symmetric)	8%	2%	3500
MFR-M15 (fluorinated asymmetric)	100% (miscible)	2%	9000
MFS-M15 (fluorinated symmetric)	100% (miscible)	3%	10,000
MCS-MXt11 (methoxy-tipped, symmetric)	100% (miscible)	100% (miscible)	10,000

* Author's work, previously unreported.

2.2. *Gas/Vapor Permeability*

In comparison to organic polymers, siloxanes possess a large free molal volume, a consequence of the length of both silicon–oxygen and silicon–carbon bonds, which allows greater permeation of small molecules. These same factors contribute to the flexibility of silicones which imposes limitations on structural properties. Hybrid polymer systems utilizing both macromer and block polymers generated by living AROP provide the balance of properties required in contact lenses, breathable films, and membranes.

2.2.1. Contact Lens/Hydrogel

Oxygen permeability is a key feature of modern contact lenses, particularly those intended for extended wear. In addition to the bloodstream, the corneal and scleral tissues of the eye take about 1/3 of their respiratory requirement directly from the air. Methacrylate-derived polymers, while possessing good optical and mechanical properties, are occlusive to oxygen transport due to the fact that the permeability of siloxanes is more than 100X greater than that of analogous hydrocarbon structures. Gaylord's pioneering material work [117,118] under the direction of Seidner led to the commercialization of rigid gas permeable contact lenses by Syntex in 1980 [119]. While these lenses satisfied the physiological requirement for extended wear, broad acceptance of silicone rigid gas permeable (RGP) lenses was not achieved, since user comfort did not match that of hydrogel lenses. Formulation and process challenges in silicone hydrogel contact lens manufacture include combining hydrophobic silicones with hydrophilic monomers while maintaining optical clarity, high water content, and high oxygen transport to the eye in the resulting hydrogels. The first successful silicone hydrogel lens, based on technology disclosed by Vanderlaan, was introduced to the marketplace by Vistakon in 2004 [102]. Monofunctional siloxane macromers polymerized by living anionic polymerization routes are key materials in such silicone hydrogel contact lens formulations: these monodisperse monofunctional materials are used as comonomers with hydrophilic hydrogel monomers such as hydroxyethylmethacrylate (HEMA) and dimethylacrylamide (DMA). The earliest report of soft silicone hydrogel lenses generated from siloxane macromers and HEMA that displayed acceptable performance utilized group transfer polymerization [120]. One material requirement of a silioxane macromer selected for such a formulation is a minimal solubility in the hydrophilic hydrogel monomers. It has been reported that, after curing the silicone hydrogel contact lens reactive monomer mix, an optically clear co-continuous silicone and hydrogel phase membrane suitable for contact lens use is formed [121]. Incorporating low molecular weight (~1000 g mol^{-1}) siloxane macromers into silicone hydrogel formulations enhances oxygen permeability in the resulting contact lenses, achieving superior oxygen transport to the eye compared to hard rigid gas permeable (RGP) and soft HEMA contact lens technologies (Figure 4) [122].

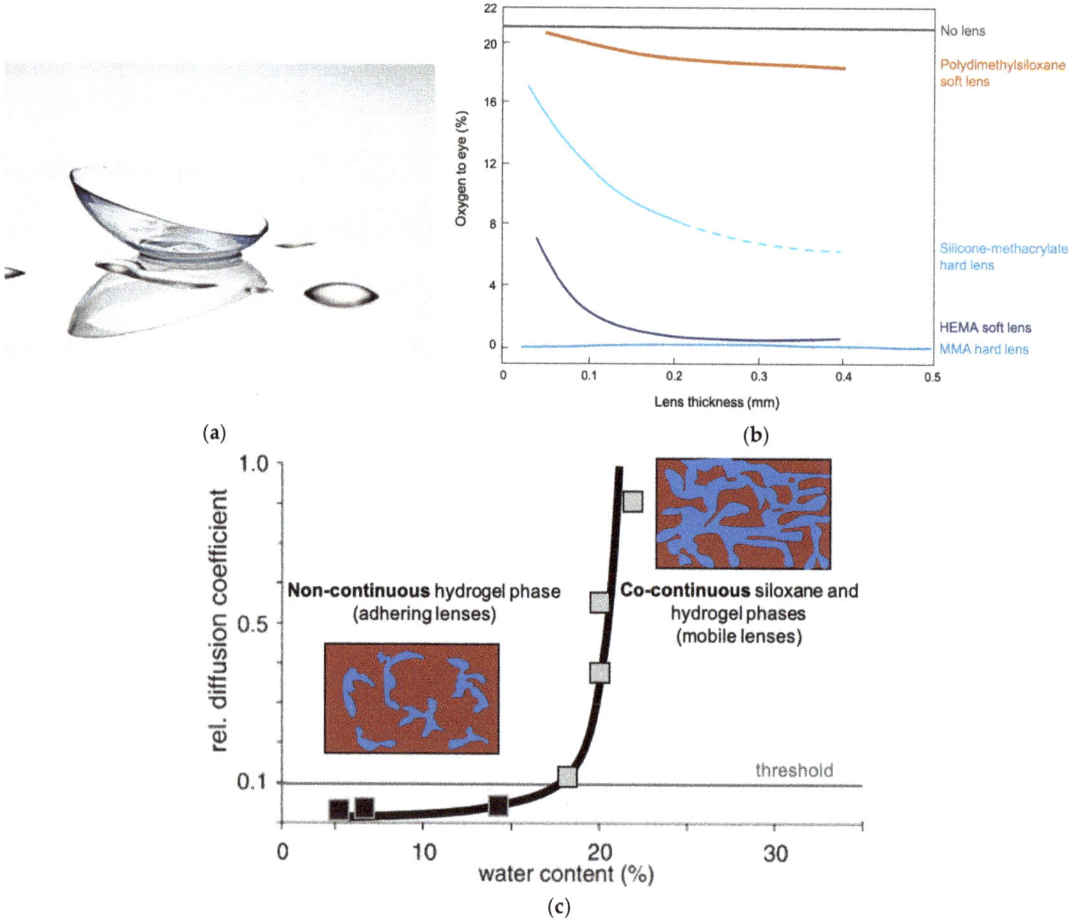

Figure 4. (**a**) Silicone hydrogel lenses incorporate low MW macromers into otherwise hydrophilic matrices based on hydroxyethylmethacrylate or dimethylacrylamide; (**b**) comparison of contact lens materials and oxygen transport to the cornea (redrawn with permission from Ref. [122] Copyright 1983 the Technomic Publication). (**c**) Hydrogel phase structure as a determinant of lens comfort (redrawn with permission from Ref. [121] Copyright 2001 Elsevier Science).

During the contact lens molding process, extraction procedures are employed to remove undesirable impurities. Water insoluble impurities, for example, decrease optical clarity and leach out of the contact lens, causing negative ocular effects such as stinging [92]. Unfortunately, the hydrophobic nature of these impurities makes conventional water extraction procedures insufficient. Silicone hydrogel extraction procedures using alcohols have been reported, but possess drawbacks including increased manufacturing costs, organic solvent waste handling concerns, and potential eye irritation from residual solvent. Living anionic polymerization enables the synthesis of monodisperse monofunctional siloxane macromers that are largely free of impurities. Indeed, newer generation silicone hydrogel formulations employ siloxane macromers of sufficient purity to rely solely on aqueous extraction and hydration steps post-molding. Table 7 gives an overview of the different generations of silicone hydrogel technologies. Generation 1 silicone hydrogel contact lenses did not use siloxane macromers and relied on tris(trimethylsilyl)-silylpropylmethacrylate (TRIS) and difunctional silicone crosslinkers as the oxygen transport enhancers in the formulation. TRIS made control of the lenses' mechanical properties difficult, resulting in higher modulus lenses that were not comfortable for the wearer [121]. Generation 2 and

3 silicone hydrogel formulations incorporated siloxane macromers as oxygen transport enhancers, which led to greater formulation flexibility and control of the final contact lens material properties (e.g., Dk and modulus) [100,101,123].

Table 7. Silicone Hydrogel Contact Lens Technology.

Technology Generation	Proprietary Name	Material	Manufacturer	Launch Date	Water Content (%)	Dk	O_2 Transport Enhancer	Hydrophilic Component *	Refs.
1	PureVision	Balafilcon A	Bausch & Lomb	1998	36	91	TRIS, Difunctional siloxane	NVP amino acid monomer	
	Focus Night & Day	Lotrfilcon A	CIBA Vision	1999	24	140	TRIS, Difunctional fluorosiloxane	DMA	Nicolson [124]
	AirOptix	Lotrafilcon B	CIBA Vision	1999	33	110	TRIS, Difunctional fluorosiloxane	DMA	Nicolson [124]
2	Biofinity	Cofilcon A	Cooper vision	2007	48	128	Monofunctional siloxane macromer	VMA	Ueyama [125] Iwata [126]
	Acuvue OASYS	Senfilcon A	JJVC	2004	38	103	SiGMA monomer, monofunctional siloxane macromer (MCR-M11)	DMA, HEMA, PVP	Vanderlaan [101,123]
	PremuiO	Amofilcon A	Menicon	2007	40	129	Difunctional fluorosiloxane macromer	DMA	
	Acuvue Advance	Gayfilcon A	JJVC	2004	47	60	SiGMA monomer, monofunctional siloxane macromer (MCR-M11)	DMA, HEMA, PVP	Vanderlaan [101,123]
3	1-DAY Acuvue TruEYE	Narfilcon A	JJVC	2008	46	100	SiGMA monomer, monofunctional siloxane macromers (OH-mPDMS, MCR-M11)	DMA, HEMA, PVP	Vanderlaan [101,123]
	Clariti 1-DAY	Somofilcon	Cooper vision	2014	56	60	Monofunctional siloxane macromer, difunctional siloxane	HEMA	Broad [100]
	Avaira	Enfilcon	Cooper vision	2011	46	100	Difunctional fluorosiloxane	VMA	
	Ultra	Samfilcon A	Bausch & Lomb	2014	46	114	TRIS, Difunctional siloxane, monofunctional carbosiloxane macromer	Unknown	Awasti [55,127]
	DAILIES TOTAL1	Delfilcon A	Alcon	2013	33	140	Unknown	DMPC, DMA	
	MyDay	Stenfilcon A	Cooper vision	2013	54	80	Difunctional PEG-PDMS copolymer, monofunctional siloxane macromer	VMA	
	Acuvue Vita	Senfilcon C	JJVC	2016	41	103	SiGMA monomer, monofunctional siloxane macromer (MCR-M11)	DMA, HEMA, PVP	

* DMA—Dimethylacrylamide; DMPC—(1,2-dimyristoyl-sn-glycero-3-phosphocholine; HEMA—Hydroxyethylmethacrylate; NVP—N-Vinylpyrrolidone; PVP—Polyvinylpyrrolidone; SiGMA—3-(3-Methacryloxy-2-hydroxypropoxy)propylbis(trimethylsiloxy)methylsilane; TRIS—Methacryloxypropyltris(trimethylsiloxy)silane; VMA—N-Vinylmethylacetamide.

Hydrogels comprised of macromers with both polycarbosiloxane units and poly(trifluoropropylmethyl)siloxane units have been reported. The macromers are synthesized by substituting D_3 with other ring-strained monomers in a living polymerization process. Awasti describes a monomethacrylate functional polycarbosiloxane synthesized from a 2,2,5,5-tetramethyl-2,5-disila-1-oxacyclopentane monomer as resistant to hydrolysis and therefore suitable for use in silicone hydrogel soft contact lens design [55,128]. Monomethacrylate functional poly(trifluoropropylmethyl)siloxane macromers possess increased polarity, improving their miscibility with hydrophilic monomers and potentially reducing the non-specific binding of proteins on soft contact lens surfaces [59,70].

Low molecular weight symmetric siloxane macromers with hydrophilic termini have increased miscibility with hydrophilic monomers compared to asymmetric siloxane macromers, as the smaller siloxane block size of the symmetric architecture limits phase separated domain formation while still maintaining the oxygen transport-enhancing benefits of asymmetric analogs. The hydrophilic termini reported by Kimble include methoxypropyl and hydroxypropyl groups [59,60]. α-Methacrylate functional, ω-polyalkyleneoxide siloxane macromers have also been described; however, their block copolymer structure results in microphase separation, rendering this macromer structure unsuitable for the production of optically clear hydrogel films [129].

One approach for improving contact lens comfort is to utilize a plasma treatment to improve hydrogel wettability and then apply hydrophilic terpolymers derived from combinations of silicone macromers and hydrophilic monomers such as diethylazetidinium methacrylate ester chloride salt, thereby providing permanent wettability [130].

In a separate but related area, siloxane-based interocular lenses (IOL) provide new opportunities for macromers. Hydride terminated macromers are utilized to control the modulus of cured elastomers used in IOLs [131]. Interestingly, silicone macromers are broadly described as components in laser adjustable IOLs that enable a post-insertion change in dioptric power by altering the refraction of the lens material [132].

Independent of contact lens development, optically clear films were derived from copolymers of styrene, ethylene glycol dimethacrylate (EGDMA), and siloxane-urethane-methacrylate (SiUMA). The SiUMA monomer was synthesized from carbinol-terminated siloxane macromers reacted with isocyanatoethylmethacrylate [133].

2.2.2. Controlled Atmospheric Packaging—Modified Atmosphere Packaging

An appreciation of the role that gases play in maintaining the freshness of meats, fruits, and vegetables during transport has created a role for polymer films with controlled permeability. Optimal oxygen concentration is associated with the appearance of "red" meat and "green" vegetables. In the case of meat, deoxymyoglobin, which is purple, forms when metabolic, diffusion, and other processes deplete oxygen availability. Over time, however, oxymyoglobin is oxidized to amber-brown methemoglobin, which is associated with a lack of freshness [134]. Ethylene is associated with fruit ripening and abscission. Gas permeable films derived from acrylate terpolymers act as overlayers to porous structures, maintaining the oxygen transmission of these structures while providing a barrier to microbial infiltration. Unlike in the case of contact lenses, transparency is usually not required for packaging applications. The use of higher molecular weight monomethacrylate macromers in packaging is therefore acceptable, and has been reported [135]. Breathable films based on this technology have become components in commercial packaging applications offered by BreatheWay, as shown in Figure 5.

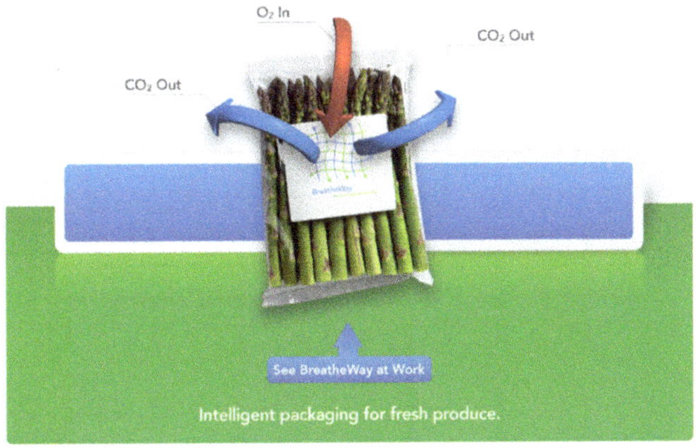

Figure 5. Breathable films and packaging (Reprinted courtesy of Breathway Corp).

2.3. Membranes

The utilization of polydimethylsiloxanes and silicon-rich homopolymers such as poly(trimethylsilylpropyne) [136] and poly(vinyltrimethylsilane) [137] for gas separation membranes has been established. The removal of volatile organic compounds from aqueous mixtures has been more effectively addressed by hybrid polymer systems in which there is formation of siloxane microphases in continuous organic phases. Reports on graft polymer and interpenetrating polymer networks were able to demonstrate selective pervaporation and removal of organics from aqueous streams [138,139]. The control of selectivity and permeation rates has been accomplished with copolymers derived from methacrylate functional macromers copolymerized with a variety of other methacrylate monomers. Uragami demonstrated that macromers with molecular weights of ~4000 Daltons formed copolymers with methyl methacrylate which, depending on comonomer content, allowed selective permeation of water or ethanol from ethanol–water solutions by varying comonomer contents [140]. In a series of reports, Urgami extended these systems to styrene copolymers and the incorporation of ionic liquids, enabling the selective pervaporation of volatile organics, including toluene and chlorinated organics, from aqueous streams [141–143].

2.4. Surface Properties/Modification

2.4.1. Dyes, Micelles and Particles for Advanced Printing, Reprographics, and Lithography

Methacrylate functional macromers have been used in both organic dye [89] and pigment-based printing applications [98,99,144]. While these technologies are quite different, the role of the silicone macromer has common features in each. In ink-jet applications, a bulk solvent-based polymerization with dye is accomplished, after which micelles are formed via the addition of water and simultaneous evaporation of solvent. In pigment-based applications, the polymer forms an encapsulant or binder. In both cases, the silicone macromer appears to serve as a hydrophobic outer surface of the micelle or pigment, thereby helping to control particle charging and, secondarily, contributing to the spread and adhesion of the micelles or particles on the substrate. Polymersomes with controlled architecture in both their overall dimensions and membrane thickness have been prepared from diblock polymers of cabinol-termined siloxane macromers and methyloxazolidine [82].

Interestingly, while the ability of styrene-PDMS deblock polymers generated via living AROP to form micelles was recognized during early development efforts [145], no recent commercial applications have been reported.

Colored fluids for electrowetting and electrofluidic applications have also been generated from highly polar dyes by reaction with aminopropyl-terminated siloxane macromers [80]. For example, the yellow dye 2-(4-carboxyphenylazo)acetoacetanalide reacted with a 1000 Mn amino-terminated macromer to form the product 4-(E)-(2,4-dioxopentan-3-yl)diazenyl-N-polydimethylsiloxane-benzamide, which was soluble in non-polar fluids including silicones. Radiation-curable films for adhesive and lithographic applications, in which aminopropyl-terminated siloxane macromers are acrylated, were reported by Leir [146]. Carbinol-terminated siloxane macromers were converted to phosphate-terminated macromers by Tao and then reacted with the surfaces of CdSe quantum dots before incorporation into bulk silicones to form electroluminescent transparent films, as shown in Figure 6 [85].

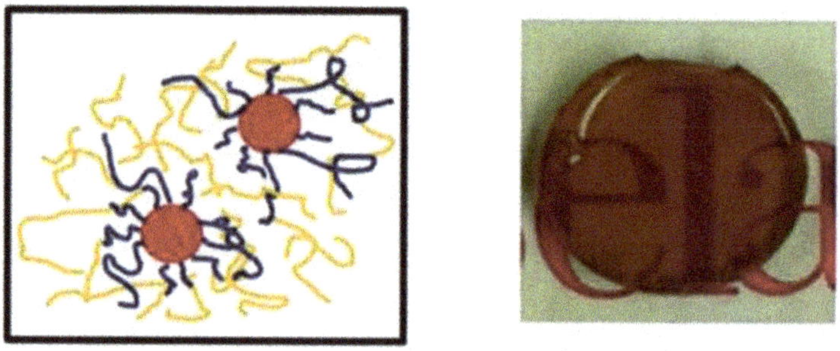

Figure 6. Electroluminescent transparent quantum dot composites (Tao) [85]. Left: depiction of macromers (blue) modifying a CdSe QD (red) in a silicone matrix; Right: specimen of composite. (Reprinted with permission from Ref. [85] Copyright 2013 the Royal Society of Chemistry).

2.4.2. Coatings Additives—Leveling Agents, Clean Surfaces and Release

Siloxane macromer block polymers are offered commercially as leveling agents to reduce waviness and orange peel in organic coatings. The effect occurs at low concentrations and is based on activity at the liquid–gas interface, in which these polymers are oriented due to limited incompatibility with the actual binder component of the coating system [147,148]. Their versatile chemistry and modular molecular structure make it possible to adjust the properties of these macromers for specific applications. For example, further improvements in levelling have been reported when a fluorinated oxetane is reacted with unsaturated termination in macromers [149].

High silicone content macromers can impart anti-graffiti properties to coated surfaces, though this often requires concentrations higher than those used in levelling applications. Macromers with a lower silicone content can reduce surface tension and improve substrate wetting without impairing recoatability. In the case of automotive coatings, other benefits include retaining the bonding characteristics of films and adhesives, while anti-blocking properties can also be achieved in decorative coatings [150,151].

One of the earliest applications of styryl and methacrylate functional macromers was for controlling release in adhesive tapes [54]. The investigators showed a correlation between molecular weight and release characteristics—e.g., in butyl methacrylate, acrylic acid, and macromer terpolymers—with macromers of low molecular weight proving ineffective but macromers with molecular weights $\geq$2000 providing control. The same general chemistry has been utilized more recently in dirt resistant coatings [93]. Release coatings, including ice-phobic coatings, have been generated from carbinol-terminated siloxane macromers by reaction with isocyanate and epoxy functional prepolymers to form amphiphilic, self-stratified thin films [152]. A reduction of marine biofouling was observed when aminopropyl-terminated macromers were incorporated into isophorone diisocyanate-derived urethanes, with low molecular weight macromers (1000 Daltons) providing more favorable results than high molecular weight macromers [81]. Methacrylate-terminated siloxane macromers with embedded hydrophilicity have also been used in this application [108]. Ultraphobic coatings—i.e., coatings exhibiting both superhydrophobic and oleophobic behavior—in which combinations of epoxy functional siloxane telechelics and macromers are reacted with linear and/or branched polyethylenimine (PEI), have been reported by Soane and Ready [94,107]. Hydride-terminated macromers have been used to modify vinyl POSS structures in slippage coatings for ice-phobic applications [91]

2.5. Coatings

2.5.1. Thin Film Silicone Coatings

The inclusion of silicone macromers into the formal polymeric structures of coatings, as distinguished from their use as additives, is a growing area of technology. These applications tend to be associated with release, lubricity, and mechanical properties related to direct physical interactions with humans. Indeed, the sensory appeal of coatings has always been an important driver of consumer applications in which positive tactile interaction is critical to acceptance [87], such as synthetic leather, textile finishes, and hair care.

Older approaches to urethane materials mainly use polydisperse telechelic carbinol-terminated siloxane polymers, in which the two identical functionalities on the termini serve to introduce siloxanes into urethanes as soft-blocks. AROP-derived siloxane macromers (oligomeric materials with functionality on one terminus) represent a newer approach in which the functionality on one terminus of the oligomer allows the formation of a brush polymer with siloxane segments as pendant, allowing the mechanical properties of the urethane backbone to be maintained (Scheme 9) [87]. These structures also demonstrate significantly greater wear resistance as well as lower friction and release properties compared to telechelic controls [153], in which siloxane is incorporated into the urethane backbone as shown in Scheme 10. Table 8 compares tribological and contact angle properties of urethane in which siloxane has been introduced as an unreacted fluid, a soft segment, and a pendant, respectively.

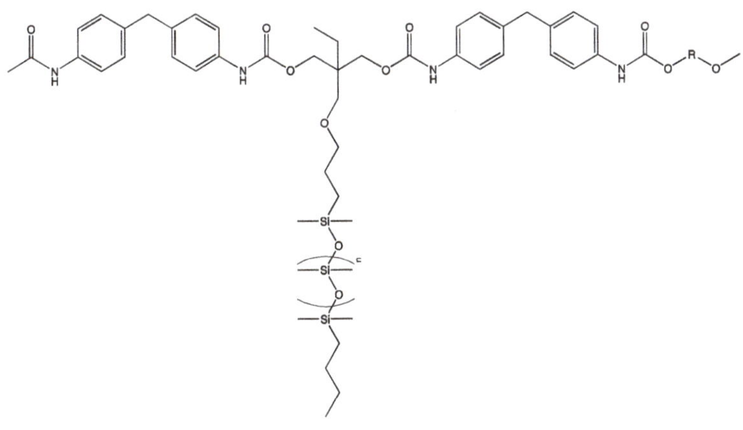

Scheme 9. Urethane Polymer with Pendant Siloxane.

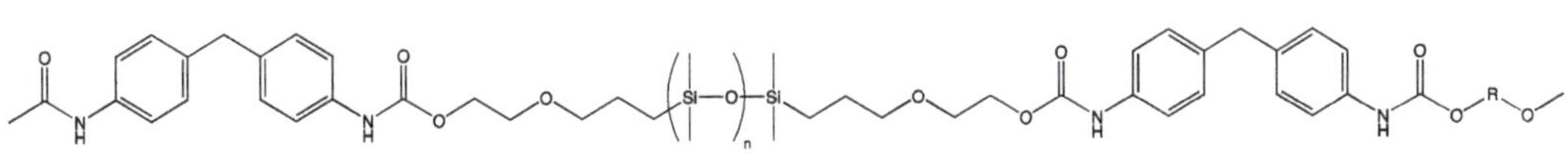

Scheme 10. Urethane-Siloxane Soft Segment Block Polymer.

In similar chemistry, self-lubricating cannulae for medical applications have been fabricated from thermoplastic urethanes derived from mono dicarbinol macromers [86].

Table 8. Comparing Contact Angle, Abrasion, and Friction in an IPDI-based Urethane.

	Control	A	B	C
Description		**Blend**	**Block Polymer**	**Pendant/Side-Chain**
Silicone Component		**Silicone Fluid**	**Telechelic Carbinol**	**Dicarbinol Terminated Asymmetric Macromer**
Abrasion Test Failure, cycles	<100	<100	<100	4200
Coefficient of Friction	0.61	0.12	0.74	0.16
Contact Angle, water before abrasion	56°	83°	72°	82°
Contact Angle, water after abrasion	55°	55°	72°	80°

Control (no siloxane); (A) introduction of 50 cSt dimethylsiloxane fluid as a non-reactive additive; (B) telechelic siloxane MW 5000 (DMS-C21) added to form a soft-block copolymer; (C) macromer MW 5000 (MCR-C62) forming a brush polymer with siloxane pendant.

2.5.2. Thick Film Silicone Coatings

Depending on their formulation, silicone coatings elicit a wide range of tactile responses. Coatings that contain low molecular weight, particularly volatile species are associated with a "silky" feel and exhibit slip. On the other hand, silicone coatings that have been depleted of low molecular weight species are associated with a "tacky" feel. Common to both these experiences is the extreme hydrophobicity of the polymer. When silicone coatings are free of low molecular-weight species, they exhibit high coefficients of friction and, due to their relatively poor mechanical properties, failure by abrasive and adhesive spalling during continuous tactile interaction. Well-defined silicones with a central vinyl functionality and discrete PEG_2 (MCS-VX15), PEG_3 (MCS-VX16—Scheme 11), or tetrahydrofurfuryl (MCS-VF14—Scheme 12) pendant end-groups can be used as comonomers in addition-cure, platinum-catalyzed 2-part silicone elastomer formulations in order to introduce hydrophilicity [61]. In such formulations, the surface tribological properties are modified by introducing a hydrodynamic lubricating layer of adsorbed water. The modified silicone elastomers retain optical clarity and mechanical performance characteristic of this class of material with up to 15 wt.% comonomer in the 2-part formulation. Contact angle measurements of deionized water on the silicone elastomer surface showed improved wettability with comonomer content: at ~3 wt.% comonomer, the elastomer surface shifts from hydrophobic (contact angle ~120°C) to hydrophilic (contact angle < 90°C). Coefficient of friction measurements for the modified silicone elastomers demonstrate increased surface lubricity with comonomer loadings (Figure 7) [112].

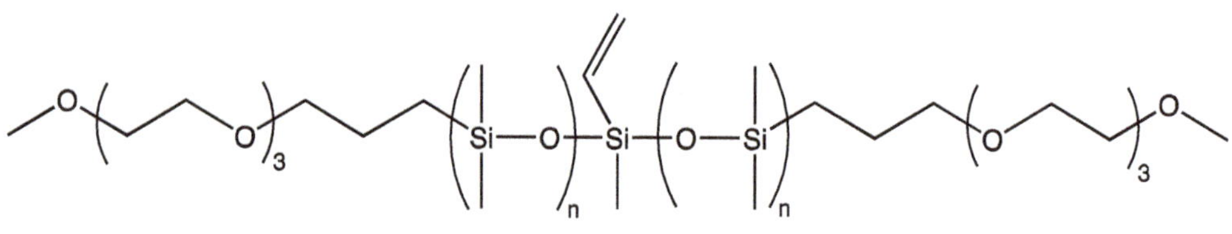

MCS-VX16

Scheme 11. Vinyl functional Siloxane with PEG Endgroups-Reactive in Two-component Pt-Cure Silicone RTVs.

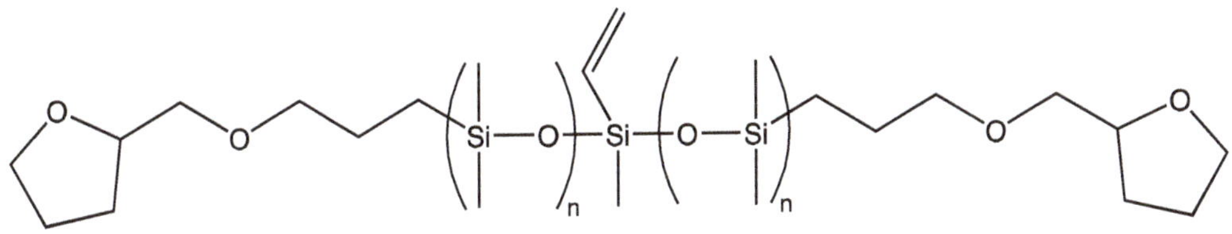

MCS-VF14

Scheme 12. Vinyl functional Siloxane with Tetrahydrofurfuryloxy Endgroups-Reactive in Two-component Pt-Cure Silicone RTVs.

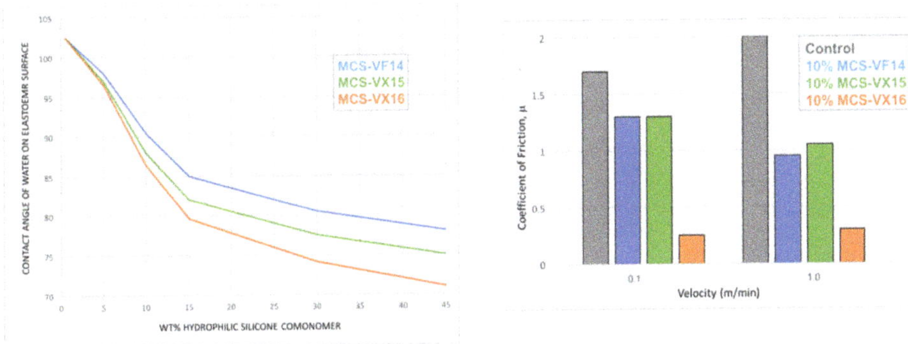

(a) (b)

Figure 7. Modification of silicone elastomers with symmetric silicone macromers. (**a**) Effect of comonomer on contact angle and hydrophobicity of silicone elastomer; (**b**) coefficient of friction of silicone elastomers modified with symmetric silicone macromers (Determined on AR-G2 Rheometer: aqueous, 37 °C, normal force 1N, velocity 1.0 rad/s (1.5 mm/s) and 1.0 rad/s (15 mm/s).)

Separately, in the field of dielectric elastomer actuators, monovinyl-terminated PDMS macromers have been used to selectively adjust the network behavior of silicone films between compliant electrodes [111].

2.6. Cosmetics and Hair Care

Hair-care formulations, including shampoos and conditioners with the ability to withstand multiple washings, require good film-forming properties with strong adhesion to the hair cuticle, but must simultaneously offer lubricity in order to provide combability. Copolymers of dimethylaminoethylmethacrylate and low molecular weight methacrylate functional silicone macromers (MW 2000) [89,90] have been used directly or in combination with other polymers [154] to provide increased lubricity and combability of hair. Dispersions of macromer-derived terpolymer particles have been reported in hydrocarbon vehicles such as isododecane, yielding film-forming compositions that are useful in eyeliners and mascaras [97]. Hydride-terminated macromers have been reacted with unsaturated terpenes and cannabidiol (CBD) (Scheme 13 to form emollient compounds with solubility in the siloxane vehicles preferred for skin care [155,156].

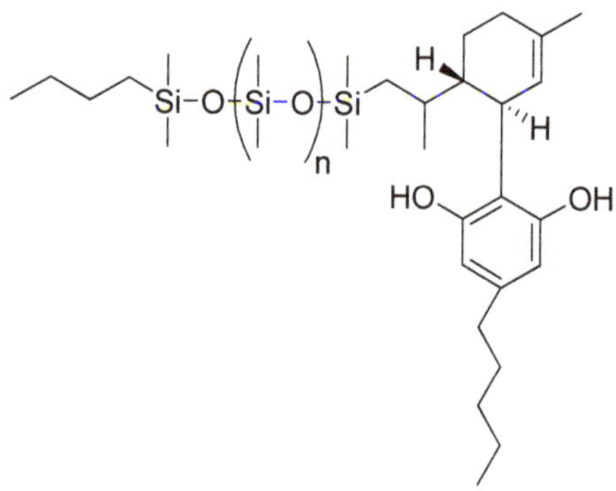

Scheme 13. CBD terminated PDMS macromer.

2.7. *Magneto-Rheological Fluids*

Well-defined biocompatible magnetic nanoparticles are of interest as materials for biomedical applications including magnetic field-directed drug delivery, biomolecule separations, and assay devices. Superparamagnetic iron oxide (Fe_3O_4) nanoparticles (SPION) sterically stabilized with PDMS macromers synthesized by Living AROP produced homogeneous hydrophobic ferrofluids that are stable against precipitation [157,158]. PDMS macromers with a tricarboxylate endgroup capable of binding to the surface of magnetite nanoparticles were synthesized by first making a trivinyl-terminated PDMS via Living AROP, followed by a thiol-ene reaction between the vinyl silane groups and mercaptoacetic acid, as depicted in Figure 8 [159,160]. Molecular effects of the PDMS tail on the stability of the PDMS-magnetite complexes were studied. Magnetic separation methods were developed to narrow the particle size distribution of the magnetite nanoparticles using tricarboxylate PDMS stabilizer while controlling the PDMS surface concentration [161]. In other studies, a monocarboxydecyl-terminated PDMS macromer (MCR-B12) was used to stabilize magnetite nanoparticles; however, the resulting ferrofluids had issues with stability and sedimentation, likely due to the lower number of carboxylate binding groups per macromer chain [88]. The magnetophoretic mobility of the magnetite-PDMS fluids was then studied in different magnetic field conditions (magnetic fields and field gradients), with the results demonstrating that the shape and speed of these droplets in viscous media can be independently manipulated [160,162].

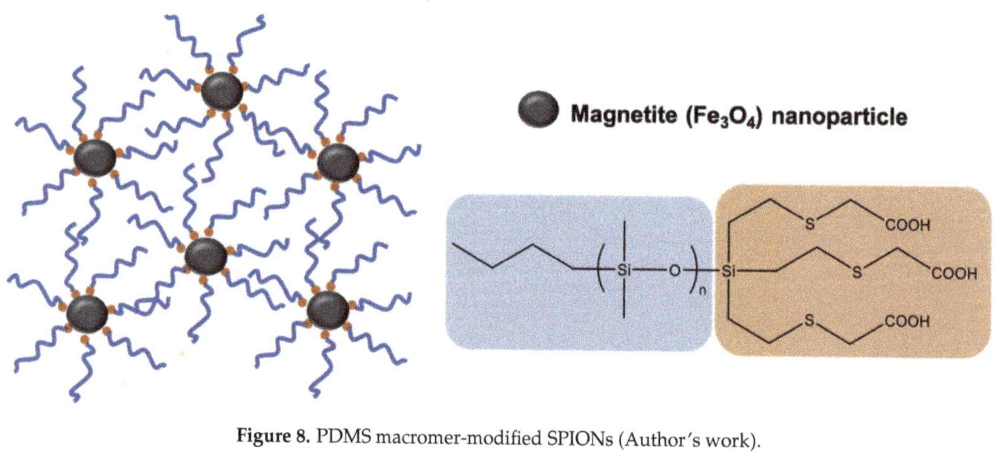

Figure 8. PDMS macromer-modified SPIONs (Author's work).

These magnetite-polydimethylsiloxane ferrofluids were proposed by Wilson as a material that could aid in the treatment of retinal detachment disorder [159]. The proposed treatment entailed inserting a pre-aligned magnet into the conjunctiva and then injecting the PDMS-based ferrofluid into the vitreous humor. As shown below in Figure 9, the ferrofluid would close the tear as it moved toward the permanent magnetite, allowing the surgeon to repair the tear.

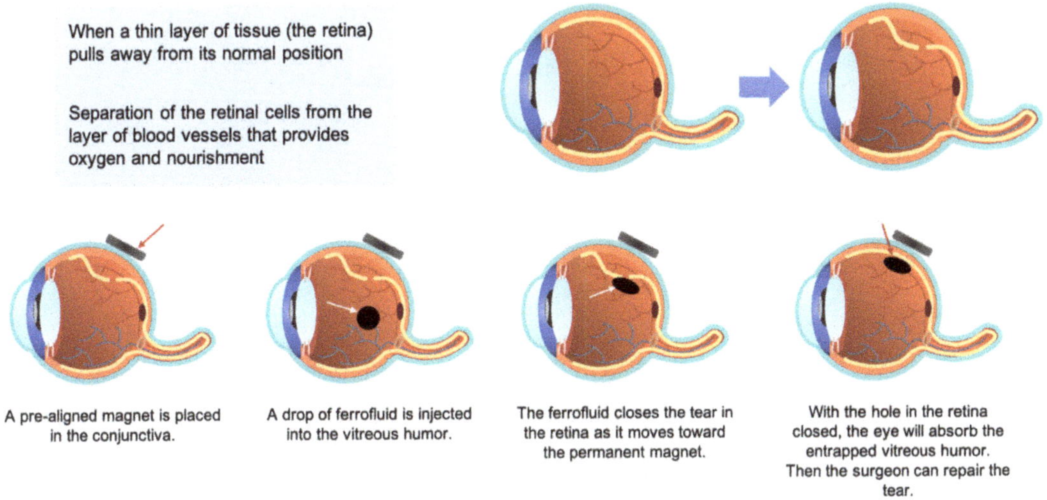

Figure 9. Retinal repair utilizing macromer-enabled magnetorheological fluids.

2.8. Bulk Macroscale Materials

Ultra-High Elongation Elastomer

An ultra-high elongation silicone elastomer has been prepared from a heterobifunctional silicone macromer compounded with reinforcing agent, achieving elongations nearing 5000%, nearly four times greater than conventional silicone elastomers. The cure mechanism of this elastomer is a step-growth polymerization of an α-vinyl-ω-hydride-terminated silicone macromer [62,113,114] via intermolecular hydrosilylation reaction, which yields a linear polymer of exceptionally high molecular weight with no apparent covalent crosslinking (Scheme 14) [73,113].

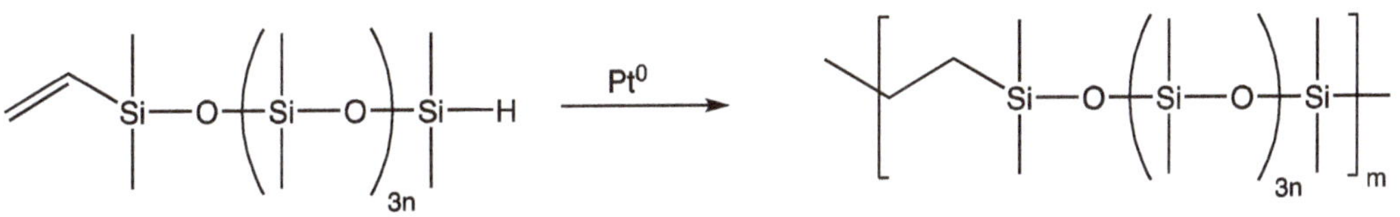

Scheme 14. Step-Growth Polymerization of Heterobifunctional α-Vinyl-ω-Hydride-terminated Silicone Macromer.

Atomistic modeling (Figure 10) of the cured silicone macromer shows the probability of knotting within a 50,000 Da segment [115], which correlates to an experimental value of critical molecular weight (M_c) for entanglement of ~42,000 Da [163]. The stress-strain curve is remarkably different from those of crosslinked silicone elastomers, as shown in Figure 11.

Figure 10. Atomistic model of a 50,000 Da segment of ultra-high elongation elastomer suggesting 1–2 knots. (Author's work).

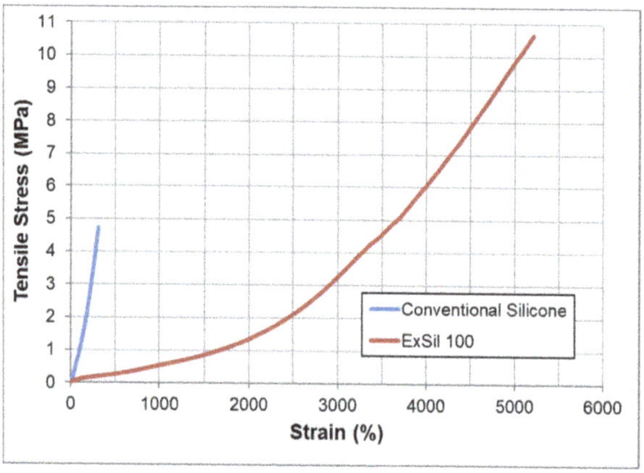

Figure 11. Mechanical properties of ultra-high elongation elastomer (ExSil 100) compared to conventional silicone elastomer.

Ultra-high elongation elastomers exhibit other behaviors not shared by conventional silicone elastomers [115]: e.g., they undergo extreme multi-axial distortions and return close to their original shape, and display remarkable tear propagation mechanisms, meaning that tear failure occurs at drastically greater elongations; recovery from minor penetration is improved below the failure limits and, finally, pseudo-self-healing behavior is also displayed, as shown in Figure 12 [116].

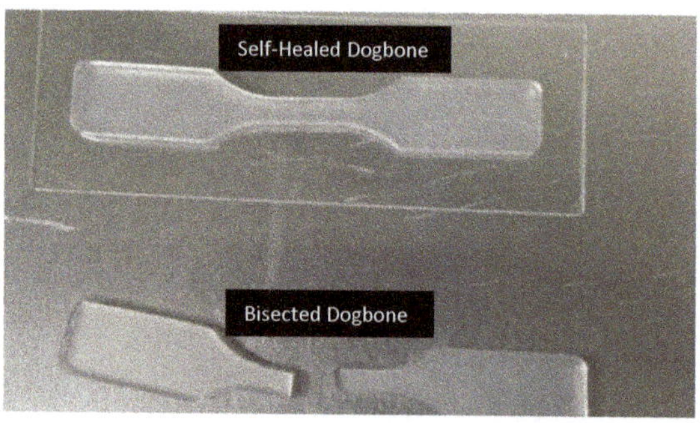

Figure 12. Pseudo-self-healing demonstration of bisected specimen of ExSil® (Author's work).

2.9. Nanoscale Morphology

2.9.1. Photoresist and Contact Printing

Templated self-assembly of a cylinder-forming poly(styrene-b-dimethylsiloxane) (PS−PDMS) diblock copolymer was first described by Saam and Fearon [51]—followed by others [164–167]—and has been investigated for nanolithography applications [44,145,168,169]. The general structure is depicted in Scheme 15. The general structure (PS−PDMS) diblock copolymer is depicted in structure 11. These copolymers undergo microphase segregation above their T_g, and the large X-polymer-solvent interaction parameter of the blocks is advantageous for achieving long-range ordering as well as for minimizing defect densities. Furthermore, the high Si content in PDMS leads to a robust oxide etch mask after two-step reactive ion etching (RIE) [170], as exemplified in Figure 13.

To address the critical needs of nano-dimensional photoresists, materials that possess dual surface properties are required. When cast or hot-pressed on a high-surface-energy substrate such as silicon, glass, or aluminum, the copolymer film forms both a lower-surface-energy component (PDMS)-enriched air/polymer interface and a higher-surface-energy component (organic block)-dominated polymer/substrate interface [171–173].

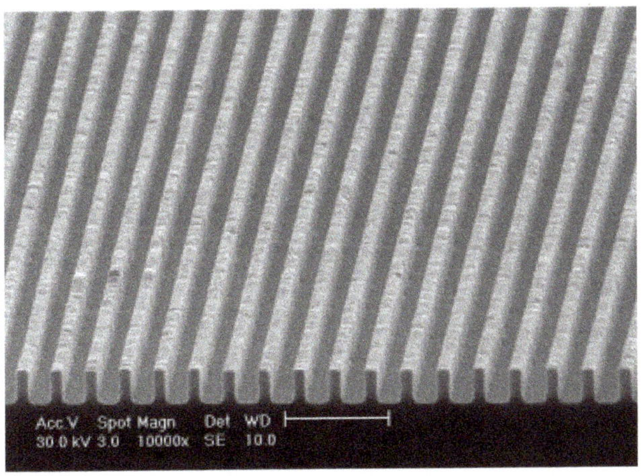

Figure 13. PS-b-PDMS after a single imprint cycle and annealing provides 60 nm features (Reprinted with permission from Ref. [173] Copyright 2007 Wiley).

Interface effects complicate the use of siloxane A-B block copolymers due to the low surface energy and wetting characteristics of the siloxane block.

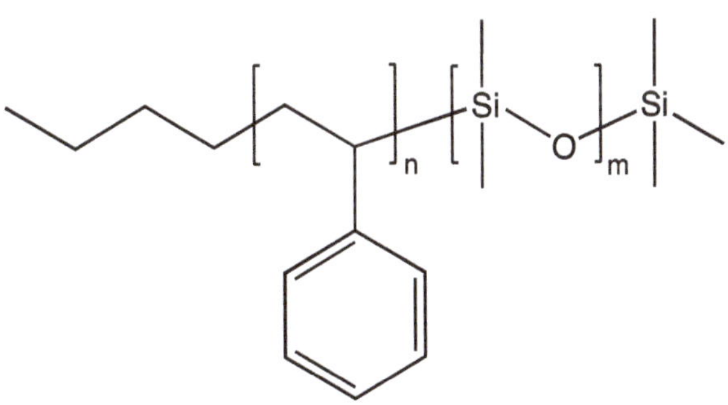

Scheme 15. Polystyrene-Polydimethylsiloxane Block Copolymer.

Silylated styrene block polymers that employ conventional radical polymerization rather than AROP have been proposed [174,175], and azidopropyl functional silicone macromers have been separately "clicked" with alkynyl-terminated ATRP-generated macromers to form PDMS-b-PMMA, producing sub-10 nm structures [174,176]. A second potential solution is star block polymers generated by AROP, which provide a mechanism through which the non-siloxane block can dominate interfacial behavior [52]. The reviewers note that similar morphologies have also been reported for polybutadiene-polydiethylsiloxane block copolymers (PBD-b-PDES) [19]; these copolymers possess less differentiation in surface energy, potentially mitigating issues with the styrenic and methacrylate systems, but have not been evaluated in lithographic systems. Similarly noteworthy in this context, while the synthesis of diblock polymers typically starts with an organic macroanion and then proceeds to a siloxane polymerization, the potential of the reverse process, in which a siloxane macromer starts and proceeds to an acrylate polymerization, was demonstrated using ATRP [177]. Other organic block copolymer polymerization examples initiated by carbinol-terminated macromers include caprolactone and trimethylene carbonate blocks [83,178]. This approach clearly expands the options for generating polymers with varying self-assembly structures.

The use of polystyrene macroanions was further elaborated by Bellas to form triblock and microarm polymers [44], and by Shefelbine to form continuous core–shell gyroid morphologies [179]. This work led to the observation of periodic double gyroid (DG) behavior, as well as a series of publications displaying an appreciation of the potential for extended novel 3D structures (Figure 14) [180–182]. As suggested by the DG topological visualizations of Thomas [183], potential structural, dielectric, charge transport, and mass transport material behavior can be controlled by the direct use of DG block polymers structures or by the removal, conversion, and/or infiltration of a DG microphase.

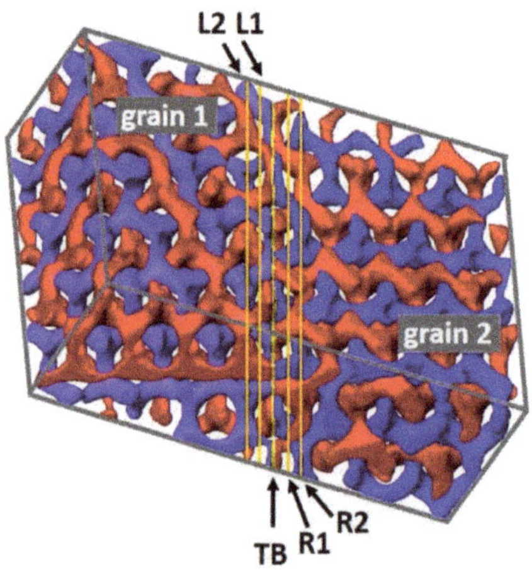

Figure 14. Reconstruction of a volume containing a Double Gyroid of PS-b-PDMS (Reprinted with permission from Ref. [183] Copyright 2021 National Academy of Sciences).

2.9.2. Biomimetic Polymers and Bottle-Brush (BB) Architecture

Recently, grafting-through polymerization and surface-initiated polymerization have led to bottle-brush polymers and particle-brush materials that have shown potential in the fabrication of biomimetic materials. These materials can be broadly considered filamentous structures that exhibit non-linear behavior under deformation, greater relaxation times, and the potential for complex non-covalent interactions leading to the formation of supermolecular structures. Siloxane macromers appear to be of particular interest in creating these structures due to both the palette of functionality and the intrinsic flexibility of the siloxane structures, which allows the length-scale of the filaments to extend beyond the primary polymer backbone or nano-feature without imposing a "hard" structural domain. Siloxane macromers allow entry to classes of materials that possess an unusually low modulus while maintaining mechanical failure properties consistent with the main polymer backbone for both methacrylate [96] and norbornene [84] functional siloxane macromers.

Separately, this recognition revealed such macromers' potential for generating biomimetic gel structures [95]. The complexity of the potential range and behavioral characteristics associated with bottle-brush polymers in terms of macromer molecular weight, graft density, and final molecular weight is readily apparent. Their tunable physical properties in both crosslinked and uncrosslinked states—based on the Dp of the main chain and grafting density of methacryloxypropyl-terminated polydimethysiloxane (Dp = 10) prepared by Grafting Through Atom Transfer Radical Polymerization—have been investigated in this context [103]. The dynamics of deformation have also been modeled [104,109,110]. A striking extension of this approach can be found in the "chameleon-like" color changes demonstrated in these systems—in which vibrant color, extreme softness, and intense strain stiffening on par with that of skin tissue have been observed as a consequence of placing a heterogeneous polymeric system under varying degrees of strain (Figure 15) [106].

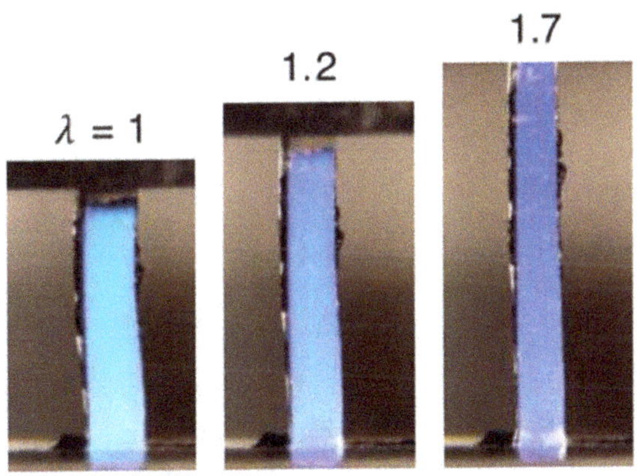

Figure 15. Color alteration from turquoise to dark blue during uniaxial stretching. (Reprinted with permission from Ref. [106] Copyright 2018 AAAS).

2.9.3. Liquid Crystal Siloxanes

Most polymers that exhibit liquid crystalline behavior have anisotropic side groups. Polydiethylsiloxane, which is distinct from both lost LCPs and its simpler homolog, polydimethylsiloxane, demonstrates mesophasic liquid crystalline behavior at ambient temperatures, as first reported by Beatty [184]. This behavior extends to poly(di-n-alkylsiloxane)s with side chains no longer than seven carbons that are also able to form a columnar mesophase. Such polymers are positionally and orientationally ordered in a two-dimensional hexagonal lattice, but without positional order along the chain. While early work was conducted with polymers of high polydispersity (>2.0), the desire to optimize liquid crystal behavior by controlling the PDI has led to the use of living AROP conditions for the polymerization of polydiethylsiloxane (PDES)—as first reported by Molenberg, who utilized lithium sec-butyldiethylsilanolate as the initiator [41,43]. The phase behavior of diethylsiloxane is shown in Figure 16.

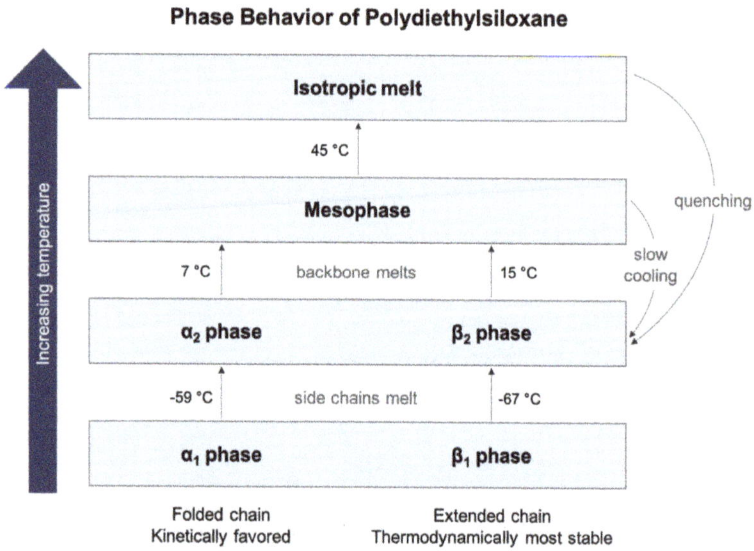

Figure 16. Liquid Crystal-Phase Behavior of Polydiethylsiloxane.

There are few reports of living AROP with higher dialkylsiloxanes, presumably because of either ineffective termination or these systems' sluggish kinetics.

Molenberg also reported on elastomeric block polymers of butadiene and diethylsiloxane that exhibited mesophase formation under tensile stress [19]. Later work with styrene-diethylsiloxane diblock polymers showed periodic nanoscale lamellar structures with compositions possessing greater than 20% styrene content, as shown in Figure 17 [42].

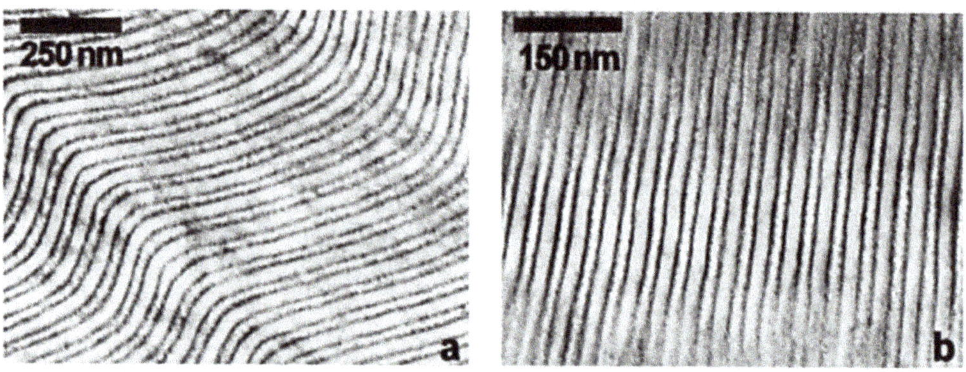

Figure 17. Bulk structures of PolyStyrene-Polydiethylsiloxane block copolymers. Bulk structures of (**a**) PS-*b*-PDES containing 37 wt% PS and (**b**) PS-*b*-PDES containing 43 wt% PS (Reprinted with permission from Ref. [19] Copyright 1998 Wiley).

The vast majority of liquid crystal polysiloxanes based on side chain substitution that have recently been reviewed are polydisperse in nature [185]. Hepenius, however, utilized AROP to form vinylmethylsiloxane copolymers from vinylpentamethylcyclotrisiloxane and then functionalized the positions with mesogenic groups: for example, by reacting 4-cyano-4′-(ω-alkenoxy)biphenyl with an excess of tetramethyldisiloxane and then reacting with the remaining Si-H group (Scheme 16) [46].

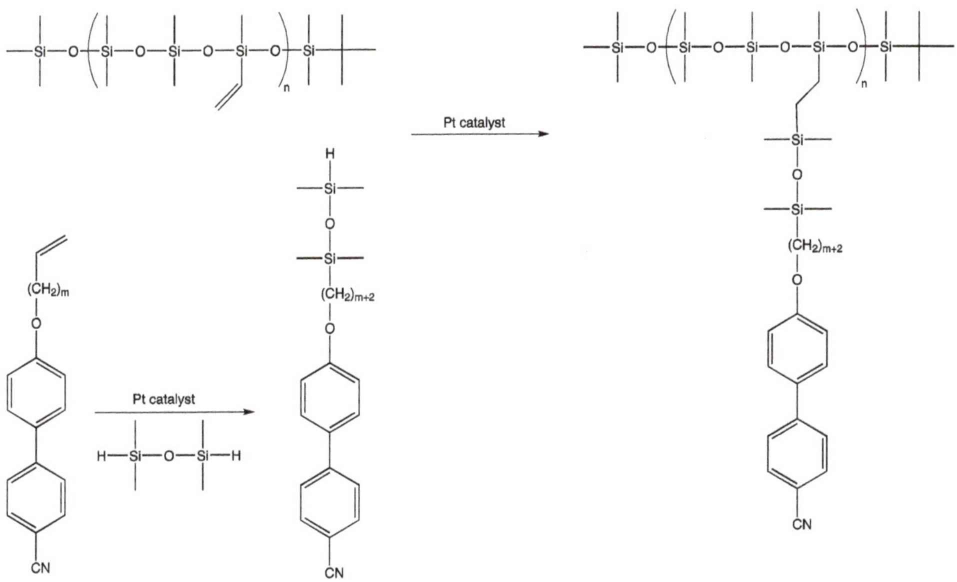

Scheme 16. Derivatization of AROP Siloxanes with mesogenic pendant groups.

3. Conclusions

The accelerating interest in living AROP-derived siloxanes has been spurred by the increasing drive to provide soft matter structures associated with tactile interaction and biocompatibility, particularly in the areas of contact lenses and biomimetic structures, and by these materials' ability enable control of nano-dimensional physical properties associated with "smart" particles and morphologies associated with self-assembly. Silicone macromers have had great impact in practical applications, primarily due to their ability to cross-over from the paradigm of inorganic siloxane chemistry to the greater paradigm of organic polymerization. Further, the elaboration of simple macromers into heterobifunctional monomers has created opportunities for polymers with new bulk phase properties, including ultra-high elongation and self-healing materials.

Funding: This research received no external funding.

Institutional Review Board Statement: Not applicable.

Informed Consent Statement: Not applicable.

Data Availability Statement: Not applicable.

Conflicts of Interest: Not applicable.

References

1. Chojnowski, J. Kinetically Controlled Siloxane Ring-Opening Polymerization. *J. Inorg. Organomet. Polym.* **1991**, *1*, 299–323. [CrossRef]
2. Johannson, O. Method of Polymerizing Cyclic Diorganosiloxanes. U.S. Patent 3,002,951, 3 October 1961.
3. McVannel, D. Method of Polymerizing Cylcotrisiloxanes II. U.S. Patent 3,294,740, 27 December 1966.
4. Frye, C.L.; Salinger, R.M.; Fearon, G.; Klosowski, J.M.; Deyoung, T. Reactions of Organolithium Reagents with Siloxane Substrates. *J. Org. Chem.* **1970**, *35*, 1308–1314. [CrossRef]
5. Lee, C.; Frye, C.; Johansson, O. Selective Polymerization of Reactive Cyclosiloxanes to Give Non-Equilibrium Molecular Weight Distributions. Monodisperse Siloxane Polymers. *Polym. Prepr.* **1969**, *10*, 1361.
6. Fessler, W.A.; Juliano, P.C. Reactivity of Solvated Lithium N-Butyldimethylsilanolate with Organosiloxane Substrates. *Ind. Eng. Chem. Prod. Res. Dev.* **1972**, *11*, 407–410. [CrossRef]

7. Saam, J.C.; Gordon, D.J.; Lindsey, S. Block Copolymers of Polydimethylsiloxane and Polystyrene. *Macromolecules* **1970**, *3*, 1–4. [CrossRef]
8. Saam, J.C.; Ward, A.H. Method for Preparing Improved Polyalphamethylstyrene-Polydiorganosiloxane Copolymers and Products Thereof. U.S. Patent 4,107,227, 15 August 1978.
9. Kawakami, Y.; Miki, Y.; Tsuda, T.; Murthy, R.; Yamashita, Y. Silicone Macromers for Graft Polymer Synthesis. *Polym. J.* **1982**, *14*, 913–917. [CrossRef]
10. Anderson, R.; Arkles, B.C.; Larson, G.L. *Silicon Compounds: Register and Review*, 4th ed.; Petrarch Systems: Bristol, PA, USA, 1987.
11. Nakano, T.; Tsuchiya, K.; Yoshimatsu, S.; Fuchigami, T. Process for Producing Diorganopolysiloxane. U.S. Patent 5,672,671, 30 September 1997.
12. Chojnowski, J. Polymerization. In *Siloxane Polymers*; Clarson, S.J., Semlyen, J.A., Eds.; Prentice Hall: Englewood Cliffs, NJ, USA, 1993; pp. 1–62.
13. Belorgey, G.; Sauvetin, G. Organosilicon Block and Graft Copolymers. In *Silicon Containing Polymers*; Jones, R.G., Ed.; Springer: Dordrecht, The Netherlands, 2000; pp. 43–78.
14. Lohmeijer, B.G.G.; Dubois, G.; Leibfart, F.; Pratt, R.C.; Nederberg, F.; Nelson, A.; Waymouth, R.M.; Wade, C.; Hedrick, J.L. Organocatalytic Living Ring-Opening Polymerization of Cyclic Carbosiloxanes. *Org. Lett.* **2006**, *8*, 4683–4686. [CrossRef]
15. Fuchise, K.; Kobayashi, T.; Sato, K.; Igarashi, M. Organocatalytic Ring-Opening Polymerization of Cyclotrisiloxanes Using Silanols as Initiators for the Precise Synthesis of Asymmetric Linear Polysiloxanes. *Polym. Chem.* **2020**, *11*, 7625–7636. [CrossRef]
16. Hunter, M.J.; Hyde, J.F.; Warrick, E.L.; Fletcher, H.J. Organo-Silicon Polymers. The Cyclic Dimethyl Siloxanes. *J. Am. Chem. Soc.* **1946**, *68*, 667–672. [CrossRef]
17. Piccoli, W.A.; Haberland, G.G.; Merker, R.L. Highly Strained Cyclic Paraffin-Siloxanes. *J. Am. Chem. Soc.* **1960**, *82*, 1883–1885. [CrossRef]
18. Pierce, O.R. New Cyclotrisiloxanes. U.S. Patent 2,979,519, 11 April 1961.
19. Molenberg, A.; Möller, M.; Soden, W.V. Block Copolymers and Networks from Polybutadiene and Polydiethylsiloxane. *Acta Polym.* **1998**, *49*, 45–51. [CrossRef]
20. Crivello, J.V.; Lee, J.L. Synthesis of Cyclic Siloxanes by the Thermal Depolymerization of Linear Poly(Siloxanes). *Chem. Mater.* **1989**, *1*, 445–451. [CrossRef]
21. Paulasaari, J.K.; Weber, W.P. Preparation of Highly Regular Poly(l-Hydrido-l,3,3,5,5-Pentamethyltrisiloxane) and Its Chemical Modification by Hydrosilylation. *Macromolecules* **1999**, *32*, 6574–6577. [CrossRef]
22. Goff, J.; Arkles, B. Hydrodosiloxanyl Substituted Cyclotrisiloxanes, Preparation Method Thereof, and Polymeric Products Derived Therefrom. U.S. Patent 10,875,968, 29 December, 2020.
23. Rózga-Wijas, K.; Chojnowski, J.; Zundel, T.; Boileau, S. Controlled Synthesis of Siloxane Copolymers Having an Organosulfur Group by Polymerization of Cyclotrisiloxanes with Mixed Units. *Macromolecules* **1996**, *29*, 2711–2720. [CrossRef]
24. Out, G.; Klok, H.-A.; Möller, M. Hexa-n-Alkylcyclotrisiloxanes-Synthesis, Melting Behaviour and Polymerization Gerhardus. *Macromol. Chem. Phys.* **1995**, *196*, 195–210. [CrossRef]
25. Fortuniak, W.; Chojnowski, J.; Sauvet, G. Controlled Synthesis of Siloxane Polymers and Siloxane-Siloxane Block Copolymers with 3-Chloropropyl Groups Pendant to the Siloxane Chain. *Macromol. Chem. Phys.* **2001**, *202*, 2306–2313. [CrossRef]
26. Lee, M.K.; Meier, D.J. Synthesis and Properties of Diarylsiloxane and (Aryl/Methyl)Siloxane Polymers: 3. New Aryl Substituents. *Polymer* **1994**, *35*, 4197–4202. [CrossRef]
27. Gädda, T.M.; Weber, W.P. Ring-Opening and Polymerization of 1-[2′-(Heptaphenylcyclotetrasiloxanyl)Ethyl]-1,3,3,5, 5-Pentamethylcyclotrisiloxane. *J. Polym. Sci. Part A Polym. Chem.* **2006**, *44*, 137–146. [CrossRef]
28. Oba, T.; Naoki, O. Acryloxypropyl-Containing Cyclotrisiloxane, Production of the Polymer Composition and Polymer Composition. Jap. Patent 2856377B2, 8 October 1993.
29. Minoru, I.; Norio, S.; Masaharu, T. Organopolysiloxane Having Cyclic Ether Group. Jap. Patent 1077346A, 24 March 1998.
30. Cai, G.P.; Weber, W.P. Synthesis of Polymethyl(Trimethylsiloxy)Siloxane by Anionic Ring-Opening Polymerization of 1,3,5-Trimethyl-1,3,5-Tris(Trimethylsiloxy)Cyclotrisiloxane. *Macromolecules* **2000**, *33*, 6310–6314. [CrossRef]
31. Suryanarayanan, B.; Peace, B.W.; Mayhan, K.G. Anionic Polymerization of a Series of Five-Membered Cyclocarbosiloxanes. *J. Polym. Sci. Part A 1 Polym. Chem.* **1974**, *12*, 1109–1123. [CrossRef]
32. Bostick, E. Ordered Organopolysiloxanes. U.S. Patent 3,337,497, 22 August 1967.
33. Pike, R. Process for Producing Cyclic Diorganosiloxanes from Diorganodihalosilanes and Alkali and Alkali Earth Metal Oxides and Carbonates. U.S. Patent 3,110,720, 12 November 1963.
34. Dobay, D. Process for Preparing Hexaethylcyclotrisiloxane. U.S. Patent 2,769,830, 6 November 1956.
35. Goff, J.; Arkles, B. Alkyl Ether Substituted Cyclotrisiloxanes and Preparation Method Thereof. U.S. Patent 10,669,294, 2 June 2020.
36. Chojnowski, J.; Cypryk, M.; Fortuniak, W.; Rozga-Wijas, K.; Scibiorek, M. Controlled Synthesis of Vinylmethylsiloxane-Dimethylsiloxane Gradient, Block and Alternate Copolymers by Anionic ROP of Cyclotrisiloxanes. *Polymer* **2002**, *43*, 1993–2001. [CrossRef]

37. Cypryk, M.; Delczyk, B.; Juhari, A.; Koynov, K. Controlled Synthesis of Trifluoropropylmethylsiloxane–Dimethylsiloxane Gradient Copolymers by Anionic ROP of Cyclotrisiloxanes. *J. Polym. Sci. Part A Polym. Chem.* **2009**, *47*, 1204–1216. [CrossRef]
38. Maschke, U.; Wagner, T. Initiation by a Monofunctional Lithium Siloxanolate. *Makromol. Chem.* **1992**, *193*, 2453–2466. [CrossRef]
39. Razzano, J.S.; Simpson, V.G. Fluorosilicone Copolymers and Process for the Preparation Thereof. U.S. Patent 3,974,120, 10 August 1976.
40. Hawkridge, A.M.; Gardella, J.A. Evaluation of Matrix-Assisted Laser Desorption Ionization Mass Spectrometry for Studying the Sec-Butyllithium and n-Butyllithium Initiated Ring-Opening Polymerization of Hexamethylcyclotrisiloxane (D3). *Am. Soc. Mass Spectrom.* **2003**, *14*, 95–101. [CrossRef]
41. Bellas, V.; Iatrou, H.; Hadjichristidis, N. Controlled Anionic Polymerization of Hexamethylcyclotrisiloxane. Model Linear and Miktoarm Star Co- and Terpolymers of Dimethylsiloxane with Styrene and Isoprene. *Macromolecules* **2000**, *33*, 6993–6997. [CrossRef]
42. Ninago, M.D.; Satti, A.J.; Ressia, J.A.; Ciolino, A.E.; Villar, M.A.; Valles, E.M. Controlled Synthesis of Poly(Dimethylsiloxane) Homopolymers Using High-Vacuum Anionic Polymerization Techniques. *J. Polym. Sci. Part A Polym. Chem.* **2009**, *47*, 4774–4783. [CrossRef]
43. Molenberg, A.; Siffrin, S.; Möller, M.; Boileau, S.; Teyssié, D. Well Defined Columnar Liquid Crystalline Polydiethylsiloxane. *Macromol. Symp.* **1996**, *102*, 199–207. [CrossRef]
44. Molenberg, A.; Möller, M. Polymerization of Cyclotrisiloxanes by Organolithium Compounds and P2-Et Base. *Macromol. Chem. Phys.* **1997**, *198*, 717–726. [CrossRef]
45. Molenberg, A.; Klok, H.A.; Möller, M.; Boileau, S.; Teyssié, D. Controlled Polymerization of Hexa-n-Alkylcyclotrisiloxanes with Long Alkyl Groups. *Macromolecules* **1997**, *30*, 792–794. [CrossRef]
46. Hempenius, M.A.; Lammertink, R.G.H.; Vancso, G.J. Well-Defined Side-Chain Liquid-Crystalline Polysiloxanes. *Macromol. Rapid Commun.* **1996**, *17*, 299–303. [CrossRef]
47. Sieburth, S.M.N.; Mu, W. Silanol Synthesis: Reaction of Hexaphenylcyclotrisiloxane with Organometallic Reagents. *J. Org. Chem.* **1993**, *58*, 7584–7586. [CrossRef]
48. Arkles, B. Silicone Molecular Weight Standards. In *Silicon Compounds: Register and Review*, 3rd ed.; Arkles, B., Peterson, W., Jr., Anderson, R., Eds.; Petrarch Systems: Bristol, TN, USA, 1984; p. 234.
49. Boileau, S. Ring Opening Polymerization of Cyclic Organosilicon Compounds: Recent Progress. *Makromol. Chem. Macromol. Symp.* **1993**, *73*, 177–181. [CrossRef]
50. Bauer, J.; Hsing, N.; Kickelbick, G. Preparation of Functionalized Block Copolymers Based on a Polysiloxane Backbone by Anionic Ring-Opening Polymerization. *J. Polym. Sci. Part A Polym. Chem.* **2002**, *40*, 1539–1551. [CrossRef]
51. Saam, J.C.; Gordon Fearon, F.W. Properties of Polystyrene-Polydimethylsiloxane Block Copolymers. *Ind. Eng. Chem. Prod. Res. Dev.* **1971**, *10*, 10–14.
52. Georgopanos, P.; Lo, T.Y.; Ho, R.M.; Avgeropoulos, A. Synthesis, Molecular Characterization and Self-Assembly of (PS-b-PDMS)n Type Linear (n = 1, 2) and Star (n = 3, 4) Block Copolymers. *Polym. Chem.* **2017**, *8*, 843–850. [CrossRef]
53. Noshay, A.; Matzner, M.; Karoly, G.; Stampa, G.B. Poly(α-Methylstyrene)-Poly (Dimethylsiloxane) Block Copolymers. *J. Appl. Polym. Sci.* **1973**, *17*, 619–628. [CrossRef]
54. Clemens, L.; Kantner, S.; Mazurek, M. Polysiloxane-Grafted Copolymer Release Coating Sheets and Adhesive Tapes. U.S. Patent 4,728,571, 1 March 1988.
55. Awasthi, A.K.; Kunzler, J.F. Ethylenically Unsaturated Polymerizable Group-Containing Polycarbosiloxane Monomers. U.S. Patent 9,039,174, 26 May 2015.
56. Cypryk, M.; Kaźmierski, K.; Fortuniak, W.; Chojnowski, J. Microstructure of the Copolymer Chain Generated by Anionic Ring-Opening Polymerization of a Model Cyclotrisiloxane with Mixed Siloxane Units. *Macromolecules* **2000**, *33*, 1536–1545. [CrossRef]
57. Leir, C. Radation-Curable Silicone Elastomers and Prssure Sensitive Adhesives. U.S. Patent 5,237,082, 17 August 1993.
58. Saho, T. Organosiloxane and Process for Preparing the Same. U.S. Patent 4,992,521, 12 February 1991.
59. Arkles, B.; Kimble, E. Asymmetric Low Molecular Wright Siloxanes with One Functional Group. U.S. Patent 8,455,599, 4 June 2013.
60. Goff, J.; Kimble, E.; Arkles, B. Living Polymerization Routes to Siloxane Macromers and Higher Order Silicone Structures. In *Progress in Silicones and Silicone-Modified Materials: ACS Symposium Series*; American Chemical Society: Washington, DC, USA, 2013; Volume 1154, pp. 59–78.
61. Goff, J.; Arkles, B.; Sulaiman, S. Modification of Silicone Elastomers Using Silicone Comonomers Containing Hydrophilic Surface Active Endgroups. *MRS Proc.* **2014**, *1626*, 518055. [CrossRef]
62. Arkles, B.; Goff, J. Dual Functional Linear Siloxanes, Step-Growth Polymers Derived Therefrom, and Methods of Preparation Thereof. U.S. Patent 8,952,118, 10 February 2015.
63. Gädda, T.M.; Nelson, A.K.; Weber, W.P. Selectivity in Anionic and Cationic Ring-Opening Polymerizations of Tetramethyl-1-(3′trifluoromethylphenyl)-1-Phenylcyclotrisiloxane and Tetramethyl-1-[3′,5′-Bis(Trifluoromethyl)Phenyl]-1-Phenylcyclotrisiloxane. *J. Polym. Sci. Part A Polym. Chem.* **2004**, *42*, 5235–5243. [CrossRef]
64. Kazama, H.; Tezuka, Y.; Imai, K. A New Bifunctional Initiator for the Living Polymerization of Hexamethylcyclotrisiloxane. *Polym. Bull.* **1989**, *21*, 31–37. [CrossRef]

65. Ogawa, T.; Suzuki, T. Branched Organopolysiloxanes. U.S. Patent 5,272,225, 21 December 1993.
66. Novozhilov, O.V.; Pavlichenko, I.V.; Demchenko, N.V.; Buzin, A.I.; Vasilenko, N.G.; Muzafarov, A.M. Multiarm Star-like Polydimethylsiloxanes Based on Dendrimers of the Sixth Generation. *Russ. Chem. Bull.* **2010**, *59*, 1909–1917. [CrossRef]
67. Vasilenko, N.G.; Rebrov, E.A.; Muzafarov, A.M.; Eßwein, B.; Striegel, B.; Möller, M. Preparation of Multi-Arm Star Polymers with Polylithiated Carbosilane Dendrimers. *Macromol. Chem. Phys.* **1998**, *199*, 889–895. [CrossRef]
68. Juliano, P.C. Substantially Transparent Polydimethylsiloxane-Polyalkylmethacrylate Compositions and Methods for Making Same. U.S. Patent 3,663,650, 16 May 1972.
69. Out, G.J.J.; Turetskii, A.A.; Snijder, M.; Möller, M.; Papkov, V.S. Model Polydiethylsiloxane Networks: 1. Synthesis and Phase Behaviour. *Polymer* **1995**, *36*, 3213–3221. [CrossRef]
70. Kimble, E.; Arkles, B. Low Molecular Weight Siloxanes with One Functional Group. U.S. Patent 7,799,888 B2, 21 September 2010.
71. Elkins, C.L.; Long, T.E. Living Anionic Polymerization of Hexamethylcyclotrisiloxane (D 3) Using Functionalized Initiation. *Macromolecules* **2004**, *37*, 6657–6659. [CrossRef]
72. Babu, J.; Sinai-Zingde, G.; Riffle, J.S. Synthesis of α, ω-Arylamine Functionalized Poly(Dimethylsiloxanes). *Polym. Prepr.* **1991**, *32*, 152–153. [CrossRef]
73. Goff, J.; Sulaiman, S.; Arkles, B.; Lewicki, J.P. Soft Materials with Recoverable Shape Factors from Extreme Distortion States. *Adv. Mater.* **2016**, *28*, 2393–2398. [CrossRef]
74. Okawa, T.; Suzuki, T. Method of Manufacturing Organopolysiloxane Having Terminal Alkenyl Groups. U.S. Patent 4,876,373, 24 October 1989.
75. Kumar, A.; Eichinger, B.E. Anionic Polymerization of Hexamethylcyclotrisiloxane with Acetylacetone Initiator to Form Telechelic Polymer. *Macromolecules* **1990**, *23*, 5358. [CrossRef]
76. Fuchise, K.; Igarashi, M.; Sato, K.; Shimada, S. Organocatalytic Controlled/Living Ring-Opening Polymerization of Cyclotrisiloxanes Initiated by Water with Strong Organic Base Catalysts. *Chem. Sci.* **2018**, *9*, 2879–2891. [CrossRef]
77. Peters, M.A.; Belu, A.M.; Linton, R.W.; Dupray, L.; Meyer, T.J.; DeSimone, J.M. Termination of Living Anionic Polymerizations Using Chlorosilane Derivatives: A General Synthetic Methodology for the Synthesis of End-Functionalized Polymers. *J. Am. Chem. Soc.* **1995**, *117*, 3380–3388. [CrossRef]
78. Belu, A.M.; DeSimone, J.M.; Linton, R.W.; Lange, G.W.; Friedman, R.M. Evaluation of Matrix-Assisted Laser Desorption Ionization Mass Spectrometry for Polymer Characterization. *J. Am. Soc. Mass Spectrom.* **1996**, *7*, 11–24. [CrossRef]
79. Arkles, B.C.; Kimble, E.; Goff, J. *Reactive Silicones: Forging New Polymer Links*; Gelest Inc.: Morrisville, PA, USA, 2016.
80. Vilner, S.; Clapp, L.; Schwartz, R. Colorants for Metalorganic Compounds and Their Application in Electronic Display Technology. U.S. Patent 9,234,101, 12 January 2016.
81. Sommer, S.; Ekin, A.; Webster, D.C.; Stafslien, S.J.; Daniels, J.; VanderWal, L.J.; Thompson, S.E.M.; Callow, M.E.; Callow, J.A. A Preliminary Study on the Properties and Fouling-Release Performance of Siloxane-Polyurethane Coatings Prepared from Poly(Dimethylsiloxane) (PDMS) Macromers. *Biofouling* **2010**, *26*, 961–972. [CrossRef] [PubMed]
82. Wu, D.; Spulber, M.; Itel, F.; Chami, M.; Pfohl, T.; Palivan, C.G.; Meier, W. Effect of Molecular Parameters on the Architecture and Membrane Properties of 3D Assemblies of Amphiphilic Copolymers. *Macromolecules* **2014**, *47*, 5060–5069. [CrossRef]
83. Viville, P.; Lazzaroni, R.; Dubois, P.; Kotzev, A.; Geerts, Y.; Borcia, G.; Pireaux, J.J. Impact of Silicone-Based Block Copolymer Surfactants on the Surface and Bulk Microscopic Organization of a Biodegradable Polymer, Poly(ε-Caprolactone). *Biomacromolecules* **2003**, *4*, 696–703. [CrossRef]
84. Reynolds, V.G.; Mukherjee, S.; Xie, R.; Levi, A.E.; Atassi, A.; Uchiyama, T.; Wang, H.; Chabinyc, M.L.; Bates, C.M. Super-Soft Solvent-Free Bottlebrush Elastomers for Touch Sensing. *Mater. Horizons* **2020**, *7*, 181–187. [CrossRef]
85. Tao, P.; Li, Y.; Siegel, R.W.; Schadler, L.S. Transparent Luminescent Silicone Nanocomposites Filled with Bimodal PDMS-Brush-Grafted CdSe Quantum Dots. *J. Mater. Chem. C* **2013**, *1*, 86–94. [CrossRef]
86. Bai, H.; Weimer, M. Self-Lubricating Medical Articles. U.S. Patent Appl. 2020/0093969, 26 March 2020.
87. Arkles, B.; Goff, J. Positive Tactile Interaction Coatings. *Paint Coat. Ind.* **2017**, *33*, 30–36.
88. Zhou, Q.; Chen, F.; Liu, Y.; Li, J. Preparation and Stability Study of Carboxyl Terminated Silicone Oil Modified Silicone-Oil-Based Ferrofluid. *J. Magn.* **2019**, *24*, 49–56. [CrossRef]
89. Hayama, K.; Narazaki, K.; Kawaguchi, S. Polymer for Hair-Care Products. U.S. Patent 5,166,276, 24 November 1992.
90. Hayama, K.; Narazaki, K.; Kawaguchi, S. Hair-Care Products Containing Copolymers Formed from Unsaturated Hydrophilic Monomers and Unsaturated Monomers and Unsaturated Monomers Having a Polysiloxane Group. U.S. Patent 5,480,634, 2 January 1996.
91. Gao, S.; Liu, B.; Peng, J.; Zhu, K.; Zhao, Y.; Li, X.; Yuan, X. Icephobic Durability of Branched PDMS Slippage Coatings Co-Cross-Linked by Functionalized POSS. *ACS Appl. Mater. Interfaces* **2019**, *11*, 4654–4666. [CrossRef]
92. Alli, A.; Zanini, D.; Ford, J.; Mahadevan, S.; Canavan, K.; Turner, D. Process for Making Ophthalmic Lenses. U.S. Patent 8,414,804, 9 April 2013.
93. Swarup, S.; Xu, S.; Vanier, R.; Endlish, E.; Simpson, D. Acrylic Polymers, Curable Film Forming Compositions Prepared Therefrom, and Method of Mitigating Dirt Build-Up on a Substrate. U.S. Patent 9,120,916, 1 September 2015.
94. Soane, D.S.; Berg, M.C.; Suddaby, N.A.; Lim, K.J.; Mowers, W.A. Ultraphobic Coating Compositions and Methods of Use. U.S. Patent Appl. 2011/0084421, 14 April 2011.

95. Daniel, W.F.M.; Burdyńska, J.; Vatankhah-Varnoosfaderani, M.; Matyjaszewski, K.; Paturej, J.; Rubinstein, M.; Dobrynin, A.V.; Sheiko, S.S. Solvent-Free, Supersoft and Superelastic Bottlebrush Melts and Networks. *Nat. Mater.* **2016**, *15*, 183–189. [CrossRef]
96. Nian, S.; Lian, H.; Gong, Z.; Zhernenkov, M.; Qin, J.; Cai, L.-H. Molecular Architecture Directs Linear-Bottlebrush-Linear Triblock Copolymers to Self-Assemble to Soft Reprocessable Elastomers. *ACS Macro Lett.* **2019**, *8*, 1528–1534. [CrossRef]
97. Portal, J.; Schultze, X.; Taupin, S. Dispersion of Polymer Particles in a Non-Aqueous Medium and Cosmetic Use Thereof. U.S. Patent 10,973,753, 13 April 2017.
98. Akers, C.E., Jr.; Franey, T.E.; Sun, J.; Butler, C. Polymeric Dispersants Used for Aqueous Pigmented Inks for Ink-Jet Printing. U.S. Patent 6,652,634 B1, 25 November 2003.
99. Tsutsumi, T.; Sawada, M. Aqueous Ink for Inkjet Printing. U.S. Patent 5,852,074, 22 December 1998.
100. Broad, R.; Gibson, I. Method of Making a Contact Lens. U.S. Patent 9,057,821 B2, 16 June 2006.
101. Vanderlaan, D.; Turner, D.; Hargiss, M.; Maiden, A.; Love, R.; Ford, J.; Molock, F.; Steffen, R.; Hill, G.; Alli, A.; et al. Soft Contact Lenses. U.S. Patent 6,943,203, 13 September 2005.
102. Azaam, A.; Vanderlaan, D.; Ford, J.; Joslin, S. Silicone Hydrogels Formed from Zero Diluent Reactive Mixtures. U.S. Patent 10,017,596 B2, 10 July 2018.
103. Martinez, M.R.; Cong, Y.; Sheiko, S.S.; Matyjaszewski, K. A Thermodynamic Roadmap for the Grafting-through Polymerization of PDMS11MA. *ACS Macro Lett.* **2020**, *9*, 1303–1309. [CrossRef]
104. Cao, Z.; Daniel, W.F.M.; Vatankhah-Varnosfaderani, M.; Sheiko, S.S.; Dobrynin, A.V. Dynamics of Bottlebrush Networks. *Macromolecules* **2016**, *49*, 8009–8017. [CrossRef]
105. Pan, Y.; Maddox, A.; Min, T.; Gonzaga, F.; Goff, J.; Arkles, B. Surface-Triggered Tandem Coupling Reactions of Cyclic Azasilanes. *Chem. Asian J.* **2017**, *12*, 1198–1203. [CrossRef] [PubMed]
106. Vatankhah-Varnosfaderani, M.; Keith, A.N.; Cong, Y.; Liang, H.; Rosenthal, M.; Sztucki, M.; Clair, C.; Magonov, S.; Ivanov, D.A.; Dobrynin, A.V.; et al. Chameleon-like Elastomers with Molecularly Encoded Strain-Adaptive Stiffening and Coloration. *Science* **2018**, *359*, 1509–1513. [CrossRef]
107. Ready, T.E.; Thomas, J.; Choi, S.; Boudjouk, P. New Technologies for the Analysis of Marine Coatings. In *New Developments in Coatings Technology ACS Symposium Series*; American Chemical Society: Washington, DC, USA, 2007; pp. 69–90.
108. Rosenberg Read, M.; Dahling, M.; Seim, M.; Oeberg Hed, K. Antifouling Composition. Eur. Patent Appl. 3,489,310 A1, 29 May 2019.
109. Keith, A.N.; Vatankhah-Varnosfaderani, M.; Clair, C.; Fahimipour, F.; Dashtimoghadam, E.; Lallam, A.; Sztucki, M.; Ivanov, D.A.; Liang, H.; Dobrynin, A.V.; et al. Bottlebrush Bridge between Soft Gels and Firm Tissues. *ACS Cent. Sci.* **2020**, *6*, 413–419. [CrossRef] [PubMed]
110. Jacobs, M.; Liang, H.; Dashtimoghadam, E.; Morgan, B.J.; Sheiko, S.S.; Dobrynin, A.V. Nonlinear Elasticity and Swelling of Comb and Bottlebrush Networks. *Macromolecules* **2019**, *52*, 5095–5101. [CrossRef]
111. Biedermann, M.; Blümke, M.; Wegener, M.; Krüger, H. Improved Actuation Strain of PDMS-Based DEA Materials Chemically Modified with Softening Agents. *Electroact. Polym. Actuators Devices* **2015**, *9430*, 94301G.
112. Goff, J.; Arkles, B. Silicone Elastomers with Intrinsic Control of Surface Hydrophilicity. *Soc. Biomater.* **2013**, *24*. Available online: http://biomaterials-org.securec7.ezhostingserver.com/abstracts/data/papers/2013/0243-000796.pdf (accessed on 2 April 2021).
113. Arkles, B.; Goff, J.D. Method for Producing Silicone Step-Growth Elastomers from Dual Functional Siloxanes. U.S. Patent 9,145,474, 29 September 2015.
114. Arkles, B.C.; Goff, J. Elastomers with Exceptional Elongation. U.S. Patent Appl. 2016/0319080, 3 November 2016.
115. Arkles, B.; Goff, J.; Sulaiman, S.; Sikorsky, A. Ultra-High Elongation Silicone Elastomers. *Rubber World* **2016**, *254*, 29.
116. Goff, J.; Sulaiman, S.; Philips, A.; Detwiler, M.; Arkles, B. Soft Tissue Compliant Silicones for Medical Devices. *Rubber World* **2018**, *258*, 22–26.
117. Gaylord, N. Method for Correcting Visual Defects, Compositions and Articles of Manufacture Useful Therein. U.S. Patent 4,120,570, 17 October 1978.
118. Gaylord, N. Oxygen-Permeable Contact Lens Composition, Methods and Article of Manufacture. U.S. Patent 3,808,178, 30 April 1974.
119. Schaeffer, J.; Beiting, J. Contact Lens Pioneers. *Rev. Optom.* **2007**, *144*, 76.
120. Spinelli, H.; Anton, W.; Seidner, L.; Coleman, H.; Ali, M.; Weintraub, L.; White, P. Silicone-Containing Polymers, Oxygen Permeable Hydrophilic Contact Lenses and Methods for Making These Lenses and Treating Patients with Visual Impairment. U.S. Patent 5,314,960, 24 May 1994.
121. Nicolson, P.C.; Vogt, J. Soft Contact Lens Polymers: An Evolution. *Biomaterials* **2001**, *22*, 3273–3283. [CrossRef]
122. Arkles, B.; Redinger, P. Silicones in Biomedical Applications. In *Biocompatible Polymers, Metals, and Composites*; Technomic: Lancaster, PA, USA, 1983; pp. 749–768.
123. Vanderlaan, D.; Nunez, I.; Hargiss, M.; Alton, M.; Williams, S. Soft Contact Lenses. U.S. Patent 5,998,498, 7 December 1999.
124. Nicolson, P.; Baron, R.; Chabreckek, P.; Court, J.; Domschke, A.; Griesser, H.; Ho, A.; Hopken, J.; Laycock, B.; Liu, Q.; et al. Extended Wear Opthalmic Lens. U.S. Patent 5,965,631, 12 October 1999.
125. Ueyama, H.; Ikawa, S.; Iwata, J. Hydrophilic Polysiloxane Macromonomer, and Product and Use of the Same. U.S. Patent 8,129,442, 6 March 2012.

126. Iwata, J.; Hoki, T.; Ikawa, S. Long Wearable Soft Contact Lens. U.S. Patent 6,867,245, 15 March 2005.
127. Awasthi, A.K.; Stanbro, K.; Kunzler, J.F. Linhardt, J. Ethylenically Unsaturated Polymerizable Group-Containing Polycarbosiloxane Monomers. U.S. Patent 8,420,711, 16 April 2013.
128. Awasthi, A.K.; Meng, F.R.; Künzler, J.F.; Linhardt, J.G.; Papagelis, P.; Oltean, G.; Myers, S.A. Ethylenically Unsaturated Polycarbosiloxanes for Novel Silicone Hydrogels: Synthesis, End-Group Analysis, Contact Lens Formulations, and Structure-Property Correlations. *Polym. Adv. Technol.* **2013**, *24*, 557–567. [CrossRef]
129. Hagiwara, M. One-Terminal Reactive Organopolysiloxane Having a Polyalkyleneoxide Group at the Omega-Terminal and a Method for the Preparation Thereof. U.S. Patent 8,404,882, 26 March 2013.
130. Qian, X.; Chang, F.; Matsuzawa, Y.; Shankar, V. Soft SIlicone Medical Devices with Durable Lubricious Coatings Thereon. U.S. Patent 10,324,233, 18 June 2019.
131. Hu, C.; Pham, D.; Lowery, M. Soft Silicone Materials for Ophthalmic Applications. U.S. Patent 8,8530,590, 10 September 2013.
132. Grubbs, R.; Sandstedt, C. Method for Modifying Power of Light Adjustable Lens. U.S. Patent 9,950,482, 24 April 2018.
133. Daimatsu, K.; Anno, Y.; Sugimoto, H.; Nakaishi, E.; Inomata, K.; Ikeda, T.; Yokoi, K. Preparation, Morphology, and Physical Properties of Transparent PSt Hybrid Materials Containing Silicone Macromonomer Kazuki. *J. Appl. Polym. Sci.* **2008**, *108*, 362–369. [CrossRef]
134. Kerry, J.P.; O'Grady, M.N.; Hogan, S.A. Past, Current and Potential Utilisation of Active and Intelligent Packaging Systems for Meat and Muscle-Based Products: A Review. *Meat Sci.* **2006**, *74*, 113–130. [CrossRef]
135. Clarke, R. Packaging of Respiring Biological Materials. U.S. Patent 8,092,848 B2, 10 January 2012.
136. Masuda, T.; Isobe, E.; Higashimura, T.; Takada, K. Poly[1-(Trimethylsilyl)-1-Propyne]: A New High Polymer Synthesized with Transition-Metal Catalysts and Characterized by Extremely High Gas Permeability. *J. Am. Chem. Soc.* **1983**, *105*, 7473–7474. [CrossRef]
137. Volkov, V.V. Free Volume Structure and Transport Properties of Glassy Polymers—Materials for Separating Membranes. *Polym. J.* **1991**, *23*, 457–466. [CrossRef]
138. Liang, L.; Dickson, J.M.; Zhu, Z.; Jiang, J.; Brook, M.A. Removal of 1,2-Dichloroethane from Aqueous Solutions with Novel Composite Polydimethylsiloxane Pervaporation Membranes. *J. Appl. Polym. Sci.* **2005**, *98*, 1477–1491. [CrossRef]
139. Uragami, T.; Sumida, I.; Miyata, T.; Shiraiwa, T.; Tamura, H.; Yajima, T. Pervaporation Characteristics in Removal of Benzene from Water through Polystyrene-Poly (Dimethylsiloxane) IPN Membranes. *Mater. Sci. Appl.* **2011**, *2*, 169–179. [CrossRef]
140. Miyata, T.; Takagi, T.; Kadota, T.; Uragami, T. Characteristics of Permeation and Separation for Aqueous Ethanol Solutions through Methyl Methacrylate-dimethylsiloxane Graft Copolymer Membranes. *Macromol. Chem. Phys.* **1995**, *196*, 1211–1220. [CrossRef]
141. Uragami, T.; Yamada, H.; Miyata, T. Removal of Dilute Volatile Organic Compounds in Water through Graft Copolymer Membranes Consisting of Poly(Alkylmethacrylate) and Poly(Dimethylsiloxane) by Pervaporation and Their Membrane Morphology. *J. Memb. Sci.* **2001**, *187*, 255–269. [CrossRef]
142. Uragami, T.; Matsuoka, Y.; Miyata, T. Permeation and Separation Characteristics in Removal of Dilute Volatile Organic Compounds from Aqueous Solutions through Copolymer Membranes Consisted of Poly(Styrene) and Poly(Dimethylsiloxane) Containing a Hydrophobic Ionic Liquid by Pervaporation. *J. Memb. Sci.* **2016**, *506*, 109–118. [CrossRef]
143. Uragami, T.; Fukuyama, E.; Miyata, T. Selective Removal of Dilute Benzene from Water by Poly(Methyl Methacrylate)-Graft-Poly(Dimethylsiloxane) Membranes Containing Hydrophobic Ionic Liquid by Pervaporation. *J. Membr. Sci.* **2016**, *510*, 131–140. [CrossRef]
144. Sun, J.; Beach, B. Chemically Prepared Toner and Process Therefor. U.S. Patent 6,991,884 B2, 31 January 2006.
145. Dawkins, J.; Taylor, G. Micelle Formation by AB Block Copolymers of Polystyrene and Poly(Dimethylsiloxane) in N-alkanes. *Chem. Macromol. Chem. Phys.* **1979**, *180*, 1737–1741. [CrossRef]
146. Gobelt, B.; Nagelsdiek, R.; Omeis, J.; Piestert, F. Method for Producing Dispersant Additives. U.S. Patent 9,573,103, 21 February 2017.
147. Haubennestel, K.; Bubat, A.; Frank, A. Levelling Agents for Surface Coatings. U.S. Patent 6,710,127, 23 March 2004.
148. Gobelt, B.; Bubat, A.; Frank, A.; Haubennestel, K. Use of Polyacrylate-Modified Polysiloxanes as Levelling Agents in Coating Compositions. U.S. Patent 7,230,051, 12 June 2007.
149. Van Der Sluis, M.; Hof, W.; Engelbrecht, L.; Onelin, S. Slip and Leveling Agent. U.S. Patent 7,915,343, 29 March 2011.
150. Nagel, C.; Heekeren, M.; Frank, A. New Silicone Structures. *Eur. Coat. J.* **2010**, *4*, 32–34.
151. BYK-3550 Technical Data Sheet. September 2015. Available online: http://www.uni-trading.com/sub/support/tds.msds/additive/BYK/BYK_etc/TDS/BYK-3550%20TDS(EN).pdf (accessed on 2 April 2021).
152. Rahimi, A.R.; Murphy, M.; Upadhyay, V.; Faiyaz, K.; Battocchi, D.; Webster, D.C. Amphiphilically Modified Self-Stratified Siloxane-Glycidyl Carbamate Coatings for Anti-Icing Applications. *J. Coat. Technol. Res.* **2020**, *18*, 83–97. [CrossRef]
153. Kozakiewicz, J. Polysiloxaneurethanes: New Polymers for Potential Coating Applications. *Prog. Org. Coat.* **1996**, *27*, 123–131. [CrossRef]
154. Bolich, R.; Torgerson, P. Hair Conditioning and Styling Compositions. U.S. Patent 5,618,524, 8 April 1997.
155. Arkles, B.; Cameron, R.; Larson, G.L. Silicones Derived from 2-Propenyl Functional Cyclic Terpenes and Methods of Preparation. U.S. Patent 8,653,294, 18 February 2014.

156. Arkles, B.C.; Goff, J.D.; Min, T.; Pan, Y.; Abel-Roberman, T. Silicon-Based Cannabadiol Derivatives and Compositions Thereof. U.S. Patent 10,933, 008 March 2021.
157. Mefford, O.T.; Vadala, M.L.; Goff, J.D.; Carroll, M.R.J.; Mejia-Ariza, R.; Caba, B.L.; Pierre, T.G.S.; Woodward, R.C.; Davis, R.M.; Riffle, J.S. Stability of Polydimethylsiloxane-Magnetite Nanoparticle Dispersions against Flocculation: Interparticle Interactions of Polydisperse Materials. *Langmuir* **2008**, *24*, 5060–5069. [CrossRef] [PubMed]
158. Miles, W.C.; Goff, J.D.; Huffstetler, P.P.; Mefford, O.T.; Riffle, J.S.; Davis, R.M. The Design of Well-Defined PDMS-Magnetite Complexes. *Polymer* **2010**, *51*, 482–491. [CrossRef]
159. Wilson, K.S.; Goff, J.D.; Riffle, J.S.; Harris, L.A.; St Pierre, T.G. Polydimethylsiloxane-Magnetite Nanoparticle Complexes and Dispersions in Polysiloxane Carrier Fluids. *Polym. Adv. Technol.* **2005**, *16*, 200–211. [CrossRef]
160. Goff, J.D. Synthesis and Characterization of Novel Polyethers and Polydimethylsiloxanes for Use in Biomaterials. Ph.D. Thesis, Virginia Tech, Blacksburg, VA, USA, 2009.
161. Riffle, J.S.; Thompson Mefford, O.; Carroll, M.R.J.; Vadala, M.L.; Goff, J.D.; Mejia-Ariza, R.; Saunders, M.; Woodward, R.C.; St Pierre, T.G.; Davis, R.M. Size Analysis of PDMS-Magnetite Nanoparticle Complexes: Experiment and Theory. *Chem. Mater.* **2008**, *20*, 2185–2191.
162. Mefford, O.T.; Woodward, R.C.; Goff, J.D.; Vadala, T.P.; St. Pierre, T.G.; Dailey, J.P.; Riffle, J.S. Field-Induced Motion of Ferrofluids through Immiscible Viscous Media: Testbed for Restorative Treatment of Retinal Detachment. *J. Magn. Magn. Mater.* **2007**, *311*, 347–353. [CrossRef]
163. Ganachaud, F.; Institut National des Sciences Appliquees de Lyon, France. Private communication, 2017.
164. Hadjichistidis, N.; Iatrou, H.; Pispas, S.; Pitsikalis, M. Anionic Polymerization: High Vacuum Techniques. *J. Polym. Sci. Part A Polym. Chem.* **2000**, *38*, 3211–3234. [CrossRef]
165. Varshney, S.K.; Khanna, D.N. Hexamethyl Cyclotrisiloxane–Styrene Block Copolymers and Their Chemical Composition. *J. Appl. Polym. Sci.* **1980**, *25*, 2501–2511. [CrossRef]
166. Dems, A.; Strobin, G. Synthesis and Properties of Well-Defined Multiblock Copolymers: Polystyrene-block-polydimethylsiloxane. *Die Makromol. Chem. Macromol. Chem. Phys.* **1991**, *2537*, 2521–2537. [CrossRef]
167. Chu, H. Morphologies of Strongly Segregated Diblock Copolymers. *Polymer.* **1995**, *36*, 1569–1575. [CrossRef]
168. Borah, D.; Cummins, C.; Rasappa, S.; Senthamaraikannan, R.; Salaun, M.; Zelsmann, M.; Liontos, G.; Ntetsikas, K.; Avgeropoulos, A.; Morris, M.A. Nanopatterning via Self-Assembly of a Lamellar-Forming Polystyrene-Block-Poly(Dimethylsiloxane) Diblock Copolymer on Topographical Substrates Fabricated by Nanoimprint Lithography. *Nanomaterials* **2018**, *8*, 32. [CrossRef] [PubMed]
169. Park, J.J.; Kim, H.H.; Kim, K.M.; Baek, S.K.; Lee, S.J.; Kim, J.H.; Choi, J.S. Method of Forming a Fine Pattern by Using Block Copolymers. U.S. Patent 9,437,452, 6 September 2016.
170. Jung, Y.S.; Ross, C.A. Orientation-Controlled Self-Assembled Nanolithography Using a Polystyrene-Polydimethylsiloxane Block Copolymer. *Nano Lett.* **2007**, *7*, 2046–2050. [CrossRef]
171. Jacoby, M. Block Copolymers for Lithography. *Chem. Eng. News* **2014**, *92*, 8–12. [CrossRef]
172. Andersen, T.H.; Tougaard, S.; Larsen, N.B.; Almdal, K.; Johannsen, I. Surface Morphology of PS-PDMS Diblock Copolymer Films. *J. Electron. Spectros. Relat. Phenom.* **2001**, *121*, 93–110. [CrossRef]
173. Choi, P.; Fu, P.F.; Guo, L.J. Siloxane Copolymers for Nanoimprint Lithography. *Adv. Funct. Mater.* **2007**, *17*, 65–70. [CrossRef]
174. Bates, C.M.; Seshimo, T.; Maher, M.J.; Durand, W.J.; Cushen, J.D.; Dean, L.M.; Blachut, G.; Ellison, C.J.; Willson, C.G. Polarity-Switching Top Coats Enable Copolymer Domains. *Science* **2012**, *338*, 775–779. [CrossRef]
175. Maher, M.J.; Bates, C.M.; Blachut, G.; Sirard, S.; Self, J.L.; Carlson, M.C.; Dean, L.M.; Cushen, J.D.; Durand, W.J.; Hayes, C.O.; et al. Interfacial Design for Block Copolymer Thin Films. *Chem. Mater.* **2014**, *26*, 1471–1479. [CrossRef]
176. Luo, Y.; Montarnal, D.; Kim, S.; Shi, W.; Barteau, K.P.; Pester, C.W.; Hustad, P.D.; Christianson, M.D.; Fredrickson, G.H.; Kramer, E.J.; et al. Poly(Dimethylsiloxane-b-Methyl Methacrylate): A Promising Candidate for Sub-10 Nm Patterning. *Macromolecules* **2015**, *48*, 3422–3430. [CrossRef]
177. Miller, P.J.; Matyjaszewski, K. Atom Transfer Radical Polymerization of (Meth)Acrylates from Poly(Dimethylsiloxane) Macroinitiators. *Macromolecules* **1999**, *32*, 8760–8767. [CrossRef]
178. Keul, H.; Müller, A.J.; Höcker, H. Preparation of Polymers with Polycarbonate Sequences and Their Depolymerization; An Example of Thermodynamic Recycling. In *Makromolekulare Chemie. Macromolecular Symposia*; Hüthig & Wepf Verlag: Basel, Switzerland, 1993; Volume 67, pp. 289–298.
179. Shefelbine, T.A.; Vigild, M.E.; Matsen, M.W.; Hajduk, D.A.; Hillmyer, M.A.; Cussler, E.L.; Bates, F.S. Core-Shell Gyroid Morphology in a Poly(Isoprene-Block-Styrene-Block- Dimethylsiloxane) Triblock Copolymer. *J. Am. Chem. Soc.* **1999**, *121*, 8457–8465. [CrossRef]
180. Politakos, N.; Ntoukas, E.; Avgeropoulos, A.; Krikorian, V.; Pate, B.D.; Thomas, E.L.; Hill, R.M. Strongly Segregated Cubic Microdomain Morphology Consistent with the Double Gyroid Phase in High Molecular Weight Diblock Copolymers of Polystyrene and Poly(Dimethylsiloxane). *J. Polym. Sci. Part B Polym. Phys.* **2009**, *47*, 2419–2427. [CrossRef]
181. Lo, T.Y.; Chao, C.C.; Ho, R.M.; Georgopanos, P.; Avgeropoulos, A.; Thomas, E.L. Phase Transitions of Polystyrene-b-Poly(Dimethylsiloxane) in Solvents of Varying Selectivity. *Macromolecules* **2013**, *46*, 7513–7524. [CrossRef]
182. Lu, K.Y.; Lo, T.Y.; Georgopanos, P.; Avgeropoulos, A.; Shi, A.C.; Ho, R.M. Orienting Silicon-Containing Block Copolymer Films with Perpendicular Cylinders via Entropy and Surface Plasma Treatment. *Macromolecules* **2017**, *50*, 9403–9410. [CrossRef]
183. Feng, X.; Zhuo, M.; Guo, H.; Thomas, E.L. Visualizing the Double-Gyroid Twin. *Proc. Natl. Acad. Sci. USA* **2021**, *118*, 1–6. [CrossRef] [PubMed]

184. Beatty, C.L.; Pechan, J.M.; Froix, M.F.; Hinman, D.D. Liquid Crystalline Type Order in Polydiethylsiloxane. *Macromolecules* **1975**, *8*, 547–551. [CrossRef]
185. Zhang, L.; Yao, W.; Gao, Y.; Zhang, C.; Yang, H. Polysiloxane-Based Side Chain Liquid Crystal Polymers: From Synthesis to Structure–Phase Transition Behavior Relationships. *Polymers* **2018**, *10*, 794. [CrossRef] [PubMed]

molecules

MDPI

Communication

Cationic Polymerization of Hexamethylcyclotrisiloxane in Excess Water

Quentin Barnes [1], Claire Longuet [2] and François Ganachaud [1,*]

1 Ingénierie des Matériaux Polymères, CNRS UMR 5223, INSA-Lyon, Univ Lyon, F-69621 Villeurbanne, France; quentin.barnes@hotmail.fr
2 IMT—Mines Ales, Polymers Hybrids and Composites (PCH), 6 Avenue De Clavières, F-30319 Alès, France; claire.longuet@mines-ales.fr
* Correspondence: francois.ganachaud@insa-lyon.fr; Tel.: +33-683-021-802

Abstract: Ring-opening ionic polymerization of cyclosiloxanes in dispersed media has long been discovered, and is nowadays both fundamentally studied and practically used. In this short communication, we show some preliminary results on the cationic ring-opening polymerization of hexamethylcyclotrisiloxane (D_3), a crystalline strained cycle, in water. Depending on the catalyst or/and surfactants used, polymers of various molar masses are prepared in a straightforward way. Emphasis is given here on experiments conducted with tris(pentafluorophenyl)borane (BCF), where high-molar polymers were generated at room temperature. In surfactant-free conditions, µm-sized droplets are stabilized by silanol end-groups of thus generated amphiphilic polymers, the latter of which precipitate in the course of reaction through chain extension. Introducing various surfactants in the recipe allows generating smaller emulsions in size with close polymerization ability, but better final colloidal stability, at the expense of low small cycles' content. A tentative mechanism is finally proposed.

Keywords: Piers-Rubinsztajn catalyst; surfactant-free polymerization; polydimethylsiloxane

Citation: Barnes, Q.; Longuet, C.; Ganachaud, F. Cationic Polymerization of Hexamethylcyclotrisiloxane in Excess Water. *Molecules* **2021**, *26*, 4402. https://doi.org/10.3390/molecules26154402

Academic Editors: Sławomir Rubinsztajn, Marek Cypryk and Wlodzimierz Stanczyk

Received: 24 June 2021
Accepted: 16 July 2021
Published: 21 July 2021

1. Introduction

Silicones are polymers of broad interest, both on industrial and academic sides. In industry, developments focus on the generation of more and more performing elastomers, widely used in numerous applications, e.g., for their thermal resistance or their innocuity [1]. In academia, a great deal of work has recently been conducted on (mostly rediscovered) chemistries to generate silicone chains with new functionalities. In the most recent research, we can cite, in particular, aza-Michael addition [2], thiol-ene click chemistry [3], and organocatalyzed polymerization and polycondensation (see, e.g., [4]).

A recurrent domain of development concerns the generation of silicone aqueous emulsions via the ring-opening polymerization of cyclosiloxanes (a recent book published by the Dow company summarizes recent comprehension of silicone dispersions [5]). Both cationic and anionic catalysts produce silicone dispersions, but the mechanism by which long polymer chains are formed differs, and is still of debate nowadays [5,6]. Basically, taking the case of octamethylcyclotetrasiloxane (D_4), polymerizations in emulsion (fast mixing in the presence of excess surfactant), in microsuspension (pre-generation of nanosized monomer droplets, typically by ultrasonication), or microemulsion (thermodynamically stable nanodroplets) follow very different pathways. In all instances, ring opening proceeds to propagate chains. Polymerization is stopped by transfer to water, but silanol chains ends can be reactivated to propagate further. Polycondensation between silanol end-groups also take place, to increase the molar masses. Side reactions that are generally observed in bulk or solution are likely prominent in water: backbiting that gives rise to a variety of cyclosiloxanes with various reactivities, and intermolecular redistributions between chains that enlarge the molar mass distribution. All the difficulty in these systems is to first understand where the different reactions take place (directly in water, at the droplet interface or inside the droplets), and second, at which paces.

Hexamethylcyclotrisiloxane is a monomer of choice when targeting silicone chains with perfectly controlled masses and functionality, as obtained by living anionic polymerization (for a very recent review, see an article just published in this Special Issue [7]). On the other hand, cationic polymerization of D_3 has been described mostly in the mid-eighties by two groups led by P. Sigwalt and J. Chojnowsky, and not further exploited (a summary of this more than 10 year's competition can be found in [8]). Using triflic acid as a superacid, the team of Sigwalt showed that water would retard the polymerization of D_3, but would not inhibit it, albeit at a high catalyst content. To our knowledge, ring-opening cationic polymerization of D_3 in water has hardly been studied. Early on, Weyenberg et al. showed, in a seminal paper, that D_4 or D_3 was easily converted into polymers in the presence of dodecylbenzenesulfonic acid [9]. They wrote that '*polymerization of hexamethylcyclotrisiloxane proceeds at a much faster rate than the cyclotetrasiloxane and, in fact, it is not necessary to pre-emulsify this monomer. Contact of even large crystals of this monomer with DBSA and water at 25 °C gave a quantitative conversion to emulsion polymer within 24 hr*'. Hemery et al. have later polymerized D_3, solubilized in toluene, in emulsion by an anionic polymerization process, where they observed fast generation of polymers with an unlikely broad distribution [10].

Tris(pentafluorophenyl)boron (acronym BCF) was discovered in the early 1960s, and was almost forgotten for 25 years before being rediscovered as a catalyst activator in metallocene-catalyzed olefin polymerization. Its strong Lewis acidity, comparable to those of $BF_3.OEt_2$, combined with its air stability and water tolerance, has made it a (co-)catalyst of choice for numerous reactions that are summarized in reviews (e.g., [11]). Since the discovery by Piers et al. that BCF catalyzes the reduction reaction of a silyl ether into alkane in the presence of an hydrogenosilane, this reaction was later patented and published to produce linear silicone chains from alkoxy- and hydrogeno-functionalized silicone molecules [12]. The so-called Piers–Rubinsztajn reaction was studied in detail, particularly in the team of Professor Chojnowsky in a series of papers explaining the mechanism of catalysis. A precision reaction could thus be performed, starting from model molecules, allowing the generation of complex branched structures with exceptional monodispersity (for a review on this, please see [13]). For the record, this Lewis acid was also proved to promote a hydrosilylation reaction, albeit in stoichiometric amount, or, more strangely, oligomerization of electron-withdrawing monomers (typically vinyl methylsulfone or acrylonitrile) onto SiH functions through a coordinated ate-type intermediate [14]. The team of Chojnowsky has shown that D_3 can be open and polymerized by tetramethyldisiloxane (L_2H) and other hydrogenosiloxanes in BCF toluene solution, but polymerization does not proceed in the absence of these molecules [15].

In this communication, we propose to show some preliminary results on the cationic polymerization of D_3 in excess water. Thanks to our deep knowledge on such processes applied to cyclosiloxanes [6] and vinyl monomers [16], we have selected a variety of Bronsted and (water-tolerant) Lewis acids. We particularly show advanced results on the polymerization of D_3 using BCF as a catalyst, and finally propose a brief discussion about a tentative polymerization mechanism.

2. Results and Discussion

2.1. First Screening

Table 1 summarizes the different results of D_3 polymerization in excess water, using different catalysts. Basically, molar masses and polydispersity, as well as the final contents of polymer, are given here, with selected SEC traces plotted in Figure 1. All the experiments were conducted at room temperature, in 10 mL vials, using a magnetic agitation (see conditions in Table 1 footnote). We did not specifically look at the colloidal state of the dispersions here, nor did we follow the kinetics of the reaction.

Table 1. First round of experiments of D_3 cationic polymerization in excess water [a].

Exp.	D_3Content (wt.%)	Catalyst	Cat.Content (wt.%)	Reaction Time (h)	M_n (kg/mol)	Đ	Conv. D_3 (%)	Yield Polym. (%)
entry 1	30	TfOH	3	12	-	-	100	0
				12 (After DEDMS addition) [b]	180	2.2	100	100
entry 2	15	DBSA	1	6	75	1.5	95	85
	30		1	6	63	1.6	90	80
entry 3	25	$YbDBS_3$ [c]	16	12	4.5	1.9	100	N.D. [e]
entry 4	25	$InDBS_3$ [d]	15	12	4.5	1.9	95	N.D. [e]

[a] Typical procedure: all ingredients are mixed at once in 3 g water into 10 mL vial and a magnetic stirrer. After drying at 90 °C, samples were recovered in toluene and analyzed by SEC; acronyms: TfOH: triflic acid; DBSA: dodecylbenzenesulfonic acid; DEDMS: diethoxydimethylsilane; [b] addition of 5 wt.% of DEDMS; [c] prepared by mixing 3 mol eq of NaDBS with 1 eq of $YbCl_3.6H_2O$; [d] prepared by mixing 3 mol eq of NaDBS with 1 eq of $InCl_3$; [e] LASCs coprecipitate with polymer, so it is not possible to calculate a yield in polymer.

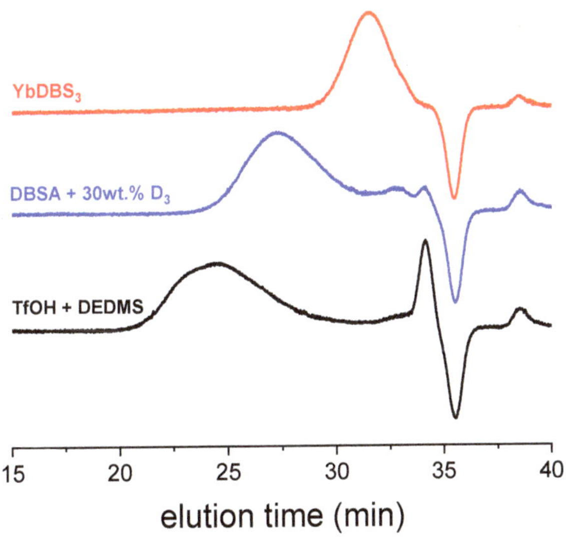

Figure 1. Selected SEC chromatograms of samples polymerized with various acids in the first set of polymeriations. Negative peak at 36 min is due to the flow marker. For the record, D_3 and D_5 were separately analyzed by SEC and came out at 34.5 and 34 min, respectively.

The first experiments conducted with triflic acid in large quantities showed that solid D_3 is rapidly consumed to give a totally transparent solution. We could not track the presence of polymer by precipitation of a sample aliquot in excess methanol, nor any other cycles that would have generated oil stains on the vial's walls. After the addition of a slight quantity of diethoxydimethylsilane (DEDMS, 0.2 eq. of initial D_3) and agitation during 12 h, we observed a precipitate at the bottom of the tube assigned to a polymer of high molar mass (M_n of 180,000 g/mol by SEC). DEDMS introduced in triflic acid aqueous solution in absence of D_3 did not produce a polymer, in agreement with a previous study [17]. We then concluded that triflic acid opens the cycle to generate water-soluble oligomers that convert into polymer by acid-catalyzed condensation between silanol and ethoxysilane groups. Note that this reaction is very different from the polycondensation reaction of bis-silanol-terminated PDMS long, hydrophobic oligomers that occurs exclusively at droplet interfaces [18].

Changing a molecular superacid to an acidic surfactant, DBSA, after only 6 h of reaction, we could detect the presence of polymers in the test tubes. SEC curves give molar masses of around 70,000 g/mol, with a larger content of small cycles at a larger D_3 content (about 10 wt.% at the end of the polymerization). Since at the time we were looking for cycle-free emulsions, we did not further explore this path; we are currently pursuing some experiments to check how fast and efficient this polymerization is.

We also tested some rare earth Lewis acids (ytterbium and indium chloride salts) combined with sodium dodecylbenzene sulfonate (NaDBS), to generate so-called Lewis acid surfactant complexes (LASCs), as reported before [16]. Even in large excesses, as tested here, these catalysts produced exclusively oligomers of molar masses around 4500 g/mol. Note that methanol also precipitated the LASC catalyst, so that it cannot be easily separated from the oligomers. According to the price of the catalysts used here, and the short oligomers produced, this alley was not pursued.

2.2. The Case of BCF

The origin of this second set of experiments comes from a study of the condensation reactions of alkoxy- and silanol-functionalized telechelic polymers in water [17]. When starting from the model molecules, tetramethyldisiloxane and dimethyldimethoxysilane, we observed, the rapid generation of cyclosiloxanes of various sizes, from D_3 to D_7, together with some polymer. The former cycle gradually disappeared with time, whereas the larger ones would accumulate in the reactor. This intriguing observation prompted us to further study the cationic ROP of D_3 in water catalyzed by BCF, in the absence of any other siloxane- or silane-based molecules. Note that we preliminary checked that BCF does not promote D_3 polymerization in toluene overnight (not shown).

A typical experiment consisted of introducing D_3 powder straightaway in a test-tube containing an aqueous solution of the catalyst, at room temperature and under magnetic agitation (see formulation in Table 2, entry 5).

Table 2. Second round of experiments of D_3 cationic polymerization in excess water [a].

Exp.	Surf. [b]	Cont. (wt.%)	State of Dispersion [c]	Reaction Time (h)	M_n (kg/mol)	Đ	Yield Polym. (%)	Cycles but D_3 (wt.% Content)
entry 5	-	-	H	12	26	2.0	56	D_5 and above (41)
				72	144	2.2	89	D_4, D_5, D_6 (7)
entry 6	Lauric acid	0.1	µE	12	51	2.4	88	D_4, D_5, D_6 (11)
entry 7	SDS	0.1	µE	12	14.4	2.2	73	D_4, D_5 (27)
entry 8	Brij 98	0.5	E	6	3	1.4	47	D_6 (3)
				12	25	2.3	65	D_4, D_5 and above (35)
entry 9	DTAB	0.1	H	12	13	1.5	39	D_4 (61)

[a] Typical procedure: all ingredients are mixed at once into a 10 mL vial equipped with a magnetic stirrer. Water = 5 g; D_3 = 0.5 g (10 wt.%); BCF = 24 mg (0.5 wt.%). Samples are precipitated in methanol, dried, recovered in toluene and analyzed by SEC. [b] Acronyms: SDS: sodium dodecyl sulfate; DTAB: dodecyl trimethyl ammonium bromide; [c] H: heterogeneous; µE: microemulsion; E: emulsion.

Even if D_3 first resides as solid chunks at the top of the water phase, it is then gently incorporated with time. A white emulsion (Figure 2a), of about 1.5 µm in size (Figure 2b), is generated, and remains almost so until the end of the reaction (we can track a slight enlargement in the size distribution with time, see Figure 2b). Zeta potential measurements give surface values of about −65 mV (Figure 2c). We suspect that the silanol groups of the oligomers/polymers protruding at the surface of the droplets stabilize them, as proposed earlier by Vincent et al. [19] and ourselves [17]. Only when the content of the silanol groups becomes too small that the polymer precipitates and deposits on the flask wall (Figure 2a).

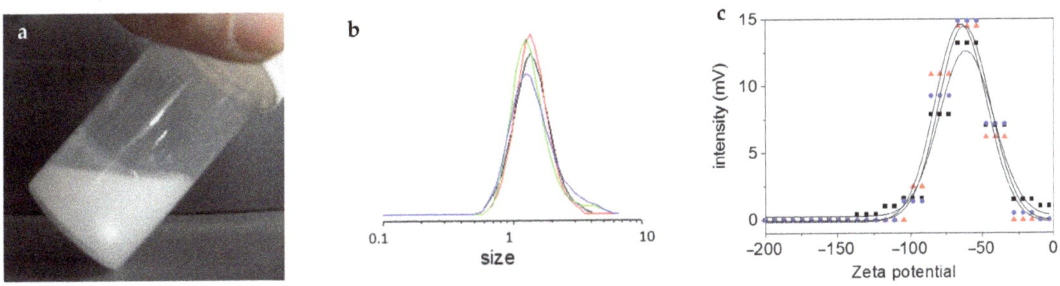

Figure 2. Colloidal state of D_3 suspension polymerization: (**a**) metastable emulsion, where polymer stains can finally be seen on the top of the wall; (**b**) average particle size at different times of reaction to highlight the emulsion stability (every 15 min in the following order: red, black, green, blue); (**c**) zeta potential of emulsion after 6 h on three different samples (average value of -65 ± 15 mV).

Typical SEC traces are given in Figure 3a, along with interpretations of it; the average molar masses are reported in Table 2 at two different reaction times. It can be seen here that polymerization of D_3 occurs quite smoothly, generating polymers of very high molar mass (typically 150,000 g/mol). This is typical of cationic polymerization in emulsion of cyclosiloxanes [6]. We were not capable of characterizing the chain-end of the polymers of such high molar masses; however, the fact that molar mass increases with time let us think that the polymers chains do not close ends here. Similar results were observed for the system starting from linear precursors [17].

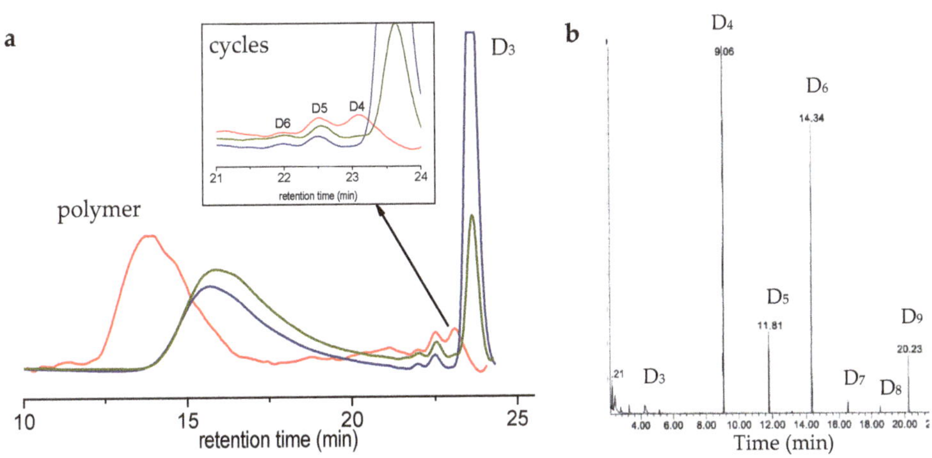

Figure 3. (**a**) SEC chromatograms of samples taken after 1/2 day (blue curve), 1 day (green olive curve) and 3 days (red curve). Zoom on the zone of small cycles is given on the inset. (**b**) GC-MS analyses of D_3/BCF system after 12 h of reaction. Abundances are relative, since peaks appear larger as molar mass of the cycle increases.

We can also track, on the SEC trace, a rapid generation of small cycles (typically D_5 and above) which contents grow slowly with time. Intermediate macrocycles are visible on the SEC trace after 3 days of reaction, showing that backbiting and intermolecular redistribution are retarded, but occur in this polymerization system (Figure 3a). To gain better insight into the course of polymerization, we have injected the intermediate sample in a GC/MS apparatus (Figure 3b). D_6 and D_9 appear together in larger proportions than other cycles, except for D_4. We observed a similar accumulation of D_{3x}^F cycles building (x = 2,3,4) in the anionic polymerization of D_3^F in miniemulsion, before a backbiting reaction occurs extensively and generates intermediate cycles (D_4^F, D_5^F …), but no macrocycles [20].

2.3. Introducing Surfactants in the Recipe

As we noted before, the simplest system described above is heterogeneous, i.e., micron-sized droplets are slowly converted into a polymer film, precipitating on the walls of the test tube as a function of time. We tried to use different surfactants to ensure a stabilization of the dispersion throughout the process, while still carrying out the polymerization. Table 2 summarizes the different trials we made, and Figure 4 shows the corresponding SEC traces.

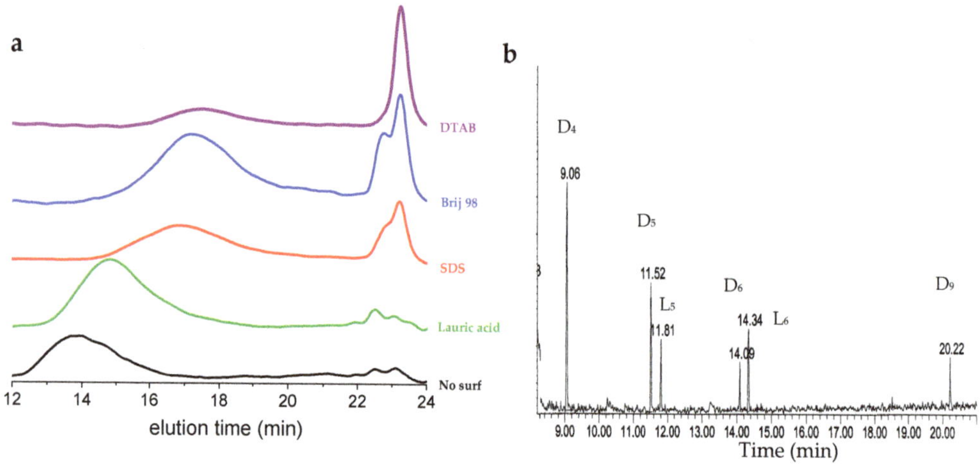

Figure 4. (**a**) SEC traces of different trials done in presence of surfactant, compared to pristine one. For polymerization conditions, see Table 2; (**b**) GC-MS analyses of D_3/BCF/Brij system after 12 h of reaction.

Using lauric acid, a translucent dispersion, typical of a microemulsion state, was obtained (average droplet size of 30 nm measured by DLS, not shown). Polymerization appeared to be quite fast, but led to the formation of larger contents of D_4 and D_5 than without a surfactant (more than 10 wt.%). Lauric acid also allowed rather large molar masses to be produced, around 55,000 g/mol, while keeping the colloidal stability of the dispersion.

With SDS, a similar microemulsion was formed, and polymerization was likely faster than without a surfactant. A polymer of a molar mass of around 15,000 g/mol was produced, together with a large load of small cycles (almost 30 wt.%). The same polymolecularity as for the other rounds was observed, typically around two. DTAB did not produce stable dispersions, and reactions led to a rapid generation of a large load of D_4 (above 60%), together with polymers of a low molar mass (13 kg/mol). The fact that only D_4 is generated here was not expected, and remains unexplained.

Not shown here is a trial with PVA, where an emulsion was formed, but polymerization did not proceed because of complexation of BCF with the alcohol groups of the dispersant. Brij 98 also complexes the catalyst through the oxygen atoms from ethylene glycol, but does not inhibit polymerization. Molar masses increase with time, together with the content of small cycles from D_4 to macro ones. Looking at the GC/MS of the sample taken after 12 h (Figure 4b), we can notice the presence of tentatively assigned linear disilanol oligomers, in addition to the same cycles observed before (D_6 and D_9). This confirms the formation of molecular intermediates before they cycle back.

2.4. Proposed Polymerization Scheme

D_3 polymerizes via a cationically catalyzed process in excess water and at room temperature. The initial screening showed that both Bronsted acids and Lewis acids catalyze the reaction, albeit at different paces and for final end results. The fact that triflic acid opens the cycle into small water-soluble oligomers, but does not convert them into polymers, seems to indicate that a condensation reaction of silanols is not likely in these conditions. This is certainly due to the absence of an interface, where this reaction generally takes place [6,18]. This also seems to confirm that, in contrast to the previously proposed emulsion polymerization of cyclosiloxane [5], small hydrophilic oligomers do not chain-extend in water. It would be worth in the future to mix together D_3, DEDMS, and a non-ionic surfactant from the first place, with a view of generating a stable latex while gaining high molar mass silicone polymer.

DBSA and BCF catalysis holds the following comparable features: fast polymerization, large molar mass polymers, and a fair load of small cycles, as expected from such cyclosiloxane cationic process. This most likely shows that BCF acts here principally as a Bronsted catalyst. Ring opening, one to two condensation steps, and back-cyclization, together with true ring-opening polymerization, take place here. When adding lauric acid to the BCF system, an acidic surfactant that is too weak to participate to the reaction, the results match perfectly with the DBSA-catalyzed ones, as follows: molar masses of typically 55 to 70 kg/mol, stable dispersions and cycle contents of 10 wt.%. It would be interesting to introduce both DBSA and BCF in the recipe, to see if synergy occurs; we plan to conduct such process soon.

Introducing other surfactants that could interact together with BCF led to more complex results. In all the cases, larger contents of cycles were observed, and polymerization was generally faster than in the absence of surfactants. Molar masses were rather low, around 15 kg/mol, which may be due to the larger content of water inside the monomer droplets, due to the presence of excess D_4 (this cycle is more polar than silicone chains). The type of dispersions depend on the content and nature of the surfactant, but this was not studied in detail here. Figure 5 summarizes the reaction taking place in this process for the exemplary case of BCF.

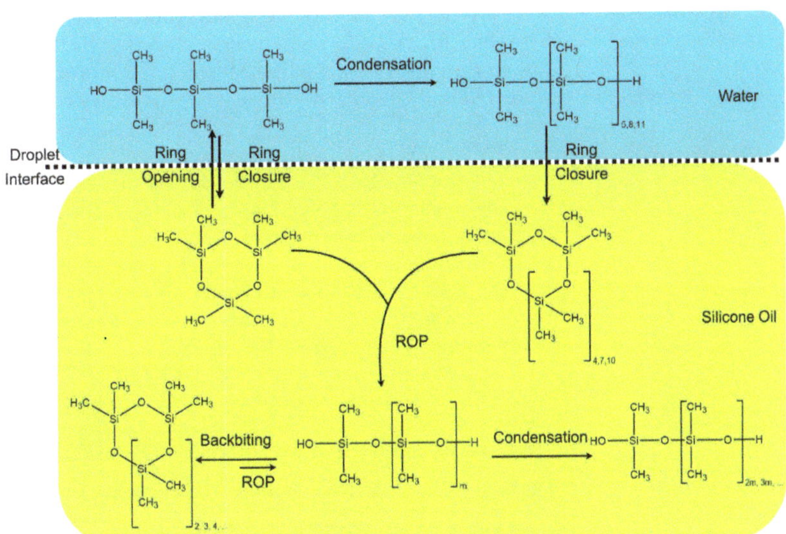

Figure 5. Proposed generic mechanistic scheme for the BCF-catalyzed D_3 cationic polymerization in excess water. Here, the catalyst was omitted for sake of clarity.

3. Materials and Methods

Hexamethylcyclotrisiloxane (D_3, 95%) was either kindly given by Bluestar Silicones (1st set of experiments) or purchased from ABCR (2nd set of experiments). Triflic acid (reagent grade, 98%), 4-DBSA (mixture of isomers, >95%), DTAB (>98%), SDS (98%) and lauric acid (≥98%) were all purchased from Sigma-Aldrich (Saint Louis, MO, USA). Brij

98 (hydroxyl titration: 50 to 65 mg KOH/g) came from ACROS organics (The Hague, Netherlands). Tris(pentafluorophenyl)borane ($B(C_6F_5)_3$, purity 97%) was obtained from Lancaster (Watd Hill, MA, USA).

Here, two series of experiments were performed at different times and locations. In the first set of experiments, size exclusion chromatography, SEC, was carried out using a Malvern Viscotek (Malvern, UK) GPC Max apparatus equipped with three Shodex columns (KF-804, -805, and -806). Detection systems were a refractive index and differential viscometry detectors. Toluene (HPLC grade, provided by Sigma-Aldrich) was eluted at 1 mL/min with diisopropylethylamine as flow marker [21]. In the second set of experiments, a Spectra Physics (Andover, MA, USA) apparatus with two PL gel columns (5 µm particles size, 300 mm length, with the following two pores sizes: one with 50 Å and one with 100 Å) and a Styragel HR2 column (7.8 mm internal diameter × 300 mm length) were used. An SP8430 differential refractometer achieved the detection. The toluene was eluted at a flow rate of 0.8 mL/min using diethylether as a flow marker. In both systems, the temperature for the SEC column set and the detector chamber was 35 °C to ensure stable baselines, high chromatographic efficiency, and consistent results. The standards used to calibrate the SEC were polystyrene standards.

Gas chromatography coupled with a mass spectrometer (GC/MS) was done on a 6890 N apparatus from Agilent Technologies (Santa Clara, CA, USA), equipped with an electrospray mass detector Agilent 5973 N and an apolar capillary column HP5-MS 30 m × 0.25 mm (stationary phase made of a film of diphenyldimethylpolysiloxane 5%, 0.25 µm). Conditions used were as follows: initial temperature 45 °C during 2 min, temperature ramp of 2 °C/min up to 50 °C then 10 °C/min up to final temperature, 250 °C, set during 10 min. Peak integration were corrected with factors inherent of each silicone species, according to the procedure published elsewhere [22].

Particle sizes were determined by dynamic diffraction of a laser beam on a Nanotrac NPA 250 device (Microtrac Inc., Montgomeryville, PA, USA), typically in a size range between 8 nm and 6.54 µm. The light dispersed by the particles entails a Doppler effect, due to Brownian motion. The Microtrac® Windows Software amplified, filtered, and mathematically treated this signal is to produce a size distribution.

4. Conclusions

To summarize, D_3 is the cyclosiloxane of choice to generate silicone dispersions of very high molar mass polymers, a priori not reachable neither by conventional polycondensation in microsuspension nor emulsification. The fact that this monomer performs polymerization in almost all the catalytic systems screened here opens large avenues in the search for the ideal polymerization process that would hopefully not produce small cycles, such as D_4 and D_5, now targeted in the 'Registration, Evaluation and Authorisation of Chemicals' (REACh) European regulation.

Author Contributions: Conceptualization, F.G.; methodology, Q.B., C.L. and F.G.; formal analysis, C.L. and F.G.; writing—original draft preparation, Q.B. and C.L.; writing—review and editing, F.G. All authors have read and agreed to the published version of the manuscript.

Funding: This research received no external funding.

Institutional Review Board Statement: Not applicable.

Data Availability Statement: The data presented in this study are available on request from the corresponding author.

Conflicts of Interest: The authors declare no conflict of interest.

Sample Availability: Samples are not available.

References

1. Köhler, T.; Gutacker, A.; Mejía, E. Industrial synthesis of reactive silicones: Reaction mechanisms and processes. *Org. Chem. Front.* **2020**, *7*, 4108–4120. [CrossRef]
2. Genest, A.; Portinha, D.; Fleury, E.; Ganachaud, F. The aza-Michael reaction as an alternative strategy to generate advanced silicon-based (macro)molecules and materials. *Prog. Polym. Sci.* **2017**, *72*, 61–110. [CrossRef]
3. Hoyle, C.; Bowman, C. Thiol–Ene Click Chemistry. *Angew. Chem. Int. Ed.* **2010**, *49*, 1540–1573. [CrossRef] [PubMed]
4. Roberts, J.M.; Belowich, M.E.; Peterson, T.H.; Bellinger, E.; Syverud, K.; Laitar, D.S.; Sidle, T. Homoconjugated Acids as Low Cyclosiloxane-Producing Silanol Polycondensation Catalysts. *ACS Omega* **2020**, *5*, 24954–24963. [CrossRef] [PubMed]
5. Liu, Y.H. *Silicone Dispersions*, 1st ed.; CRC Press: Boca Ranton, FL, USA, 2017.
6. Ganachaud, F.; Boileau, S. Siloxane-Containing Polymers. In *Handbook of Ring-Opening Polymerization*; Dubois, P., Coulembier, O., Raquez, J.-M., Eds.; Wiley: Weinheim, Germany, 2009; pp. 65–95.
7. Goff, J.; Sulaiman, S.; Arkles, B. Applications of hybrid polymers generated from living anionic ring opening polymerization. *Molecules* **2021**, *26*, 2755. [CrossRef] [PubMed]
8. Chojnowski, J.; Cypryk, M. Synthesis of Linear Polysiloxanes. In *Silicon-Containing Polymers*; Jones, R.G., Ando, W., Chojnowski, J., Eds.; Springer: Dordrecht, Germany, 2000; pp. 3–41.
9. Weyenberg, D.R.; Findlay, D.E.; Cekada, J.; Bey, A.E. Anionic Emulsion Polymerization of Siloxanes. *J. Polym. Sci. C* **1969**, *27*, 27–34. [CrossRef]
10. De Gunzbourg, A.; Maisonnier, S.; Favier, J.-C.; Maitre, C.; Masure, M.; Hémery, P. Ionic Polymerization in Aqueous Emulsion. *Macromol. Symp.* **1998**, *132*, 359–370. [CrossRef]
11. Piers, W.E.; Chivers, T. Pentafluorophenylboranes: From obscurity to applications. *Chem. Soc. Rev.* **1997**, *26*, 345–354. [CrossRef]
12. Rubinsztajn, S.; Cella, J.A. A new polycondensation process for the preparation of polysiloxane copolymers. *Macromolecules.* **2005**, *38*, 1061–1063. [CrossRef]
13. Brook, M.A.; Grande, J.B.; Ganachaud, F. New Synthetic Strategies for Structured Silicones Using $B(C_6F_5)_3$. *Adv. Polym. Sci.* **2010**, *235*, 161–183.
14. Pouget, E.; Holgado-Garcia, E.; Vasilenko, I.V.; Kostjuk, S.V.; Campagne, J.M.; Ganachaud, F. Oligomerization of electron-deficient vinyl monomers through an ate-complex mechanism: A new role for $B(C_6F_5)_3$ Lewis Acid. *Macromol. Rapid Commun.* **2009**, *30*, 1128–1132. [CrossRef] [PubMed]
15. Chojnowski, J.; Rubinsztajn, S.; Fortuniak, W.; Kurjata, J. Oligomer and Polymer Formation in Hexamethylcyclotrisiloxane (D_3)-Hydrosilane Systems Under Catalysis by tris(pentafluorophenyl)borane. *J. Inorg. Organomet. Polym. Mat.* **2007**, *17*, 173–187. [CrossRef]
16. Kostjuk, S.V.; Ganachaud, F. Cationic Polymerization of Vinyl Monomers in Aqueous Media: From Monofunctional Oligomers to Long-Lived Polymer Chains, Acc. *Chem. Res.* **2010**, *43*, 357–367. [CrossRef] [PubMed]
17. Longuet, C.; Joly-Duhamel, C.; Ganachaud, F. Copolycondensation of regular functional silane and siloxane in aqueous emulsion using $B(C_6F_5)_3$ as a catalyst. *Macromol. Chem. Phys.* **2007**, *208*, 1883–1892. [CrossRef]
18. Saam, J.C.; Huebner, D.J. Condensation Polymerization of Oligomeric Polydimethylsiloxanols in Aqueous Emulsion. *J. Polym. Sci. Polym. Chem.* **1982**, *20*, 3351–3368. [CrossRef]
19. Obey, T.M.; Vincent, B. Novel monodisperse "silicone oil"/water emulsions. *J. Colloid Interf. Sci.* **1994**, *163*, 454–463. [CrossRef]
20. Barrère, M.; Maitre, C.; Dourges, M.A.; Hémery, P. Anionic Polymerization of 1,3,5-Tris(trifluoropropylmethyl)cyclotrisiloxane (F3) in Miniemulsion. *Macromolecules* **2001**, *34*, 7276–7280. [CrossRef]
21. Barrère, M.; Ganachaud, F.; Bendejacq, D.; Dourges, M.-A.; Maitre, C.; Hémery, P. Anionic polymerization of octamethylcyclotetrasiloxane in miniemulsion II. Molar mass analyses and mechanism scheme. *Polymer* **2001**, *42*, 7239–7246. [CrossRef]
22. Steinmeyer, R.D.; Becker, M.A. *The Analytical Chemistry of Silicones*; Smith, A.L., Ed.; John Wiley & Sons, Inc.: New York, NY, USA, 1991; p. 255.

molecules

MDPI

Communication

When Attempting Chain Extension, Even Without Solvent, It Is Not Possible to Avoid Chojnowski Metathesis Giving D_3

Mengchen Liao, Yang Chen and Michael A. Brook *

Department of Chemistry and Chemical Biology, McMaster University, 1280 Main Street, W. Hamilton, ON L8S 4M1, Canada; liaom6@mcmaster.ca (M.L.); dychen@mcmaster.ca (Y.C.)
* Correspondence: mabrook@mcmaster.ca; Tel.: +1-905-525-9140 (ext. 23483)

Abstract: A simple, mild and efficient method to prepare HSi- or HOSi-telechelic, high-molecular-weight polydimethylsiloxane polymers (to 41,600 g·mol^{-1}) using the one-shot hydrolysis of M^HM^H is reported; titration of the water allowed for higher molecular weights (to 153,900 g·mol^{-1}). The "living" character of the chain extension processes was demonstrated by adding a small portion of M^HM^H and $B(C_6F_5)_3$ (BCF) to a first formed polymer, which led to a ~2-fold, second growth in molecular weight. The heterogeneous reaction reached completion in less than 30 min, much less in some cases, regardless of whether it was performed neat or 50 wt% in dry toluene; homogeneous reactions in toluene were much slower. The process does not involve traditional redistribution, as judged by the low quantities (<3%) of D_4 produced. However, it is not possible to avoid Chojnowski metathesis from M^HDDM^H giving D_3, which occurs competitively with chain extension.

Keywords: silicone polymers; Piers–Chojnowski–Rubinsztajn–Kawakami reaction; chain extension by Si-H hydrolysis; siloxane metathesis; D_3 synthesis

check for updates

Citation: Liao, M.; Chen, Y.; Brook, M.A. When Attempting Chain Extension, Even Without Solvent, It Is Not Possible to Avoid Chojnowski Metathesis Giving D_3. *Molecules* **2021**, *26*, 231. https://doi.org/10.3390/molecules26010231

Academic Editors: Sławomir Rubinsztajn, Marek Cypry and Wlodzimierz Stanczyk

Received: 11 December 2020
Accepted: 31 December 2020
Published: 5 January 2021

1. Introduction

Polydimethylsiloxane (PDMS) oils are the parent commercial silicone polymers from which almost all other silicones are derived. PDMS possesses unique properties including low T_g, good thermal stability, high optical transparency, excellent dielectric properties and excellent biocompatibility. Conventionally, two methods dominate the commercial preparation of high-molecular-weight PDMS oil: dehydration of silanol-terminated oligomers [1] and acid- or, particularly, base-catalyzed equilibration of cyclic monomers [2] (Figure 1A,B). However, each of these approaches suffers from inherent shortcomings. The former process, as with all condensation processes, slows down with increasing conversion, which makes the synthesis of high-molecular-weight polymers challenging; both processes typically lead to polymers with high dispersities $Đ_M$. In the latter case, with base-catalyzed equilibration that has an equilibrium constant near 1, the desired polymer is accompanied by the formation of large quantities of cyclic monomers, e.g., <15% for D_4 (D = ~Me_2SiO~) [3,4] (Figure 1B). Cyclooligosiloxanes, particularly D_4, have attracted different levels of concerns by regulatory agencies because of their purported environmental behaviors [5,6]; D_4 concentrations are regulated in Canada and the UK [7,8]. Hence, there is an increasing consensus that the value of silicone polymers would be increased if they contained lower cyclooligosiloxane concentrations. Note that the removal of cyclic monomers from silicone oils becomes more difficult as the MW (molecular weight) and viscosities of both cyclics and oils increase.

The competing commercial process for synthesis of high-MW silicone oils is ring-opening polymerization [9], typically initiated by anions [10–12] (Figure 1C). Polymers with high MW can result, but the process is challenged by the need to be scrupulously dry to avoid premature termination; higher MW PDMS polymers with narrow dispersity $Đ_M$ are more easily achieved when the more expensive, ring-strained monomer D_3 is utilized as a starting material instead of D_4.

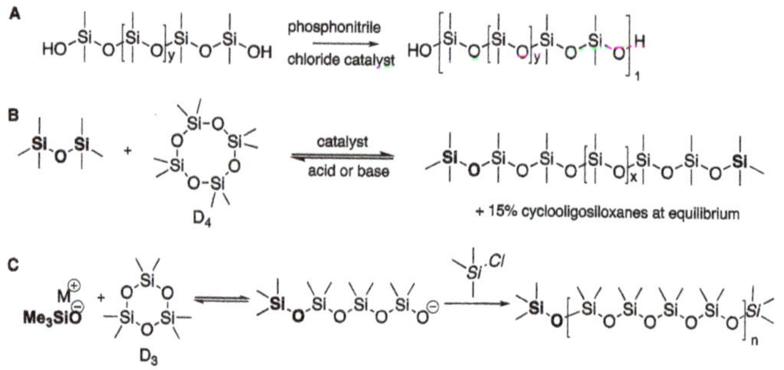

Figure 1. Traditional routes to silicone polymers: (**A**) silanol condensation; (**B**) redistribution; (**C**) ring-opening polymerization.

Hydrosilane monomers, particularly $HMeSiCl_2$ and HMe_2SiCl, are produced in a direct process [13] in concentrations that typically exceed commercial needs. As a consequence, oligomers $HMe_2SiOSiMe_2H$) (M^HM^H) and polymers $Me_3Si(OSiMeH)_nOSiMe_3$ (PHMS) based on these monomers are readily available. The compounds are potent reducing agents and, as Larson has noted, are both efficacious and inexpensive [14,15]. In organic synthesis, a benefit is the ease with which silicone products, after reduction, are readily separated from the desired organic product(s).

Tris(pentafluorophenyl)borane $B(C_6F_5)_3$ (BCF), a metal-free, water-tolerant [16] and thermally stable (up to 270 °C) compound [17], is renowned for its effectiveness as a co-initiator for industrial olefin polymerization [18]. In his extensive and elegant studies of reduction of carbonyl groups, Piers showed that it was also a potent catalyst for hydrosilane reductions [17]. We used M^HM^H in the presence of BCF to reduce the sulfur crosslinks in automobile rubber tires, permitting reuse of the organic constituents [19]. When using $HSiEt_3$ in the presence of BCF, Piers et al. originally reported that over-reduction of carbonyl groups led first to silyl ethers and then to alkanes plus disiloxanes.

Rubinsztajn and Cella recognized that the Piers reduction was a new route to silicones, and that was first patented and then published in the open literature [20]. Chojnowski et al. have made several seminal contributions to our understanding of the mechanism of this process [21,22]. It is worth noting that growing optically active siloxanes from silanols, rather than alkoxysilanes, using $B(C_6F_5)_3$ was pioneered by Kawakami [23]. In retrospect, perhaps we should have named the reaction the Piers–Chojnowski–Rubinsztajn–Kawakami (PCRK) reaction, rather than the PR reaction [24], and will do so for this paper.

The PCRK reaction is a particularly convenient route to synthesize silicone polymers. One simply chooses the number of alkoxysilane or silanol substituents, or water, required for the synthesis of a given linear or branched monomer, and then adds the appropriate mono-, di- or oligofunctional HSi-containing molecules in the presence of BCF [25]. It is thus possible to create linear polymers, including block-copolymers, simply by combining telechelic HSi + HOSi silicones or HSi silicones + water [26–28]. We have previously exploited this method to reliably introduce branches along linear silicone backbones [29] including, in the limit, highly branched dendron-like structures [30], including MDTQ resins (D = $Me_2SiO_{2/2}$) [31] (Figure 2).

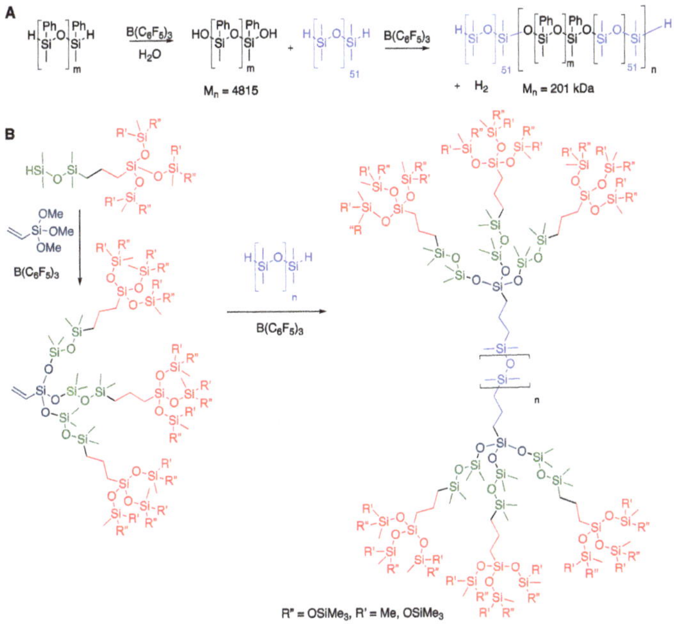

Figure 2. (**A**) Polymerization of telechelic HSi-silicones and water to give block copolymers. (**B**) Formation of highly branched silicones using the PCRK (Piers–Chojnowski–Rubinsztajn–Kawakami) reaction.

Tetramethyldisiloxane (M^HM^H) is an inexpensive, atom-efficient hydrosilane that has been selected as a silane source for a number of reactions [14,32,33]. We note that, at the time of writing, M^HM^H is slightly more expensive that non-functional D_3 and much more expensive than D_4 monomers. Attempts to produce high-molecular-weight PDMS oil from M^HM^H and H_2O in aqueous media, a traditional PCRK reaction (mole ratio, [OH]/[SiH] = 56) was made by the group of Ganachaud who reported formation of an elastomer; it was concluded that the reaction was not readily controllable [32]. Chojnowski et al. showed under anhydrous Schlenk line conditions that the oligomerization of M^HM^H (Figure 3H–J) in the presence of a BCF catalyst led to HSi-terminated oligomers and D_3—Chojnowski metathesis [22], however, with large quantities of D_3 that were produced M^HDDM^H; secondary copolymerization of D_3 and M^HM^H with activation with $B(C_6F_5)_3$ could be used to lead to higher molecular weight polymers [33].

Neither the Ganachaud nor Chojnowski outcomes with M^HM^H matched our experience of HSi-terminated silicones in the presence of water and $B(C_6F_5)_3$, which led smoothly to high-molecular-weight PDMS oils [26] (Figure 2). We hypothesized that differences arose because of the quantity of available water and BCF and, perhaps, other experimental conditions. Herein, we report a simple and mild process for the formation of HOSi- or HSi-terminated high-molecular-weight PDMS oil by the hydrolysis of M^HM^H in a kinetic process that generates relatively low quantities of D_4 (<3%). However, dilution with good solvents for silicone enhances the fraction of D_3 produced alongside the polymer.

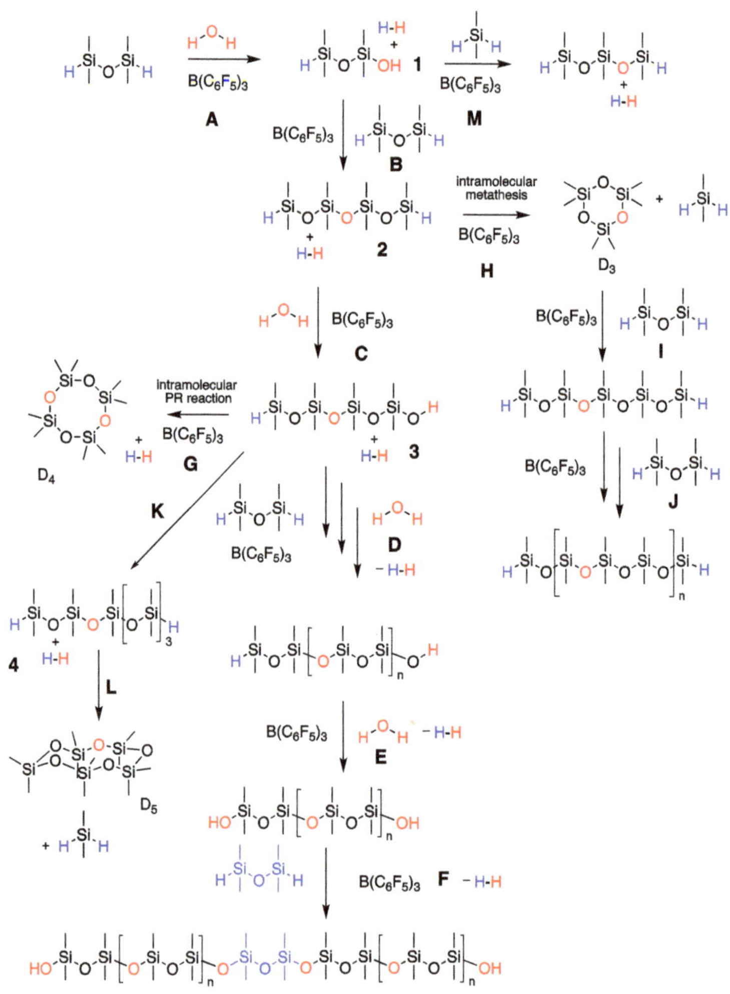

Figure 3. Proposed reactions for cyclooligosiloxane and polymer formation from M^HM^H. Hydrolysis of SiH compounds to give hydroxy-capped (**A**) dimer; (**C**) tetramer; (**D**,**E**) polymers. Chain extension to give HSi-terminated (**B**) tetramer; (**F**) polymer; (**K**) hexamer. Chojnowski metathesis to give (**H**) D_3; (**K**) D_5; (**L**) Me_2SiH_2; (**G**) Cyclization to give D4. (**I**,**J**). Metathesis polymerization to higher polymers. (**M**) Chain homologation between silanols and Me_2SiH_2.

2. Results

The reaction between M^HM^H and water leads first to disiloxanol **1** and then to the tetrasiloxane formation **2** (Figure 3A,B); both reactions are rapid and lead to the concomitant formation of hydrogen at the water droplet interface. *Note: caution must be taken, as this can lead to a pressure spike of this flammable gas*. Repetition of these processes lead to PDMS polymer initially terminated with SiH and then SiOH groups (Figure 3C–E). The processes were followed using ^{1}H-NMR (particularly the relationship between integrated peak areas of Si-*H* (~4.7 ppm vs. Si-CH_3 ~0.1–0.2 ppm [34,35])) and ^{29}Si-NMR spectra that allow clear differentiation between cyclic silicone monomers, linear silicone polymers and HSi-terminated linear materials [35], as well as gel permeation chromatography (GPC) and Fourier-transform infrared spectroscopy (FTIR; Supplementary Material, SM).

The objective of the research was to identify simple, efficient processes that would lead to linear PDMS, optionally terminated with SiH groups. Preliminary experiments examined the impact of reaction parameters, including catalyst concentration, solvent and, in particular, the effect of homogeneity between water and silicone phases during the course of hydrolysis. In addition, the impact of "one-shot" addition of reagents was compared to a titration in which water and/or additional BCF were added when reaction ceased.

Simply mixing M^HM^H, $B(C_6F_5)_3$ stock solution in dry toluene with bulk water led to linear silicones (entries 1–3, Table 1). Unsurprisingly, reactions were faster with more catalyst, but satisfactory rates were already achieved with only 0.02 mol% of this not inexpensive catalyst (<30 min, see below). Adding water in excess to the stoichiometric requirement did not lead to an improved outcome: cyclics were formed as a byproduct (see below, Supplementary Materials). Dilution with solvent, initially dichloromethane (containing 72.5 ppm water), demonstrated that much higher molecular weight polymers were accessible with even lower levels of catalyst (entries 4–6, Table 1); however, D_4 was also a byproduct of these reactions (Supplementary Materials). Thus, if high molecular weight is most desired, using small amounts of organic solvents is beneficial (entries 5, 6, Table 1); if suppression of cyclics is key, use slightly more BCF (entry 1, Table 1). Note that the use of a completely homogeneous reaction was disadvantageous for practical and kinetic reasons. For example, the reaction of 1 g of M^HM^H would require 14.3 mL of "wet" toluene (saturated with sufficient water to complete SiH hydrolysis), so scale up would not be practicable. When we did attempt the reaction on a small scale under these conditions, it proceeded for only a small extent over 5 h and then remained unchanged even after 24 h (see Supplementary Materials).

Table 1. High-molecular-weight PDMS (polydimethylsiloxane) preparation using hydrolysis.

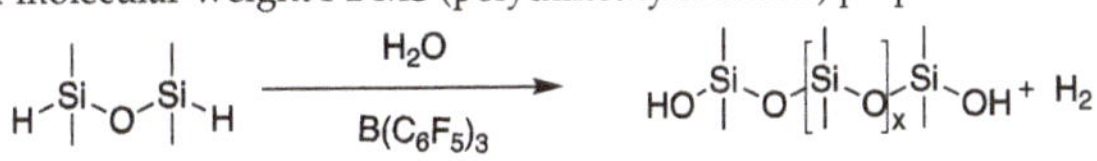

optional - dry toluene, wet DCM, addition in portions

Entry [a]	[BCF]/ [SiH] mol%	[OH]/ [SiH] [a]	Addition [b]	M^HM^H/DCM [c] (g·mL^{-1})	Cyclics % [d]	PDMS % [d]	M_n (g·mol^{-1})	M_w (g·mol^{-1})	$Đ_M$	Mass Bal [e]
1	0.1	1	O	-	2	98	43,700	129,200	2.97	70.1
2	0.02	1	O	-	26	74	52,700	105,400	2.00	71.3
3	0.02	3.2	O	-	5.9	94	41,600	101,500	2.44	70.6
4	0.008	1	P	-	5.9	92	20,700	32,500	1.57	79.3
5	0.006	0.73	P	1.3	12.9	87	111,400	221,500	1.99	95.6
6	0.004	0.75	P	1	5.7	94	152,500	314,000	2.06	82.9
				$M^HM^H/PhCH_3$ [c] (g·mL^{-1})						
7	0.02	1	O	-	18.9	81.2	20,500	37,400	1.82	73.8 [f]
8	0.02	1	O	1.53	47.9	52.1	40,000	63,100	1.58	89.1 [g]

[a] Stoichiometric addition of bulk water except for entries 5, 6. Reactions were quenched between 2.5 and 22 h (SM). Most reactions were complete within 30 min (see below), conditions under which the catalyst remains highly effective [16]. [b] O = one-shot reaction; P = reagents were added portion by portion, each time bubble evolution ceased. [c] DCM contained 72.5 ppm water. Toluene was dry. Note: experiments in Table 1 were completed in DCM to optimize reaction conditions. We then switched to the greener solvent toluene. [d] Fraction of D units in $D_3 + D_4 + D_5$ or PDMS based on integration in ^{29}Si-NMR. [e] Mass % of non-volatiles. [f] Cold trap (volatiles) contained 23.9% (total mass balance 97.7%), which consisted of M^HM^H 50%, Me_2SiH_2 42%. [g] Cold trap contained 10.5% (total mass balance 99.6%), which consisted of M^HM^H 66%, Me_2SiH_2 26%.

2.1. High-Molecular-Weight PDMS Preparation Using Hydrolysis

The preparation of telechelic PDMS polymers from M^HM^H terminated with either SiOH or SiH groups was straightforward by adding water in a one-shot process in the presence of 0.02% $B(C_6F_5)_3$. The two-phase reaction of water/silicone was surprisingly rapid; reaction times of less than 30 min led to polymers and concomitant formation of

H_2. The hydrolysis/condensation reactions in 50 wt% toluene were yet more rapid, as judged by the rate of build of the polymer molecular weight (entries 1–9, Table 2). We have not determined if the hydrolytic processes in toluene/water involve only homogeneous or a mixture of homogeneous/heterogeneous steps. The efficiency of chain extension will decrease as the living polymer increases in size, particularly above the entanglement limit of about 29,000 g·mol^{-1} (note: in the literature, reported entanglement limits ranges from about 15,000–35,000 g·mol^{-1}. Here we use data from the seminal study of Mrozek et al. [36]). This was clearly observed here, as the final M_n were 31–45 kg·mol^{-1} (entry 9, Table 2); complete consumption of SiH groups at higher conversion terminates polymerization by forming HOSi-terminated polymers. Re-initiation of such "dead" polymers is straightforward; however, addition of small quantities of M^HM^H caps the SiOH groups leading to dimerization or higher homologues of the existing polymer (entry 10, Table 2, Figure 3F). That is, a beneficial consequence of this process is that if HSi-terminated polymers are desired, one need only add excess M^HM^H to cap (Silicone-Me_2SiOH → Silicone-$(Me_2SiO)_2Me_2SiH$) and, if desired, grow the polymers before quenching the catalyst. This observation demonstrates that the process is living (entry 10, Table 2) [26]. Note that any residual water will compete for the silanol chain ends such that, if high molecular weights are desired, it is advantageous before capping to remove using distillation the small quantities of low molecular weight materials present, including residual water.

Table 2. Molecular weight versus reaction time using hydrolysis of M^HM^H.

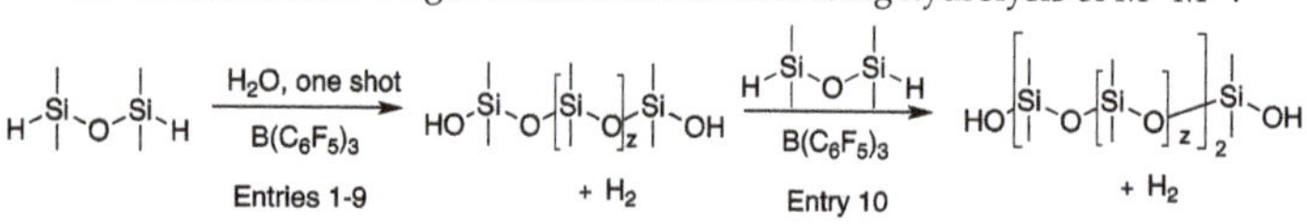

		Neat				H₂O/Toluene [a]			
Entry	Time(min)	M_n	MW	$Đ_M$	MW [b]	M_n	MW	$Đ_M$	MW [b]
1	0	-	-	-	134	-	-	-	134
2	2	-	-	-	240	-	-	-	250
3	4	-	-	-	300	-	-	-	310
4	6	-	-	-	520 [c]	-	-	-	330
5	8	-	-	-	310	-	-	-	440
6	10	-	-	-	330	-	-	-	1030
7	30	21,600	54,000	2.50	-	39,000	69,700	1.79	-
8	60	21,800	63,000	2.90	- [a]	45,300	97,100	2.14	-
9	180	31,200	82,900	2.66	- [a]	45,500	79,900	1.76	-
		Chain	Extension [d]						
		Starting	Polymer			Product	Polymer		
10		21,300	33,200	1.56	→	41,900	137,000	3.27	

[a] Reactions performed with 50 wt% M^HM^H dry toluene + liquid water. [b] Calculated based on Si*H* peak area versus SiC*H*$_3$ (integrated to 100) region in the ^{1}H-NMR. [c] The observation at 6 min. is considered an outlier. [d] Starting polymer entry 4, Table 1 contained no cyclics, and none formed by ^{29}Si-NMR; mass balance 99.9%.

2.2. *Managing Cyclics*

Our objective was to develop simple, practicable polymer syntheses that avoided the need for inert gas blankets; a septum with a bubbler was used to control pressure. One of the challenges presented by M^HM^H is its high volatility, which was problematic with or without solvents. The evolution of cyclics was followed by a combination of gravimetric analysis for volatile products (trapped in a cold trap that permitted cogenerated H_2 to escape) and, for the polymerization mixture, ^{29}Si-NMR, which is particularly sensitive to subtle differences in the chemical environment of D units; it is straightforward to distinguish D_3 (−8.3 ppm), D_4 (−19.1 ppm) and D_5 (−21.5 ppm) from D units in linear polymers (−21.6 ppm) [37]. M^HM^H was the main constituent captured in a cold finger, with small amounts of M^HDM^H, Me_2SiH_2 and D_3 (entries 7,8, Table 1).

The hydrolysis/condensation of M^HM^H in dilute, homogenous toluene solution was very slow. However, two-phase reactions (water/silicone or water/silicone+toluene) were very rapid reactions and completed in <30 min. Reactions can occur at the interface or within the organic fluid. In the case of the water/silicone mixture, the polymer yield was near 80%, with competing growth in D_3 and small amounts of D_5; D_4 was only inefficiently formed (Figure 4A). By contrast, even with the small dilution provided in the 50% water/silicone+toluene system, much less polymer was formed at the expense of D_3 and D_5 production, again with little D_4 (Figure 4B).

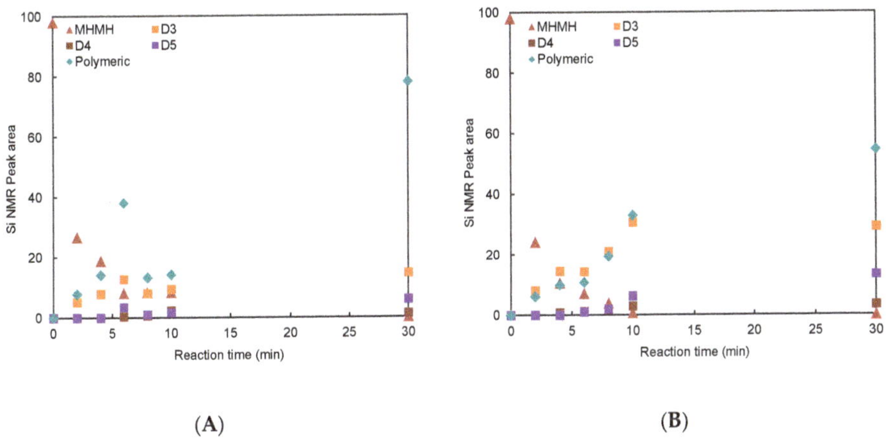

Figure 4. (**A**) Conversion to polymer without solvent. (**B**) Conversion to polymer in 50 wt% toluene. Rate of reaction (loss of HSi) during hydrolysis of M^HM^H was monitored using ^{1}H-NMR and silicone constituents using ^{29}Si-NMR. Integrations assumed identical sensitivity for all D units and are normalized to 100%.

Cyclic monomers are important starting materials for silicone synthesis. As noted above, the ring strain in D_3 makes it an attractive starting material for ring-opening polymerizations. On the other hand, there is an interest for a variety of reasons in making cyclic-free silicones, particularly from an inexpensive starting material like M^HM^H that polymerizes so rapidly. We had hoped to be able to fine tune the polymerization to create high molecular polymers in the absence of cyclics.

After dimerization, $M^HM^H \rightarrow M^HDDM^H$ **2**, the tetramer can undergo hydrolysis to give **3** then further chain extension, or cyclization to give D_4 (Figure 3D vs. Figure 3G). The generation of D_4 during polymerization under equilibrating conditions is favored both enthalpically and entropically, possessing virtually zero ring strain, and the SiO bond strength in cyclics is similar to those in linear chains and larger number of molecules than in linear polymers [1]. In neither of the cases examined was D_4 a significant product (Figure 4, Table 1). Thus, under these conditions of chain extension, **3** outcompetes cyclization (Figure 3D,K vs. Figure 3G). Adding small amounts of toluene led to an increase in D_4 product only from 1% to 3%.

More problematic, with respect to cyclics, was the formation of D_3 and, to a lesser extent, D_5. The most concentrated (neat) solution produced 13% D_3, but in the diluted sample 33% of the product mixture was D_3, in addition to linear polymer (entries 7,8, Table 1). These data show that, even in the presence of water, the Chojnowski metathesis to give D_3 and Me_2SiH_2 from **2** is highly competitive with hydrolysis (Figure 3H). The presence of D_5 in such high quantities is consistent with chain extension from **3** to **4** and then a different Chojnowski metathesis leading to D_5 and Me_2SiH_2 (e.g., Figure 3K,L. We thank a referee for this suggestion). It is believed that the formation of M^HDM^H, found in the cold trap, can be ascribed to reactions with Me_2SiH_2 with disiloxanes in the presence of $B(C_6F_5)_3$ (Figure 3M).

Previous experience with polymer chain extension of HSi-telechelic polymers with water did not lead to the formation of cyclics. Under these conditions with low catalyst concentrations (0.02%) (BCF/H_2O and $B(C_6F_5)_3{\cdot}OH_2$ [38]), redistribution reactions are similarly not efficient. While the reactions here were complete in 30 min, there was no change in the cyclics profile between 30–180 min (SM); redistribution would favor the formation of D_4 which, of the cyclics characterized, was formed in the lowest concentration. Under equilibration in the absence of solvents, the normal concentration of D_4 is ~15% [4]. Similarly, redistribution cannot explain the higher concentrations of D_3 and D_5.

The formation of cyclics vs. linears is therefore a consequence of the kinetics of intra- vs. intermolecular reactions within a silicone/organic solvent, or at the water interface. When done in 50wt% toluene, more cyclics, particularly D_3, were produced than when the only solvent for the silicone was M^HM^H and silicone products themselves, i.e., than when the system was more concentrated (Figure 4B vs. Figure 4A). This shows the power of Chojnowski metathesis. The conversion of **2** → D_3 (and **4** → D_5) effectively competes with hydrolysis of **2** → **3** when the reaction is slightly diluted, and is an important reaction even when done neat. The formation and then disappearance in the ^{29}Si-NMR of a maximum of 1.4% of Me_2SiH_2 was observed based on D-units (for full details of all intermediates from Figure 4, see Supplementary Materials). M^HDM^H, observed in the cold trap, could arise from reaction of Me_2SiH_2 with **3** and there may be additional homologation reactions that consume this reactive material. We cannot distinguish between polymerization involving hydrolysis/condensation (Figure 3B–E) vs. the reaction between D_3 and hydrosiloxanes, as shown by Chojnowski et al. (Figure 3I,J) [33]. However, the rapid rate and high molecular weight support the hydrolytic/condensation process as dominant.

Molecular weight is, of course, an important determinant of polymer properties. One advantage of this process is the relatively narrow dispersities achieved in the higher molecular weight polymers (Table 2). They do not match dispersities of polymers formed from ring-opening polymerization of D_3 but, apart from the challenge of managing hydrogen generation, are easier to perform. In part, the objectives of the work were met. It was possible to obtain rapid polymerization of M^HM^H with water to get medium-molecular-weight linear silicones terminated, if desired, with SiH groups. High concentrations facilitated the rates of polymerization. The process is living and any premature suppression of polymerization by complete hydrolysis to SiOH groups can be overcome by the addition of small amounts of M^HM^H, preferably after removing any residual water, a process that allows much longer polymers to form. However, unlike the hydrolytic process with starting materials that have a higher MW (e.g., DP > 6) [26], polymerization competed with undesired cyclic formation. Distinctions between these in previous studies are mostly related to concentrations. The important precedent work of Ganachaud showed that with very high water and higher catalyst concentrations, complex equilibration reactions occur under less controlled conditions [32]. On the other hand, with exceptionally low concentrations of water and catalysts, Chojnowski et al. showed metathesis of **2** dominates the process; one can elect to capture and then use D_3 for polymerization [21,22]. This work demonstrates the power of Chojnowski metathesis. Even without a solvent, efficient formation of D_3 is observed, up to 33%, made worse even by small amounts of solvent.

3. Experimental Section

3.1. Materials

Tetramethyldisiloxane (M^HM^H) was purchased from Gelest (Morrisville, PA, USA) and dried over molecular sieves (obtained from Sigma Aldrich (Oakville, ON, Canada), Molecular sieves, 4 Å beads, 8–12 mesh) before use. Ultrapure water (18 MΩ-cm) was obtained from Barnstead Easy pure RF (https://www.barnstead-water.com/). The $B(C_6F_5)_3$ catalyst was provided by Alfa Aesar (Tewksbury, MA, USA). Toluene (Caledon, ON, Canada) was dried over an activated alumina column (2.3 ppm water present). "Wet"

DCM (72.5 ppm water present) and "wet" toluene (466.6 ppm water present) were purchased from Caledon (Georgetown, ON, Canada).

3.2. Methods

^{1}H- and ^{29}Si-NMR spectra were recorded on a Bruker Advance 600 MHz nuclear magnetic resonance spectrometer using deuterated solvent chloroform-*d* (Milton, ON, Canada). Chromium(III) acetylacetonate (3×10^{-3} mol·L^{-1}) was added as a relaxation reagent for some of these measurements.

Gel permeation chromatography was carried out using a Viscotek GPC Max (VE 2001 GPC Solvent/Sample Module) (no longer manufactured). The system was equipped with a Viscotek VE 3580 RI Detector, a Viscotek 270 Dual Detector, and a PolyAnalytik SupeRes PAS-101 (8 mm Å~30 cm) column with a single pore, styrene-divinylbenzene gel, 6 µm particle size (London, Canada). It was additionally calibrated using a single narrow polydispersity polystyrene standard (93 kDa) from Polymer Laboratories. Toluene was used as the eluent at a flow rate of 1.0 mL·min^{-1}.

FTIR data were collected on a Nicolet 6700 FTIR using Thermo Electron's OMNIC software (version 8.0).

GC-MS analyses were performed using an Agilent 6890 N gas chromatograph (Santa Clara, CA, USA) equipped with a DB-17ht column (30 m × 0.25 mm i.d. × 0.15 µm film, J & W Scientific), and coupled to an Agilent 5973 MSD single quadruple mass spectrometer (Santa Clara, CA, USA). One microliter of sample was injected using Agilent 7683 autosampler (Santa Clara, CA, USA) using an injector temperature was 250 °C and a carrier gas (helium) flow of 0.8 mL/min. The transfer line was set to 280 °C and the MS source temperature was 230 °C. The column temperature started at 40 °C, raised to 70 °C at 5 °C/min, raised to 95 °C at 10 °C/min, raised to 300 °C at 40 °C/min, and was then held at 300 °C for 8 min for a total run time of 21.73 min. Full scan mass spectra between m/z 50 and 800 were acquired, with the MS detector turned off between 2.0–2.8 min for solvent. Please note that the sample is dissolved in propyl formate.

Water concentrations were determined using Karl Fischer titrations (Mettler Toledo DL39 Coulometer) (Columbus, OH, USA) with a one-component system containing a Hydranal Composite solution.

3.3. Experimental Procedures for High-Molecular-Weight PDMS Preparation Using Hydrolysis

The following paragraphs refer to entries in Table 1.

3.3.1. One-Shot Addition

$B(C_6F_5)_3$ stock solution in dry toluene.

To a pre-dried 25.0 mL vial added $B(C_6F_5)_3$ (0.052 g, 0.01 mmol) catalyst together with dry toluene (1.0156 mL, 0.881 g) to prepare a $B(C_6F_5)_3$ stock solution in dry toluene (0.01 M).

entry 1 ([BCF]/[SiH] = 0.02 mol%; [OH]/[SiH] = 1; one-shot) (Table 1):

To a pre-dried 100.0 mL round-bottomed flask we added tetramethyldisiloxane (M^HM^H) (134 g·mol^{-1}, 3.13 g, 0.02 mol) and distilled water (18 g·mol^{-1}, 0.402 mL, 0.02 mol) which was then capped with a septum with a needle with a bubbler open to atmosphere to balance the pressure. The mixture was stirred for 5–10 min prior to the addition of $B(C_6F_5)_3$ stock solution in dry toluene (0.01 M, 0.094 mL, 0.0094 mmol). The $B(C_6F_5)_3$ stock solution was added by an Eppendorf pipette into the flask through opening the septa. Once the $B(C_6F_5)_3$ stock solution was added, immediately, vigorous bubbling inside of the round bottle flask was observed. The mixture was stirred at room temperature for 3 h and the reaction was quenched by alumina for 5 h and the residual water droplet was removed by adding sodium sulfate.

entry 2 ([BCF]/[SiH] = 0.02 mol%; [OH]/[SiH] = 3.2; one-shot) (Table 1):

To a pre-dried 100.0 mL round-bottomed flask we added tetramethyldisiloxane (M^HM^H) (134 g·mol^{-1}, 5.01 g, 0.037 mol) and distilled water (18 g·mol^{-1}, 2.144 mL,

0.119 mol) which was then capped with a septum with a needle with a bubbler open to atmosphere to balance the pressure. The mixture was stirred for 5–10 min prior to the addition of $B(C_6F_5)_3$ stock solution in dry toluene (0.01 M, 0.149 mL, 0.0149 mmol). The $B(C_6F_5)_3$ stock solution was added by an Eppendorf pipette into the flask through opening the septa. Once the $B(C_6F_5)_3$ stock solution was added, immediately, vigorous bubbling inside of the round bottle flask was observed. The mixture was stirred at room temperature for 3 h and the reaction was quenched by alumina for 5 h and the residual water droplet was removed by adding sodium sulfate.

entry 3 ([BCF]/[SiH] = 0.1 mol%; [OH]/[SiH] = 1; one-shot) (Table 1):

To a pre-dried 100.0 mL round-bottomed flask we added tetramethyldisiloxane (M^HM^H) (134 g·mol^{-1}, 1.00 g, 0.0074 mol) and distilled water (18 g·mol^{-1}, 0.134 mL, 0.0074 mol) which was then capped with a septum with a needle with a bubbler open to atmosphere to balance the pressure. The mixture was stirred for 5–10 min prior to the addition of $B(C_6F_5)_3$ stock solution in dry toluene (0.01 M, 0.150 mL, 0.015 mmol). The $B(C_6F_5)_3$ stock solution was added by an Eppendorf pipette into the flask through opening the septa. Once the $B(C_6F_5)_3$ stock solution was added, immediately, vigorous bubbling inside of the round bottle flask was observed. The mixture was stirred at room temperature for 3 h and the reaction was quenched by alumina for 5 h and the residual water droplet was removed by adding sodium sulfate. The mixture was stirred at room temperature for 3 h and the reaction was quenched by alumina. The product was collected by filtration through Celite under reduced pressure.

3.3.2. Portion by Portion Addition

entry 4 (overall [BCF]/[SiH] = 0.008 mol%; [OH]/[SiH] = 1; portion by portion adding of $B(C_6F_5)_3$ catalyst, neat) (Table 1):

To a pre-dried 100.0 mL round-bottomed flask we added tetramethyldisiloxane (M^HM^H) (134 g·mol^{-1}, 5.01 g, 0.037 mol) and the distilled water (18 g·mol^{-1}, 0.67 mL, 0.037 mol) which was then capped with a septum with a needle with a bubbler open to atmosphere to balance the pressure. The mixture was stirred for 5–10 min prior to the addition of $B(C_6F_5)_3$ stock solution in dry toluene (0.001 M, 0.149 mL, 0.00149 mmol, [BCF]/[SiH] = 0.002 mol%, first portion of catalyst). The $B(C_6F_5)_3$ stock solution was added by an Eppendorf pipette into the flask through opening the septa. Bubbles immediately formed once the catalyst was added. After no bubble formation was observed (throughout the reaction, water droplets always existed, indicating sufficient amount of water), once again, $B(C_6F_5)_3$ stock solution in dry toluene (0.001 M, 0.149 mL, 0.00149 mmol, [BCF]/[SiH] = 0.002 mol%, second portion of catalyst) was added and vigorous bubbling occurred again. The procedure of adding $B(C_6F_5)_3$ stock solution in dry toluene (0.001 M, 0.149 mL, 0.00149 mmol, [BCF]/[SiH] = 0.002 mol%) repeated until no bubbles were observed even after the addition of a fresh portion of catalyst; overall, four portions of catalyst were added (0.008 mol%). The mixture was stirred at room temperature for 22 h and the reaction was quenched by alumina for 5 h and the residual water droplet was removed by adding sodium sulfate.

entry 5 (overall [BCF]/[SiH] = 0.006 mol%; [OH]/[SiH] = 0.73; portion by portion, 1.3 g·mL^{-1}, mass of M^HM^H versus volume of DCM) (Table 1):

To a pre-dried 100.0 mL round-bottomed flask we added tetramethyldisiloxane (M^HM^H) (134 g·mol^{-1}, 4.12 g, 0.031 mol) and "wet" DCM (3 mL, 72.5 ppm water, 0.016 mmol of water present) which was then capped with a septum with a needle with a bubbler open to atmosphere to balance the pressure. The mixture was stirred for 5–10 min prior to the addition of $B(C_6F_5)_3$ stock solution in dry toluene (0.001 M, 0.119 mL, 0.0012 mmol, [BCF]/[SiH] = 0.002 mol%, first portion of catalyst) followed by a quick addition of the distilled water (18 g·mol^{-1}, 0.074 mL, 0.0041 mol, [OH]/[SiH] = 0.13, first portion). The $B(C_6F_5)_3$ stock solution and water were added by an Eppendorf pipette into the flask through opening the septa and a vigorous bubble formation was observed immediately after the addition of water. Here is how we decided to whether add catalyst or water:

First, we added the distilled water (18 g·mol^{-1}, 0.055 mL, 0.0031 mol, [OH]/[SiH] = 0.1, second portion). (i) If bubble formation was observed, the indicated the cease of bubbles originated from insufficient amount of water. (ii) If no bubble formation was observed after the addition of the water indicating the catalyst was insufficient, then $B(C_6F_5)_3$ stock solution in dry toluene (0.001 M, 0.119 mL, 0.0012 mmol, [BCF]/[SiH] = 0.002 mol%) was added. The above-mentioned procedure repeated until no bubbles formed regardless of the addition of catalyst or water; overall, [BCF]/[SiH] = 0.006 mol%; [OH]/[SiH] = 0.73. The mixture was stirred at room temperature for 5.6 h and the reaction was quenched by alumina for 5 h and the residual water droplet was removed by adding sodium sulfate.

entry 6 (overall [BCF]/[SiH] = 0.004 mol%; [OH]/[SiH] = 0.75; portion by portion, 1.0 g·mL^{-1}, mass of MHMH versus volume of DCM) (Table 1):

To a pre-dried 100.0 mL round-bottomed flask we added tetramethyldisiloxane (M^HM^H) (134 g·mol^{-1}, 3.05 g, 0.023 mol) and "wet" DCM (3 mL, 72.5 ppm water, 0.016 mmol of water present) which was then capped with a septum with a needle with a bubbler open to the atmosphere to balance the pressure. The mixture was stirred for 5–10 min prior to the addition of $B(C_6F_5)_3$ stock solution in dry toluene (0.001 M, 0.092 mL, 0.00092 mmol, [BCF]/[SiH] = 0.002 mol%, first portion of catalyst) followed by a quick addition of the distilled water (18 g·mol^{-1}, 0.062 mL, 0.0034 mol, [OH]/[SiH] = 0.15, first portion). The $B(C_6F_5)_3$ stock solution and water were added by an Eppendorf pipette into the flask through opening the septa and a vigorous bubble formation was observed immediately after the addition of water. Here is how we decided to whether add catalyst or water: First, we added the distilled water (18 g·mol^{-1}, 0.041 mL, 0.0023 mol, [OH]/[SiH] = 0.1, second portion). If bubble formation was observed, this indicated the cessation of bubbles originated from an insufficient amount of water. If no bubble formation was observed after the addition of the water, indicating the catalyst was insufficient, then $B(C_6F_5)_3$ stock solution in dry toluene (0.001 M, 0.092 mL, 0.00092 mmol, [BCF]/[SiH] = 0.002 mol%) was added. The above-mentioned procedure was repeated until no bubbles formed regardless of the addition of catalyst or water; overall, [BCF]/[SiH] = 0.004 mol%; [OH]/[SiH] = 0.75. The mixture was stirred at room temperature for 2.6 h and the reaction was quenched by alumina for 5 h and the residual water droplet was removed by adding sodium sulfate.

3.3.3. Capturing the Volatiles, Entries 7, 8

To a pre-dried 100.0 mL round-bottomed flask, tetramethyldisiloxane (M^HM^H) (134 g·mol^{-1}, 10.38 g, 77.5 mmol) with distilled water (18 g·mol^{-1}, 1.34 mL, 74.4 mmol)(entry 7, Table 1) was added. To the mixture $B(C_6F_5)_3$ stock solution in dry toluene (0.1 M, 0.298 mL, 0.0298 mmol) was added. The cold trap condenser was submerged in dry ice/acetone with the addition of acetone and was directly connected to the round-bottomed flask through 6 mm O.D. Pyrex tubing. The mixture in the round bottle flask was stirred at room temperature for 3h and the reaction was quenched by alumina and the residual water droplet was removed by adding sodium sulfate with 5h quenching time. Before the work up procedures, the mixtures in the round bottle flask were weighed and characterized by ^{1}H-, ^{29}Si-NMR (SR) followed by filtration through Celite under reduced pressure. The same process was repeated for the experiment that also included toluene (entry 8, Table 1): M^HM^H (134 g·mol^{-1}, 10.32 g, 77.0 mmol) with distilled water (18 g·mol^{-1}, 1.34 mL, 74.4 mmol), dry toluene (6.73 mL, 7.76 g) and $B(C_6F_5)_3$ stock solution in dry toluene (0.1 M, 0.298 mL, 0.0298 mmol).

3.4. *Experimental Procedure for Controlled Growth of Linear PDMS Using M^HM^H and Water*

The following paragraphs refer to Table 2.

3.4.1. In Neat Water, or in Water/Toluene Mixtures

To a series of pre-dried 25.0 mL vials we added M^HM^H (134 g·mol^{-1}, 1.04 g, 7.8 mmol) with distilled water (18 g·mol^{-1}, 0.134 mL, 7.4 mmol). To the mixtures, $B(C_6F_5)_3$ stock solution in dry toluene (0.1 M, 0.03 mL, 0.003 mmol, respectively) was added and stirred at room temperature. At time points from 2–180 min alumina was added to quench the reaction at different time intervals together with ~0.5 g of sodium sulfate to remove the residual water. Before the work up procedures, the products were characterized by ^{1}H-, ^{29}Si-NMR (SR) followed by filtration through Celite under reduced pressure. The same process was repeated for the experiments that also included toluene: M^HM^H (134 g·mol^{-1}, 1.04 g, 7.8 mmol) with distilled water (18 g·mol^{-1}, 0.134 mL, 7.4 mmol), dry toluene (0.677 mL, 0.585 g) and $B(C_6F_5)_3$ stock solution in dry toluene (0.1 M, 0.03 mL, 0.003 mmol), respectively, the mixture was then stirred at room temperature (entries 1–9, Table 2).

3.4.2. Chain Extension

To a 10.0 mL round-bottomed flask we added tetramethyldisiloxane (M^HM^H) (134 g·mol^{-1}, 0.015 mL, 0.011 g, 0.085 mmol) with pre-prepared PDMS oil terminated with silanol (21,300 g·mol^{-1}, 0.90 g, 0.042 mmol, entry 4, Table 1). To be noted here, to the pre-prepared silanol PDMS we performed Kugelrohr distillation at 100 °C under vacuum (635 mmHg) for 60 min before running the GPC to remove the residual water and volatile small molecules. To the mixture, $B(C_6F_5)_3$ stock solution in dry toluene (0.01 M, 0.0086 mL, 0.086 μmol) was added and stirred at room temperature. After 3 h, alumina was added to the mixture to quench the reaction. The product was collected by filtration through Celite under reduced pressure (Entry 10, Table 2).

4. Conclusions

The simple, efficient and mild process of hydrolyzing M^HM^H leads to high-molecular-weight PDMS (maximum 153,900 g·mol^{-1}) in a controllable and "living" manner. The reactions can be completed in as short a time as 30 min, in two-phase reactions using either bulk water with neat M^HM^H or M^HM^H in dry toluene (50 wt%). Under either set of conditions, however, low D_4 (1–3%) content is produced and higher values of D_5 (2–10%) and particularly D_3 (13–33%) are produced. These outcomes are attributed to the particular efficiency of Chojnowski metathesis in organic solvents, even with high water content.

Supplementary Materials: Figures S1 and S2: kinetic plots of reaction progress with and without toluene solvent, ^{1}H-NMR and ^{29}Si-NMR and IR characterization of each datapoint on Figure S1, Tables S1 and S2, GC and NMR data for the cold trap with and without toluene solvent (entries 7, 8, Table 1), NMR data for polymer (entry 10, Table 2).

Author Contributions: The original concept of this research was proposed by M.A.B. and Y.C. The research was performed by M.L., M.A.B. and M.L. shared the writing activities. All authors have read and agreed to the published version of the manuscript.

Funding: This research was funded by the Natural Sciences and Engineering Research Council of Canada.

Data Availability Statement: Data is contained within the article or supplementary material.

Acknowledgments: We gratefully acknowledge the financial support of the Natural Sciences and Engineering Research Council of Canada.

Conflicts of Interest: The authors declare no conflict of interest.

Dedication: Julian Chojnowski is a father of modern silicone polymer chemistry. His scientific prowess and excellence is matched by his generosity. When I asked to use his name in what became the "Piers–Rubinsztajn reaction", he wanted instead to have his colleagues to take all the credit. We acknowledge with gratitude his personal and scientific contributions to our ability to undertake silicone-based science and, for this paper, rename the reaction the Piers–Chojnowski–Rubinsztajn–Kawakami (PCRK) reaction. The metathesis reaction should have his name.

Sample Availability: Samples of the compounds are not available from the authors.

References

1. Chojnowski, J.; Cypryk, M. Synthesis of Linear Polysiloxanes. In *Silicon-Containing Polymers: The Science and Technology of Their Synthesis*; Jones, R.G., Ando, W., Chojnowski, J., Eds.; Springer: Berlin/Heidelberg, Germany, 2000; pp. 1–42.
2. Kantor, S.W.; Grubb, W.T.; Osthoff, R.C. The Mechanism of the Acid- and Base-catalyzed Equilibration of Siloxanes. *J. Am. Chem. Soc.* **1954**, *76*, 5190–5197. [CrossRef]
3. Chojnowski, J. Polymerization. In *Siloxane Polymers*; Clarson, S.J., Semlyen, J.A., Eds.; Prentice Hall: Englewood Cliffs, NJ, USA, 1993; pp. 1–71.
4. Wright, P.V.; Semlyen, J.A. Equilibrium ring concentrations and the statistical conformations of polymer chains: Part 3. Substituent effects in polysiloxane systems. *Polymer* **1970**, *11*, 462–471. [CrossRef]
5. Franzen, A.; Greene, T.; Van Landingham, C.; Gentry, R. Toxicology of octamethylcyclotetrasiloxane (D4). *Toxic. Lett.* **2017**, *279*, 2–22. [CrossRef] [PubMed]
6. Wang, D.-G.; Steer, H.; Tait, T.; Williams, Z.; Pacepavicius, G.; Young, T.; Ng, T.; Smyth, S.A.; Kinsman, L.; Alaee, M. Concentrations of cyclic volatile methylsiloxanes in biosolid amended soil, influent, effluent, receiving water, and sediment of wastewater treatment plants in Canada. *Chemosphere* **2013**, *93*, 766–773. [CrossRef]
7. European Chemical Agency. Restriction proposal for siloxanes (D4, D5, D6) in personal care products. *ECHA 2020*. Available online: http://echa.europa.eu/-/echa-s-committees-conclude-on-five-restrictions (accessed on 3 January 2021).
8. Environment Canada Regulation on D4. Available online: www.canada.ca/en/environment-climate-change/services/management-toxic-substances/list-canadian-environmental-protection-act/slioxane-d4.html (accessed on 3 January 2021).
9. Peters, M.A.; Belu, A.M.; Linton, R.W.; Dupray, L.; Meyer, T.J.; DeSimone, J.M. Termination of Living Anionic Polymerizations Using Chlorosilane Derivatives: A General Synthetic Methodology for the Synthesis of End-Functionalized Polymers. *J. Am. Chem. Soc.* **1995**, *117*, 3380–3388. [CrossRef]
10. Frye, C.L.; Salinger, R.M.; Fearon, F.W.G.; Klosowski, J.M.; DeYoung, T. Reactions of organolithium reagents with siloxane substrates. *J. Org. Chem.* **1970**, *35*, 1308–1314. [CrossRef]
11. Maschke, U.; Wagner, T.; Coqueret, X. Synthesis of high-molecular-weight poly(dimethylsiloxane) of uniform size by anionic polymerization, 1. Initiation by a monofunctional lithium siloxanolate. *Makromol. Chem.* **1992**, *193*, 2453–2466. [CrossRef]
12. Fuchise, K.; Igarashi, M.; Sato, K.; Shimada, S. Organocatalytic controlled/living ring-opening polymerization of cyclotrisiloxanes initiated by water with strong organic base catalysts. *Chem. Sci.* **2018**, *9*, 2879–2891. [CrossRef]
13. Brook, M.A. *Silicon in Organic, Organometallic, and Polymer Chemistry*; Wiley: New York, NY, USA, 2000.
14. Pesti, J.; Larson, G.L. Tetramethyldisiloxane: A Practical Organosilane Reducing Agent. *Org. Proc. Res. Devel.* **2016**, *20*, 1164–1181. [CrossRef]
15. Larson, G.L. Silicon-Based Reducing Agents. *Gelest*. 2004. Available online: https://www.gelest.com/wp-content/uploads/Silicon-Based_Reducing_Agents.pdf (accessed on 3 January 2021).
16. Schneider, A.F.; Chen, Y.; Brook, M.A. Trace water affects tris(pentafluorophenyl)borane catalytic activity in the Piers–Rubinsztajn reaction. *Dalton Trans.* **2019**, *48*, 13599–13606. [CrossRef]
17. Piers, W.E. The Chemistry of Perfluoroaryl Boranes. In *Advances in Organometallic Chemistry*; Academic Press: Cambridge, MA, USA, 2004; Volume 52, pp. 1–76.
18. Jacobsen, H.; Berke, H.; Doring, S.; Kehr, G.; Erker, G.; Frohlich, R.; Meyer, O. Lewis acid properties of tris(pentafluorophenyl)borane. Structure and bonding in L-B(C_6F_5)$_3$ complexes. *Organometallics* **1999**, *18*, 1724–1735. [CrossRef]
19. Zheng, S.; Liao, M.; Chen, Y.; Brook, M.A. Dissolving used rubber tires. *Green Chem.* **2020**, *22*, 94–102. [CrossRef]
20. Rubinsztajn, S.; Cella, J.A. A New Polycondensation Process for the Preparation of Polysiloxane Copolymers. *Macromolecules* **2005**, *38*, 1061–1063. [CrossRef]
21. Chojnowski, J.; Rubinsztajn, S.; Cella, J.A.; Fortuniak, W.; Cypryk, M.; Kurjata, J.; Kaźmierski, K. Mechanism of the B(C6F5)3-Catalyzed Reaction of Silyl Hydrides with Alkoxysilanes. Kinetic and Spectroscopic Studies. *Organometallics* **2005**, *24*, 6077–6084. [CrossRef]
22. Chojnowski, J.; Fortuniak, W.; Kurjata, J.; Rubinsztajn, S.; Cella, J.A. Oligomerization of Hydrosiloxanes in the Presence of Tris(pentafluorophenyl)borane. *Macromolecules* **2006**, *39*, 3802–3807. [CrossRef]
23. Zhou, D.; Kawakami, Y. Tris(pentafluorophenyl)borane as a Superior Catalyst in the Synthesis of Optically Active SiO-Containing Polymers. *Macromolecules* **2005**, *38*, 6902–6908. [CrossRef]
24. Brook, M.A.; Grande, J.B.; Ganachaud, F. New Synthetic Strategies for Structured Silicones Using B(C_6F_5)$_3$. In *Silicon Polymers*; Muzafarov, A.M., Ed.; Springer: Berlin/Heidelberg, Germany, 2011; pp. 161–183. [CrossRef]
25. Brook, M.A. New Control Over Silicone Synthesis using SiH Chemistry: The Piers–Rubinsztajn Reaction. *Chem. Eur. J.* **2018**, *24*, 8458–8469. [CrossRef]
26. Liao, M.; Schneider, A.F.; Laengert, S.E.; Gale, C.B.; Chen, Y.; Brook, M.A. Living synthesis of silicone polymers controlled by humidity. *Eur. Polym. J.* **2018**, *107*, 287–293. [CrossRef]
27. Schneider, A.F.; Brook, M.A. High-Throughput Synthesis and Characterization of Aryl Silicones Using the Piers-Rubinsztajn Reaction. *Chem. Eur. J.* **2019**, *25*, 15367–15374. [CrossRef]
28. Schneider, A.F.; Lu, E.K.; Lu, G.; Brook, M.A. Facile synthesis of phenyl-rich functional siloxanes from simple silanes. *J. Polym. Sci.* **2020**, *58*, 3095–3106. [CrossRef]

29. Morgan, J.; Chen, T.; Hayes, R.; Dickie, T.; Urlich, T.; Brook, M.A. Facile synthesis of dendron-branched silicone polymers. *Polym. Chem.* **2017**, *8*, 2743–2746. [CrossRef]
30. Thompson, D.B.; Brook, M.A. Rapid Assembly of Complex 3D Siloxane Architectures. *J. Am. Chem. Soc.* **2008**, *130*, 32–33. [CrossRef] [PubMed]
31. Zheng, S.; Liang, S.; Chen, Y.; Brook, M.A. Hyperbranched Silicone MDTQ Tack Promoters. *Molecules* **2019**, *24*, 4133. [CrossRef] [PubMed]
32. Longuet, C.; Joly-Duhamel, C.; Ganachaud, F. Copolycondensation of regular functional silane and siloxane in aqueous emulsion using $B(C_6F_5)_3$ as a catalyst. *Macromol. Chem. Phys.* **2007**, *208*, 1883–1892. [CrossRef]
33. Chojnowski, J.; Rubinsztajn, S.; Fortuniak, W.; Kurjata, J. Oligomer and Polymer Formation in Hexamethylcyclotrisiloxane (D3) -Hydrosilane Systems Under Catalysis by tris(pentafluorophenyl)borane. *J. Inorg. Organomet. Polym. Mat.* **2007**, *17*, 173–187. [CrossRef]
34. Babonneau, F.; Gualandris, V.; Maquet, J.; Massiot, D.; Janicke, M.T.; Chmelka, B.F. Newly Applied Two-Dimensional Solid-State NMR Correlation Techniques for the Characterization of Organically Modified Silicates. *J. Sol-Gel Sci. Technol.* **2000**, *19*, 113–117. [CrossRef]
35. Taylor, R.B.; Parhboo, B.; Fillmore, D.M.; Fillmore, D.M. Nuclear Magnetic Resonance Spectroscopy. In *Analysis of Silicones*; Smith, A.L., Ed.; Wiley: Hoboken, NJ, USA, 1991; Chapter 12; pp. 347–420.
36. Mrozek, R.A.; Cole, P.J.; Otim, K.J.; Shull, K.R.; Lenhart, J.L. Influence of solvent size on the mechanical properties and rheology of polydimethylsiloxane-based polymeric gels. *Polymer* **2011**, *52*, 3422–3430. [CrossRef]
37. Williams, E.A. NMR Spectroscopy of Organosilicon Compounds. In *The Chemistry of Organic Silicon Compounds*; Patai, S., Rappoport, Z., Eds.; John Wiley & Sons: Chichester, UK, 1989; Volume 1, p. 511.
38. Rabanzo-Castillo, K.M.; Kumar, V.B.; Söhnel, T.; Leitao, E.M. Catalytic Synthesis of Oligosiloxanes Mediated by an Air Stable Catalyst, $(C_6F_5)_3B(OH_2)$. *Front. Chem.* **2020**, *8*, 477. [CrossRef]

molecules

MDPI

Article

Multiarm Star-Shaped Polydimethylsiloxanes with a Dendritic Branching Center

Pavel A. Tikhonov [1], Nataliya G. Vasilenko [1,*], Marat O. Gallyamov [2,3], Georgii V. Cherkaev [1], Viktor G. Vasil'ev [2], Nina V. Demchenko [1], Mikhail I. Buzin [2], Sergey G. Vasil'ev [4] and Aziz M. Muzafarov [1,2]

1 Enikolopov Institute of Synthetic Polymeric Materials, Russian Academy of Sciences, 117393 Moscow, Russia; tikhonpa@ispm.ru (P.A.T.); georgij.cherkaev@gmail.com (G.V.C.); nika1946@mail.ru (N.V.D.); aziz@ispm.ru (A.M.M.)
2 Nesmeyanov Institute of Organoelement Compounds, Russian Academy of Sciences, 119991 Moscow, Russia; glm@polly.phys.msu.ru (M.O.G.); viktor@ineos.ac.ru (V.G.V.); buzin@ineos.ac.ru (M.I.B.)
3 Faculty of Physics, Lomonosov Moscow State University, 119991 Moscow, Russia
4 Institute of Problems of Chemical Physics, Russian Academy of Sciences, Academician Semeonov av. 1, Chernogolovka, 142432 Moscow, Russia; viesssw@mail.ru
* Correspondence: n-vasilenko@mail.ru; Tel.: +79166269133

Abstract: New multiarm stars have been synthesized based on polylithium derivatives of high-generation carbosilane dendrimers. In the synthesis of multiarm stars based on the eighth-generation dendrimer, steric hindrances were observed even during the synthesis of a polylithium initiator. Subsequently, this led to chain transfer reactions between growing arms, as well as other side effects. As a result, dense nanogel formations with a higher tendency of ordering than in classical objects of this type were isolated from the reaction mixture. The study of the rheology of multiarm stars based on sixth-generation dendrimers made it possible to determine the activation energies of viscous flow in these objects, which makes it possible to consider them as objects with a macromolecular nature and a reptation flow mechanism.

Keywords: polylithium initiator; multiarm stars; anionic polymerization of hexamethylcyclotrisiloxane; carbosilane dendrimers

Citation: Tikhonov, P.A.; Vasilenko, N.G.; Gallyamov, M.O.; Cherkaev, G.V.; Vasil'ev, V.G.; Demchenko, N.V.; Buzin, M.I.; Vasil'ev, S.G.; Muzafarov, A.M. Multiarm Star-Shaped Polydimethylsiloxanes with a Dendritic Branching Center. *Molecules* **2021**, *26*, 3280. https://doi.org/10.3390/molecules26113280

Academic Editors: Sławomir Rubinsztajn, Marek Cypryk and Wlodzimierz Stanczyk

Received: 5 May 2021
Accepted: 26 May 2021
Published: 29 May 2021

1. Introduction

The anionic polymerization of organocyclosiloxanes, especially in the nonequilibrium variant, makes it possible not only to obtain silicone rubbers, whose production is currently estimated in the hundreds of thousands of tons, but also different macromolecular systems of various designs [1–10]. We dedicate this work to Chojnowski, whose large contribution to this area, including in terms of the synthesis of various multiarm systems, cannot be overestimated [11–13].

Star-shaped polydimethylsiloxane polymers have a long history; the first works on obtaining three- and four-armed systems were carried out in the 1960s in Kraus' [14] and Andrianov's [7] laboratories. At that time, researchers regarded them primarily as tools for the formation of defect-free polymer networks, but this approach was postponed for a long time due to the imperfection of the synthetic methods for such structures. The modern approach to the synthesis and research of star-shaped polydimethylsiloxanes was much more widespread and prepared. In this case, the real transition to precise multiarm stars occurred because of the appearance of dendrimers, with their practically unlimited and, at the same time, definite number of functional groups [15,16]. The work on synthesizing such structures was carried out by Roovers [17], Muzafarov [4], and Gnanou [18] and was largely initiated by the theoretical studies by Daoud and Cotton [19]. This time, the researchers were interested in the unusual rheology of stars, i.e., a decrease of their solution and melt viscosities upon an increase in the number of arms. Roovers and Hadjichristidis synthesized the first star-shaped structures using carbosilane dendrimers as branching centers, and they found that the high functionality of the initial branching center is a necessary but insufficient condition as the problem of the complete substitution of the center active groups by oligomeric arms in the case of their large number is not easy. Various strategies for multiarm star synthesis have been developed: convergent (arm-first) and divergent (core-first), as well as a combination of these two approaches [20]. The structures obtained with their help had properties so different from classical polymers that it was proposed to refer to them as macromolecular nanoobjects, or macromolecular particles, along with dendrimers, dense molecular brushes, and nanogels [21,22]. The boundary between low-arm stars that belong to branched polymers and multiarm stars can be determined using the α coefficient of the Mark–Kuhn–Houwink equation $[\eta] = K_\eta M^\alpha$. Its magnitude and, more importantly, the absence of the dependence of its magnitude on the molecular weight of the arm within the experimentally obtained values make it possible to unambiguously refer to the new objects as one or another classification group [23–26]. If a star-shaped polymer has a number of arms $f > 30$, the value of the α coefficient decreases with the number of arms to very low values for such systems (at $f > 100$, $\alpha = 0.06$) [24], which characterizes the object as a rigid and compact globule. The nature of these specific polymeric macromolecules determines the peculiarities of their behavior. A number of papers [27,28] have distinguish two limiting cases in star-shaped systems: the first type is stars with a small number of long arms and a low-volume core that remains permeable to elements of the neighboring stars and do not become arranged, and their terminal dynamics are mainly controlled by the relaxation of the arms. The second type is stars with a large number of short arms and a core that is impermeable to the elements of the neighboring stars, which show a clear arrangement and where the slowest dynamics are controlled by structural rearrangements. The first group can be regarded as a typical polymer system, but the second one demonstrates analogies with colloids. The studies were carried out on examples of polyisoprene and polybutadiene stars.

This work is an extension of the studies on star-shaped polydimethylsiloxanes (PDMSs) with a different number of arms [5,29–31]. It was previously shown that PDMS stars with a number of arms $f > 30$, based on an examination of their diluted solutions, refer to dense globular objects with $\alpha = 0.06$–0.15. A study on PDMS star rheology showed the Newtonian nature of the flow in systems with $f = 8$ and 32, and a pseudoplastic flow at $f = 128$ [31]. In this case, the viscous flow activation energy value E_{act} of all the synthesized samples, regardless of the number of arms, insignificantly differed from the E_{act} of linear PDMS [32], which indicated the polymeric nature of the objects—even a 128-arm star with an arm length of ~60 links did not show signs of colloidal behavior. To further study the properties of multiarm PDMS systems, a series of model 128-arm objects based on a sixth-generation carbosilane dendrimer with different arm lengths were synthesized and studied, and an attempt was made to synthesize a 512-arm PDMS star based on dendrimer G8.

2. Results and Discussion

The possibility of obtaining multiarm stars with a known number of arms and their adjustable and known length was determined by the use of the nonequilibrium anionic polymerization of hexamethylcyclotrisiloxane with a regularly structured polylithium dendritic initiator (Scheme 1). The polyfunctional initiator was a derivative of a carbosilane dendrimer with a protective hydrocarbon layer to prevent the intermolecular aggregation of lithium atoms [31].

The shielding outer layer consisting of methyldidodecylsilyl groups was created by the hydrosilylation of didodecylmethylsilane (DDMS) and dendrimer **1**, leaving half of the outer allyl groups unreacted (Figures S1–S4). Based on the synthesized DDMS derivative **2**, we obtained a polylithium anionic polymerization macroinitiator **3** via reaction with n-butyllithium and TMEDA in a hexane solution. The lithiation reaction was monitored by the appearance of lithium carbanion signals on the ^{1}H NMR spectra, along with disappearance signals from allyl groups, using diffusion filtration to suppress hexane signals. Based on the absence of signals from allyl groups in the ^{1}H NMR spectra (Figure S5), it is fair to say that the conversion of the carbanion formation reaction in the case of DDMS derivatives of fourth- and sixth-generation dendrimers was close to 100%, which allows, within the NMR technique error, one to assert with a high degree of certainty the presence of active lithium centers in the structure of the macroinitiator, in an amount similar to the number of allyl groups in the dendrimer-derived DDMS. Earlier, the GPC technique showed [29] that the organolithium product of lithiation terminated by trimethylchlorosilane is monomodal and has a narrow polydispersity, which proves the complete absence of intermolecular crosslinking reactions.

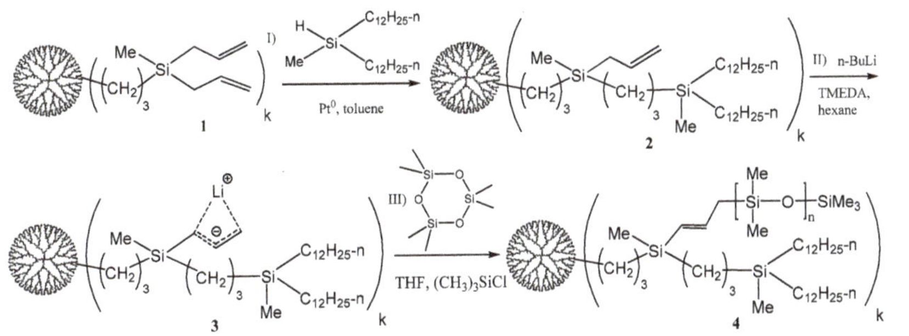

Scheme 1. Synthesis of star-shaped PDMS based on carbosilane dendrimer, where k = 32, 128, and 512.

The polymerization of hexamethylcyclotrisiloxane using a synthesized polyfunctional initiator with different monomer/initiator ratios (Table S1) yielded a series of 128-arm star-shaped PDMSs, with monomodal narrow-dispersed molecular weight distributions according to GPC data (Figure 1). The presence of a certain number of didodecylsilyl group protons on the branching center in the structure of the synthesized compounds made it possible to calculate the number average molecular weight of the macromolecule and the arm length using the ^{1}H NMR spectra data (Figures S6–S9, Formula S1). The results of calculations are presented in Table 1.

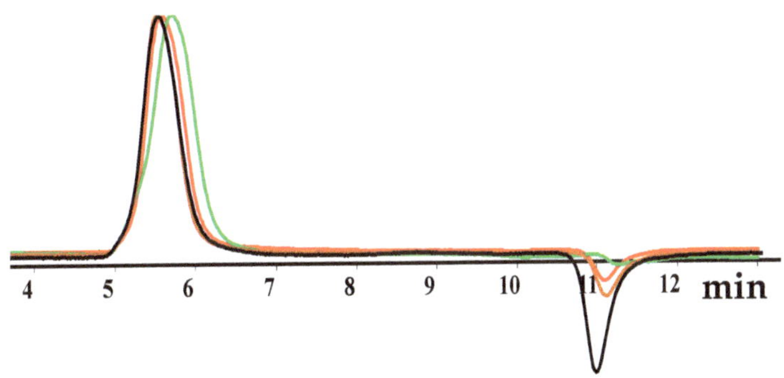

Figure 1. GPC curves of star-shaped PDMSs with number of arms f = 128 and different arm lengths.

In the text, samples of star-shaped polydimethylsiloxanes (PDMS) are designated as St-arms number-arm length, for example, St-128-114.

The light scattering method confirmed the monomodality and narrow dispersion of the obtained objects according to the GPC data; as an example, the obtained curves for a specimen St-128-33 are shown in Figure 2.

Table 1. Characteristics of the obtained star-shaped 128-arm PDMSs.

Sample	PDI	MW (According to ^{1}H NMR)	E_a of the Viscous Flow, kJ/mol	Viscosity (at 20 °C, Shear Rate 10 s^{-1})	E_a, kJ/mol (Self-diffusion, Melt)	D, 10^{-13} m^2/s (20 °C, Melt)	D, 10^{-11} m^2/s (25 °C, 0.9% Solution in Toluene)	Hydrodynamic Radius R_h, nm (25 °C, 0.9% Solution in Toluene)	E_a, kJ/mol (Self-diffusion, 0.9% Solution in Toluene)
St-128-33	1.12	402,700	17.7	4.04	15.1	4.32	8.97	4.06	9.58
St-128-59	1.19	639,500	18.5	4.39	14.6	3.47	8.97	4.42	8.78
St-128-87	1.07	914,200	18.5	3.11	15.1	3.37	9.37	4.24	9.29
St-28-114	1.07	1,169,900	16.1	2.56	13.2	3.87	8.59	4.62	8.59

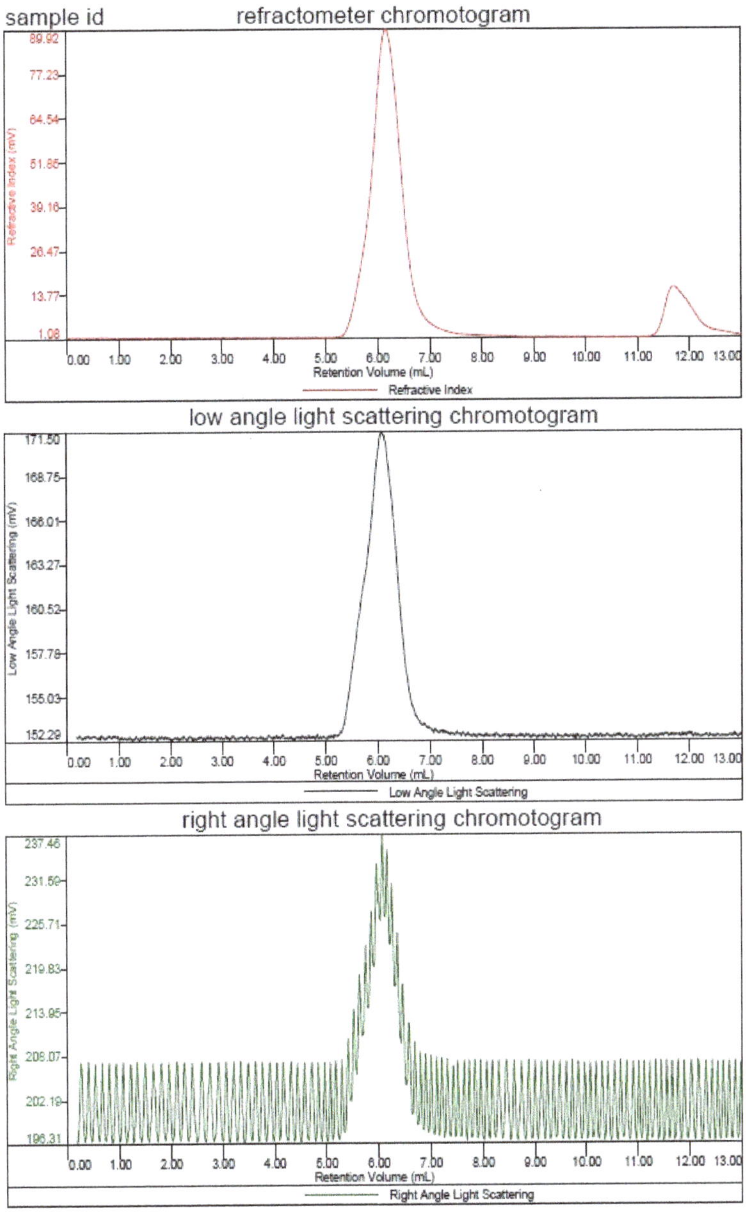

Figure 2. Static light scattering data obtained for St-128-33.

Thus, the synthesis of the 32-arm PDMS star as a comparison object and a series of 128-arm PDMS stars with different arm lengths was successfully carried out; however, the use of this approach for the synthesis of star-shaped 512-arm PDMSs with an eighth-generation carbosilane dendrimer as the initial compound was unsuccessful. While the structure of the initial allyl functional G8 dendrimer, according to GPC and ^{1}H NMR spectroscopy data, corresponded to the theoretical one, the reaction of the didodecylmethylsilane hydrosilylation of half of the allyl groups did not yield the target conversion of the hydrosilyl groups above ~60%. Even with the toughening of the reaction conditions and a significant increase in the reaction time, only the increased migration of the double bond was observed (Figure 3).

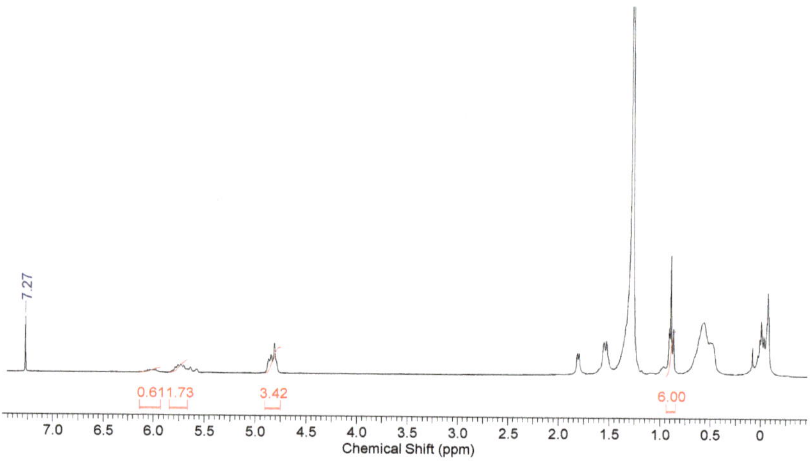

Figure 3. ^{1}H NMR spectrum of the purified DDMS derivative of the eighth-generation carbosilane dendrimer.

The GPC of the reaction mixture showed the presence of a low-molecular-weight fraction, probably as a result of the presence of side reactions of hydrosilanes with the toughening of the process conditions (Figure 4).

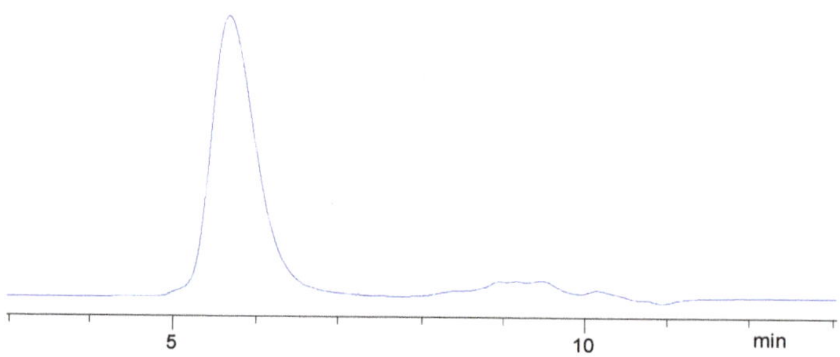

Figure 4. GPC curve of the 8th generation dendrimer DDMS derivative reaction mixture.

It can be assumed that, in the case of a G8 dendrimer, the outer layer is packed so densely that the accessibility of all unsaturated groups is impaired, and the steric factor does not allow the required number of hydrosilyl group additions to occur.

The lithiation reaction also proceeded slowly and not to the same high conversion as in the case of dendrimer G6. The introduction of an additional amount of n-BuLi and the continuation of the reaction for 92 h did not result in a target conversion of allyl groups, which noticeably differentiates the processes on the eighth-generation dendrimer from similar reactions using previous objects of this series with less dense outer layers (Figure S10).

Thus, it was found that the processes on dendrimer G8, both the silylation of half of the allyl groups and lithiation of the remaining ones, were much more difficult, and they proceeded to a lower depth than in previous generations. In this case, the tougher reaction conditions, as well as an increase in their duration, did not lead to an increase in the conversion of functionalities, which indicates the presence of inaccessible groups, apparently due to steric hindrances in the especially dense outer sphere of the eighth-generation dendrimer, as there were no other differences from processes on the previous generations. Quite naturally, the use of such an initiator, which possessed unequally active centers from 500 to 700 in the anionic polymerization of hexamethylcyclotrisiloxane, led to the formation of a star-shaped product with obvious polydispersity both in the number of arms and in their length. Product fractionation in a preparative chromatograph and fraction composition analysis by ^{1}H NMR spectroscopy showed that the content of the dimethylsiloxane component was ~10 times as high in the low-molecular-weight fraction compared to the high-molecular-weight fraction.

Thus, it can be stated that the sixth-generation carbosilane dendrimer is a kind of threshold, the last object in the series of carbosilane dendrimers with an outer layer density sufficient for full accessibility and equivalence of all terminal functional groups. It should be noted that, already in this case, the silylation processes were slower and under tougher conditions than in the previous generations; however, as a result, their target conversion was achieved. The structure of the eighth-generation carbosilane dendrimer, apparently due to steric hindrances and the appearance of inaccessible groups, did not allow target conversion of the terminal functionalities, which prevented the obtaining of a regularly structured polyfunctional macroinitiator based on it.

Earlier, we showed [31] that low-arm star-shaped PDMSs with f = 8 and 32 were Newtonian fluids, similar to linear PDMS, but an increase in the number of arms (f = 128 in our case) led to the appearance of pseudoplastic properties. The obtained flow curves of the synthesized series of 128-arm PDMS showed (Figures S12–S15) that, while the specimens with an arm length of up to 100 dimethylsiloxane units were shear thinning liquids, an increase in the arm length to n = 114 again demonstrates a flow pattern close to Newtonian, that is, the rheological properties were again determined by the behavior of the lateral PDMS arms. Moreover, based on repeated measurements of the flow curve upon cooling to room temperature after heating, it can be seen that the original curve at 20° did not coincide with the curve obtained after cooling the specimens, which may indicate the presence of slow relaxation processes in the star-shaped PDMS specimens.

Based on the obtained flow curves, viscous flow activation energy values were calculated for all 128-arm star-shaped PDMSs using the Arrhenius Equation (1):

$$\eta = Ae^{Ea/RT} \tag{1}$$

The values obtained in this series insignificantly differ from the E_{act} of linear PDMSs [32], and at a large arm length n = 114, their values completely coincide, which characterizes even such multiarm high-density structures within the considered arm lengths as polymeric objects rather than colloidal particles (Table 1).

A structural investigation of the synthesized series of specimens by means of PFG NMR in diluted solutions of toluene confirmed their narrow dispersion. The determined self-diffusion E_a of 128-arm PDMS stars in melts are presented in Table 1. Diffusion attenuation was obtained as the dependence of the stimulated echo amplitude (A) on the amplitude of the magnetic field gradient (g). The self-diffusion coefficient (SDC) was determined from the diffusion attenuation (Equation (2):, according to the expression [33]:

$$A = A_0 \exp\left(-\gamma^2 g^2 \delta^2 t_d D\right) \; (*) \tag{2}$$

where γ is the gyromagnetic ratio, δ is the gradient pulse duration, $t_d = (\Delta - \delta/3)$ is the effective diffusion time, Δ is the interval between the gradient pulses, and D is the self-diffusion coefficient. For diluted star-shaped PDMS solutions in toluene, the diffusion attenuations had an exponential form (2), which confirms the narrow dispersion of the molecular weight distribution. The SDC values were of the order of 1×10^{-10} m^2/s. A complex nonexponential attenuation form was observed for the 128-arm star-shaped PDMS melts. The gradient pulse duration δ and the diffusion time t_d were 3 and 99 ms, respectively, in the experiments used to determine the activation energy of the self-diffusion (Figure 5). The attenuations were approximated by two terms of the form (2) with relative contributions p_1 and p_2 ($p_1 + p_2 = 1$), from which an average SDC was determined Equation (3):

$$D = p_1 D_1 + p_2 D_2 \quad (3)$$

The values of p_1 and p_2 were selected individually for each specimen but were kept constant at different temperatures. Such an approach allows for a good approximation of at least a fifty-fold decay of the stimulated echo amplitude. The p_1 values corresponding to the higher self-diffusion coefficient were in the range of 0.3–0.7. The dependence of the average coefficient on temperature is shown in Figure 5. The self-diffusion activation energy values were determined using approximation by Arrhenius dependences of the form Equation (4) :

$$D = D_0 e^{-Ea/RT} \quad (4)$$

in the corresponding linear coordinates. The results are shown by solid lines in Figure 5. The greatest differences between D_1 and D_2 SDCs, as well as between the values of p_1 and p_2, were observed for the specimen with the highest molecular weight (St-128-114). For this sample, a series of experiments with varying diffusion time t_d was performed at 25 °C. Changes in the shape of diffusion attenuation were not observed for t_d in the range of 100–500 ms. Thus, the restricted in-cage motion, which is characteristic of the behavior of colloidal suspensions [26], was not observed on the considered timescale. On the other hand, the appearance of nonexponential diffusion attenuations in melts contrasted the exponential attenuations in diluted solutions, indicating the presence of dynamical heterogeneities. On the whole, the activation energies of multiarm star-shaped PDMSs were close to each other. These values are slightly lower than the viscous flow activation energies and are also lower than the value of ~18.8 kJ/mol determined for the self-diffusion of linear PDMSs [34].

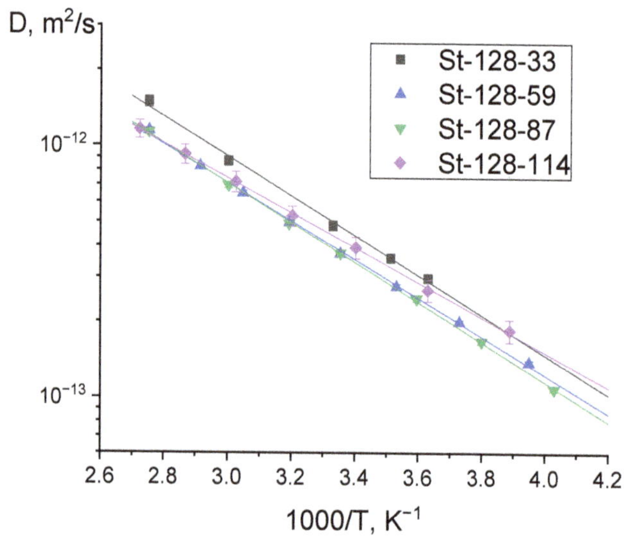

Figure 5. Dependence of the average self-diffusion coefficient of 128-arm PDMSs with different arm lengths on temperature.

The synthesized star-shaped polymers were studied via DSC.

It was found that the DSC curves (Figure 6) showed thermal effects characteristic of a linear PDMS [35]. The jump in heat capacity at the glass transition temperature was expressed very weakly, as well as the exothermic cold crystallization effect (Table 2). The glass transition temperature of the synthesized stars was generally lower than that of a linear PDMS, similar to those observed in other star-shaped systems [36,37]. The melting peak of the crystalline phase was bimodal. The heat of fusion was lower than that observed for a linear PDMS with a comparable molecular weight, as was previously observed [22–24]. Up to a temperature of 170 °C, no heat effects were observed above the melting of the PDMS crystalline phase.

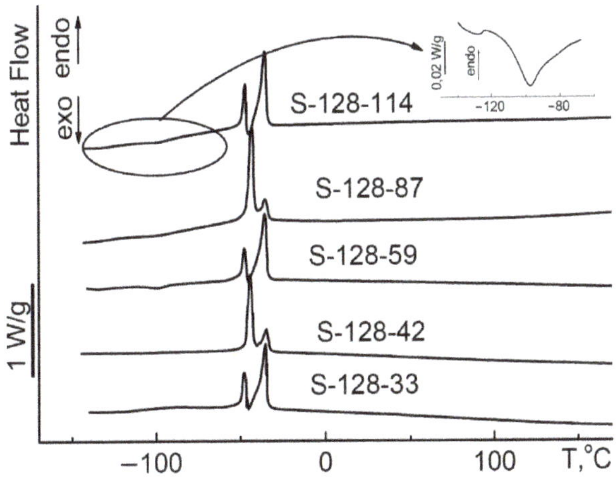

Figure 6. DSC curves of star-shaped PDMSs with different arm lengths.

Table 2. Thermal characteristics of 128-arm PDMSs with various arm lengths.

Sample	T_g, °C	T_{cc}, °C	ΔH_{cc}, J/g	T_m, °C	ΔH_m, J/g
St-128-33	−126	−97	0.1	−48/−35	27.8
St-128-59	−128	−99	2.0	−47/−36	27.5
St-128-87	−121	−97	4.0	−42/−36	30.5
St-128-114	−125	−98	3.3	−49/−35	31.7

The results of examining the properties of multiarm stars have led to conflicting outcomes. On the one hand, we have received answers to key questions as to which system these objects are closer to—the colloidal or polymeric one. On the other hand, the DLS and GPC data indicated the monodisperse nature of the objects obtained, but the study of self-diffusion in melt indicated the presence of objects with different diffusion coefficients, while, in dilute solutions, we also see monodisperse objects by self-diffusion measurements. The impeccability of the synthetic approach also turns out to have its limits. In addition, while side processes during the synthesis of multiarm stars based on an eighth-generation dendrimer were expected, certain doubts arose: they were possibly also observed to a lesser extent when using lower generations. Indeed, in other cases, the formation of inhomogeneities was also reported during the drying of star specimens. Most of these doubts were dispelled at the stage of structure identification. As was already noted, specimens of sixth-generation dendrimers of different consistency always gave an identical monomodal GPC pattern and equally adequate NMR spectra. The unexpected diffusion coefficient results were attributed to the increased aggregation tendency of multiarm objects. Similar to dendrimers [38], while dried, multiarm stars form metastable aggregates that are stable under normal conditions. They become completely destroyed when dissolved; therefore, we see monomodal narrowly dispersed signals on GPC or DLS. The results of examining these objects by atomic force microscopy techniques gave rich food for thought (Figure 7).

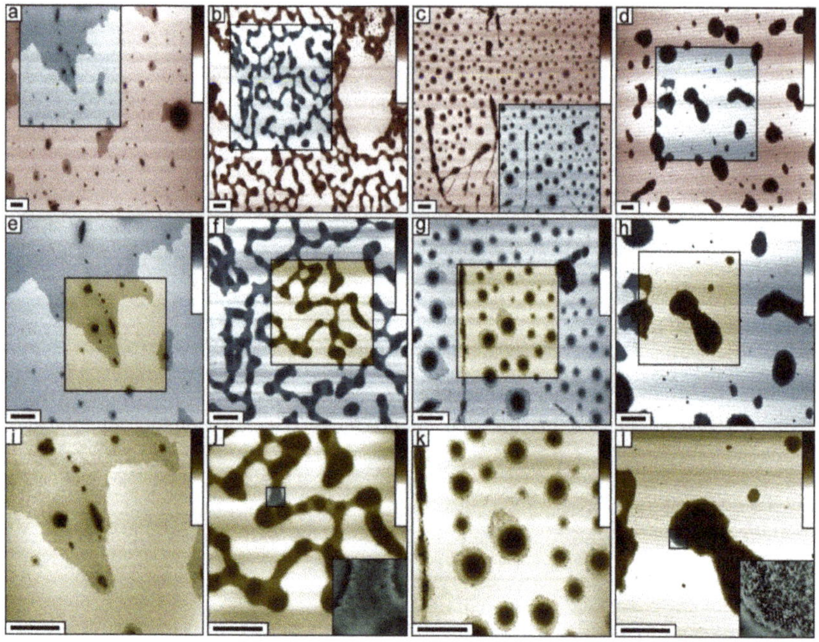

Figure 7. AFM results. From left to right: St-32-230 (G4 dendrimer as object of comparison) (**a**,**e**,**i**); St-128-33 (G6 dendrimer) (**b**,**f**,**j**); St-128-114 (G6 dendrimer) (**c**,**g**,**k**); St ~190-10 (G8 dendrimer) (**d**,**h**,**l**). Height scale, 30 nm. Scale line, 500 nm. Frame sizes in rows from top to bottom are 8 × 8 μm (**a**–**d**), 4 × 4 μm (**e**–**h**), and 2 × 2 μm (**i**–**l**). The inset images show a size of 190 × 190 nm (St-128-33 j, St ~190-10 l). The areas in rectangles have been rescanned and are shown in the figures below on a larger scale.

The AFM results showed that small stars (St-32-230, G4 dendrimer) are prone to aggregation: we saw a continuous layer of aggregated stars and individual aggregates in the figures in the first vertical column (a, e, i) with a sequential increase in resolution. The transition to objects obtained on the sixth-generation dendrimer (St-128-33 and St-128-114) led to a change in the aggregate nature. In the case of long arms (c, g, k), they were presented in the form of dense multilayer spherical objects and, at the same time, individual particles were observed, which, in terms of their size, can be attributed to single stars. Their diffusion characteristics were significantly lower than those of the lighter analogs from the first column, and in the process of rapid contraction of the solution, some of them "stray from the herd". Thus, the AFM data to a certain extent correlate with the results of self-diffusion studies.

However, of course, the most spectacular shots were obtained on the most unsuccessful specimen with a G8 dendrimer as a nucleus (St ~190-10). In this case, as described above, all the processes did not follow the general pattern; there was a cross-transition of the chain to the neighboring arms, and free linear PDMS oligomers and compacted systems were formed, which, of course, can be attributed to multiarm stars only conditionally. On the other hand, thanks to the stability of the core and the compaction of the shells, we see that after fractionation, these objects were prone to the formation of ordered layers with a crystal-like structure. The period of the visualized two-dimensional lattice (shown in the inserts) was 9.2 ± 0.6 nm for specimen St ~ 190-10 (G8 dendrimer, compacted siloxane shell) and 7.0 ± 0.4 nm for specimen St-128-33 (G6 dendrimer). Obviously, these numbers determine the lateral size of the corresponding individual polymeric object in the closely packed layer.

3. Materials and Methods

In this work, the following reagents and excipients were used: the carbosilane dendrimers G4, G6, and G8 with allyl functional groups in the shell, previously obtained using the technique from [39]; methyldichlorosilane 97% distilled with a reflux condenser manufactured by ABCR ((Karlsruhe, Germany); distilled allyl chloride 98% manufactured by Acros (Geel, Belgium); hexane 97.76% manufactured by Ruskhim.ru (Moscow, Russia), dried over calcium hydride and 3 Å molecular sieves; analytical-grade toluene manufactured by Khimpromtorg (Reutov, Russia), dried over sodium with benzophenone and 3 Å molecular sieves; tetrahydrofuran (THF) 99.8% manufactured by Ruskhim.ru, dried over sodium with benzophenone and 3 Å molecular sieves; platinum(0)-1,3-divinyl,1,1,3,3-tetramethyldisiloxane complex–2% solution in xylene (Karstedt's catalyst, Sigma-Aldrich, St. Louis, Missouri, MO, USA); 1-bromododecane 98% manufactured by Acros, magnesium shavings 99% manufactured by Fluka (Sigma-Aldrich, St. Louis, Missouri, MO, USA); hexamethylcyclotrisiloxane (D_3) 95% manufactured by ABCR, dried and distilled over calcium hydride; n-butyllithium 1.6 M solution in hexane manufactured by Acros; tetramethylethylenediamine (TMEDA) (99%, Acros), dried over 3 Å molecular sieves; petroleum ether distilled on a rotary evaporator manufactured by Ruskhim.ru; silica gel manufactured by Khimmed (Moscow, Russia), 0.040–0.063 mm; anhydrous sodium sulfate manufactured by Komponent-reaktiv (Moscow, Russia); saturated ammonium chloride solution manufactured by MCD (Moscow, Russia).

GPC analysis was carried out on the following chromatographic system: a STAYER s.2 high-pressure pump (Akvilon, Moscow, Russia), a Smartline RI 2300 refractometer, and a JETSTREAM 2 PLUS thermostat (KNAUER, Berlin, Germany). The thermostat temperature was 40 °C (±0.1 °C). The eluents were THF and toluene +2% THF; the flow rate was 1.0 mL/min. Columns 300 × 7.8 mm, Phenogel sorbent (Phenomenex, Los Angeles, CA, USA), 5 μm with pores ranging from 10^3 to 10^5 Å.

^{1}H NMR spectra were recorded on a Bruker WP250 SY spectrometer (Billerica, Massachusetts, MA, USA), $CDCl_3$ solvent, as well as on a BrukerAvance AV300 spectrometer (Billerica, Massachusetts, MA, USA), tetramethyl silane external standard, n-hexane and $CDCl_3$ solvent. The spectra were processed on a computer while using the ACD/ChemSketch software, version 2020.1.2, Advanced Chemistry Development, Inc., (Toronto, ON, Canada).

Self-diffusion studies were carried out via pulsed magnetic field gradient NMR (PFG NMR),) on ^{1}H nuclei using the stimulated echo technique [32] on a Bruker Avance III–400 spectrometer (Billerica, Massachusetts, MA, USA). The spectrometer is equipped with a gradient system that allows one to obtain magnetic field gradient pulses with a maximum amplitude of 30 T/m. Diffusion attenuations were obtained as a function of the gradient amplitude at constant time intervals in the pulse sequence. The temperature was controlled within ±0.2 °C. The samples of neat star-shaped PDMS (melts) and 0.9%-by-weight solution in toluene (diluted solutions) were investigated.

Rheological characteristics were measured on an Anton Paar MCR 302 rheometer (Graz Austria) in constant shear rate mode with a cone-plane measuring unit (d = 25 mm).

Differential scanning calorimetry (DSC) studies were carried out on a DSC-822e instrument (Mettler-Toledo, Greifensee, Switzerland) in an argon atmosphere at a heating rate of 10 °C/min.

AFM experiments were carried out on a multimode scanning probe microscope with a NanoScope-IIIa controller and an MMAFM-2 scanning head in tapping mode (Digital Instruments, Santa Barbara, CA, USA). We used silicon cantilevers manufactured by Nanoworld (Neuchatel, Switzerland) with a resonant frequency of about 300 kHz. The recording density was 512 × 512 points at a horizontal-sweep frequency of 1 Hz. The recorded images were analyzed using the Nanoscope (Digital Instruments, Tonavanda, New York, NY, USA) and Femtoscan (ATC, Moscow, Russia) software. For examination, 1 μL of polymer object solution in hexane at a concentration of c = 1 g/L was applied to the surface of freshly cleaved mica and was dried in air.

4. Experimental Part

The polylithium initiator was synthesized using 4th, 6th, and 8th generation carbosilane dendrimers with an outer hydrocarbon layer according to the technique described in [4].

Multiarm star-shaped polydimethylsiloxanes were obtained by the anionic polymerization of hexamethylcyclotrisiloxane using a polylithium initiator in an amount corresponding to the given arm length in a 10% hexane solution with a THF polymerization promoter in a glove box under an inert atmosphere. The polymerization time ranged from 14 to 24 h, and the product was terminated afterward with trimethylchlorosilane. The resulting polymer was purified by extraction, filtration, and preparative gel permeation chromatography.

5. Conclusions

A series of 128-arm star-shaped PDMSs with arm lengths from 33 to 114 siloxane units were obtained using the core-first approach with a branching center based on carbosilane dendrimer G6 having a didodecylmethylsilyl layer that prevents the intermolecular aggregation of lithium atoms. It was found that the G6 dendrimer is the last generation where the packing density of the outer layer allows the existence of equally accessible and equally active terminal functional groups. When carrying out similar reactions of the terminal groups of a carbosilane dendrimer G8, the processes proceed with a noticeably lower conversion, and tougher reaction conditions, as well as an increase in their duration, lead to a number of side processes, such as the transition of the chain to neighboring arms, and unequal development of the polymerization process from the beginning. These results give us reason to say that the proposed approach for the formation of multiarm stars has limits for the formation of regular structures, and the eight generation is outside of this limit.

The rheology of the synthesized series almost completely coincides with the behavior of a linear PDMS with respect to viscous flow E_{act} when their viscosity characteristics differ by several orders of magnitude. In this case, specimens of 128-arm star-shaped PDMS with short arms, which are shear thinning liquids, fall out of the state of Newtonian fluids, usual for PDMS. All this characterizes even such multiarm high-density structures within the considered arm lengths as polymeric objects rather than colloidal particles. DSC data show the presence of glass transition and crystallization processes in the synthesized objects, completely identical to a linear PDMS.

Supplementary Materials: The following are available online. Figure S1: GPC curve of DDMS-derivative of carbosilane dendrimer of 4th generation, Figure S2: GPC curve of DDMS-derivative of carbosilane dendrimer of 6th generation, Figure S3: 1H NMR spectrum of DDMS-derivative of carbosilane dendrimer of 4th generation, Figure S4: 1H NMR spectrum of DDMS-derivative of carbosilane dendrimer of 6th generation, Figure S5: 1H NMR spectra recorded during lithiation of the 6th generation DDMS dendrimers without (left) and with (right) using diffusion filtration after 20 (a) and 44 h (b), Figure S6: 1H NMR spectrum of St-128-33, Figure S7: 1H NMR spectrum of St-128-59, Figure S8: 1H NMR spectrum of St-128-87, Figure S9: 1H NMR spectrum of St-128-114, Figure S10: 1H NMR spectra of the product of lithiation with n-butyllithium of the DDMS-derivative of carbosilane dendrimer G8 using diffusion filtration after 68 (1) and 92 h (2) since the start of the reaction, Figure S11: GPC curve of DDMS-derivative of the carbosilane dendrimer G8, Figure S12: Flow curves of St-128-33 at temperatures from 20 to 120 °C and re-measured at 20 °C after cooling from 120 °C, Figure S13: Flow curves of St-128-59 at temperatures from 20 to 120 °C and re-measured at 20 °C after cooling from 120 °C, Figure S14: Flow curves of St-128-87 at temperatures from 20 to 120 °C and re-measured at 20 °C after cooling from 120 °C, Figure S15: Flow curves of St-128-114 at temperatures from 20 to 120 °C and re-measured at 20 °C after cooling from 120 °C, Table S1: Polymerization parameters of 128-arm PDMS.

Author Contributions: Conceptualization—A.M.M.; methodology—N.G.V., synthesis and purification—P.A.T.; validation—G.V.C. (NMR) and N.V.D. (GPC); investigation—V.G.V. (Rheology), M.O.G. (AFM), M.I.B. (DSC), and S.G.V. (PFG NMR). All authors have read and agreed to the published version of the manuscript.

Funding: This research was funded by Russian Science Foundation grant NO 19-13-00340 (Synthesis). In studies of molecular weight distribution, NMR spectra were recorded with the support of the Ministry of Science and Higher Education of the Russian Federation (№ 0086-2019-0005) using scientific equipment of the Shared Facility Center "Center for Polymer Research" ISPM RAS.

Institutional Review Board Statement: Not applicable.

Informed Consent Statement: Not applicable.

Data Availability Statement: The data presented in this study are available in supplementary material.

Acknowledgments: Rheological measurements were provided in the laboratory of Polymer Physic INEOS RAS with financial support from the Ministry of Science and Higher Education of the Russian Federation. DSC results were obtained using the equipment of the Educational and Scientific Centre of Functional and Nanomaterials, Moscow Pedagogical State University with financial support from the Ministry of Science and Higher Education of the Russian Federation. The self-diffusion measurements were performed using the equipment of the Multi-User Analytical Center of IPCP RAS and Research Resource Center of Science Center in Chernogolovka RAS.

Conflicts of Interest: The authors declare no conflict of interest. The authors declare that they have no known competing financial interests or personal relationships that could have appeared to influence the work reported in this paper.

Sample Availability: Samples of the compounds are available from the authors.

References

1. Zhu, W.; Zhong, M.; Li, W.; Dong, H.; Matyjaszewski, K. Clickable stars by combination of AROP and aqueous AGET ATRP. *Macromolecules* **2011**, *44*, 1920–1926. [CrossRef]
2. Mathers, R.T.; Magenau, A.J.D.; Schroder, K.; Matyjaszewski, K. Overview of controlled/living polymerization methods of vinyl monomers. In *Monitoring Polymerization Reactions: From Fundamentals to Applications*, 1st ed.; Reed, W.F., Alb, A.M., Eds.; John Wiley & Sons, Inc.: New York, NY, USA, 2014. [CrossRef]
3. Matyjaszewski, K.; Miller, P.J.; Kickelbick, G.; Nakagawa, Y.; Diamanti, S.; Pacis, C. Organic-inorganic hybrid polymers from atom transfer radical polymerization and poly(dimethylsiloxane). In *Silicones and Silicone-Modified Materials*; American Chemical Society: Washington, DC, USA, 2000; Volume 729, pp. 270–283. [CrossRef]
4. Vasilenko, N.G.; Rebrov, E.A.; Muzafarov, A.M.; Esswein, B.; Striegel, B.; Moller, M. Preparation of multi-arm star polymers with polylithiated carbosilane dendrimers. *Macromol. Chem. Phys.* **1998**, *199*, 889–895. [CrossRef]
5. Vasilenko, N.G.; Ingat'eva, G.M.; Myakushev, V.D.; Rebrov, E.A.; Moller, M.; Muzafarov, A.M. Functional multiarms star-like polydimethylsiloxanes. *Dokl. Chem.* **2001**, *377*, 348–352. [CrossRef]
6. Tverdokhlebova, I.I.; Sipyagina, M.A.; Stepanova, Y.Y.; Golub, A.S.; Larina, T.A.; Pavlova, S.A. Synthesis and properties of star-shaped polymethylphenyl siloxane. *Polym. Sci. USSR* **1981**, *23*, 1562–1569. [CrossRef]
7. Andrianov, K.A.; Pavlova, S.A.; Tverdokhlebova, I.I. Trifunctional star-shaped poly(dimethylsiloxane). *Vysokomol. Soedin. Ser. B* **1968**, *10*, 16–18.
8. Molenberg, A.; Klok, H.A.; Möller, M.; Boileau, S.; Teyssié, D. Controlled polymerization of hexa-n-alkylcyclotrisiloxanes with long alkyl groups. *Macromolecules* **1997**, *30*, 792–794. [CrossRef]
9. Hubert, S.; Hemery, P.; Boileau, S. Anionic polymerization of cyclosiloxanes with cryptates as counterions: New results. *Makromol. Chem., Macromol. Symp.* **1986**, *6*, 247–252. [CrossRef]
10. Molenberg, A.; Sheiko, S.; Moller, M. A block copolymer from polystyrene and columnar liquid-crystalline poly(diethylsiloxane). *Macromolecules* **1996**, *29*, 3397–3400. [CrossRef]
11. Chojnowski, J. Kinetically controlled siloxane ring-opening polymerization. *J. Inorg. Organomet. Polym.* **1991**, *630*, 299–323. [CrossRef]
12. Fortuniak, W.; Chojnowski, J.; Sauvet, G. Controlled synthesis of siloxane polymers and siloxane-siloxane block copolymers with 3-chloropropyl groups pendant to the siloxane chain. *Macromol. Chem. Phys.* **2001**, *202*, 2306–2313. [CrossRef]
13. Chojnowski, J.; Cypryk, M.; Fortuniak, W.; Rozga-Wijas, K.; Scibiorek, M. Controlled synthesis of vinylmethylsiloxane-dimethylsiloxane gradient, block and alternate copolymers by anionic ROP of cyclotrisiloxanes. *Polymer* **2002**, *43*, 1993–2001. [CrossRef]
14. Kraus, G.; Gruver, J.T. Rheological properties of multichain polybutadienes. *J. Polym. Sci. A Polym. Chem.* **1965**, *3*, 105–122. [CrossRef]
15. Tomalia, D.A.; Frechet, J.M.J. Discovery of dendrimers and dendritic polymers: A brief historical perspective. *J. Polym. Sci. A Polym. Chem.* **2002**, *40*, 2719–2728. [CrossRef]
16. Newkome, G.R.; Moorefield, C.N.; Vögtle, F. *Dendritic Molecules: Concepts, Syntheses, Perspectives*; Wiley-VCH: Weinheim, Germany, 1996. [CrossRef]
17. Roovers, J.; Zhouj, L.-L.; Toporowski, P.M.; Hadjichristidis, N. Regular Star Polymers with 64 and 128 Arms. Models for Polymeric Micelles. *Macromolecules* **1993**, *26*, 4324–4331. [CrossRef]

18. Cloutet, E.; Fillaut, J.L.; Astruc, D.; Gnanou, Y. Newly designed star-shaped polystyrene: Synthesis and characterization. *Macromolecules* **1998**, *31*, 6748–6755. [CrossRef]
19. Daoud, M.; Cotton, J.P. Star shaped polymers: A model for the conformation and its concentration dependence. *J. Physique* **1982**, *43*, 531–538. [CrossRef]
20. Ren, J.M.; McKenzie, T.G.; Fu, Q. Star Polymers. *Chem. Rev.* **2016**, *116*, 6743–6836. [CrossRef]
21. Kim, C.; Hong, J. Carbosilane and carbosiloxane dendrimers. *Molecules* **2009**, *14*, 3719–3730. [CrossRef]
22. Muzafarov, A.M.; Tatarinova, E.A.; Vasilenko, N.V.; Ignat'eva, G.M. Organosilicon dendrimers and irregular hyperbranched polymers. *Elsevier* **2017**, 323–382. [CrossRef]
23. Ito, K.; Tomi, Y.; Kawaguchi, S. Poly(ethylene oxide) macromonomers. 10. characterization and solution properties of the regular comb polymers with polystyrene main chains and poly(ethylene oxide) side chains. *Macromolecules* **1992**, *25*, 1534–1538. [CrossRef]
24. Drohmann, C.; Gorbatsevich, O.B.; Muzafarov, A.M.; Moller, M. Detection of molecular weight and structure of hyperbranched polymers. *Am. Chem. Soc. Polym. Prepr. Div. Polym. Chem.* **2000**, *41*, 959–960.
25. Goh, T.K.; Coventry, K.D.; Blencowe, A.; Qiao, G.G. Rheology of core cross-linked star polymers. *Polymer* **2008**, *49*, 5095–5104. [CrossRef]
26. Filippov, A.; Amirova, A.I.; Kirila, T.; Belyaeva, E.V.; Sheremetyeva, N.A.; Muzafarov, A.M. Influence of branching regularity on the behavior of hyperbranched polymers in solution. *Polym. Int.* **2015**, *64*, 780–786. [CrossRef]
27. Vlassopoulos, D.; Fytas, G.; Pakula, T.; Roovers, J. Multiarm star polymers dynamics. *J. Phys. Condens. Matter* **2001**, *13*, R855–R876. [CrossRef]
28. Pakula, T.; Vlassopoulos, D.; Fytas, G.; Roovers, J. Structure and dynamics of melts of multiarm polymer stars. *Macromolecules* **1998**, *31*, 8931–8940. [CrossRef]
29. Novozhilov, O.V.; Pavlichenko, I.V.; Demchenko, N.V.; Buzin, A.I.; Vasilenko, N.G.; Muzafarov, A.M. Multiarm star-like polydimethylsiloxanes based on dendrimers of the sixth generation. *Russ. Chem. Bull.* **2010**, *59*, 1909–1917. [CrossRef]
30. Novozhilov, O.V.; Vasilenko, N.G.; Buzin, M.I.; Scherbina, M.A.; Chvalun, S.N.; Muzafarov, A.M. Structuring of star-like multiarm polydimethylsiloxanes. *Russ. Chem. Bull.* **2011**, *60*, 1019–1021. [CrossRef]
31. Tikhonov, P.A.; Vasilenko, N.G.; Cherkaev, G.V.; Vasil'ev, V.G.; Demchenko, N.V.; Tatarinova, E.A.; Muzafarov, A.M. Synthesis and rheological properties of star-shaped polydimethylsiloxanes based on carbosilane dendrimers. *Mendeleev Commun.* **2019**, *29*, 625–627. [CrossRef]
32. Kataoka, T.; Ueda, S. Viscosity-molecular weight relationship for polydimethylsiloxane. *J. Polymer Sci. B* **1966**, *4*, 317. [CrossRef]
33. Tanner, J.E. Use of the stimulated echo in NMR diffusion studies. *J. Chem. Phys.* **1970**, *52*, 2523–2526. [CrossRef]
34. McCall, D.W.; Anderson, E.W. Self-diffusion in linear polydimethyl siloxane liquids. *Appl. Phys. Lett.* **1965**, *7*, 153–154. [CrossRef]
35. Clarson, S.J.; Dodgson, K.; Semlyen, J.A. Studies of cyclic and linear poly(dimethylsiloxanes): 19. Glass transition temperatures and crystallization behavior. *Polymer* **1985**, *26*, 930. [CrossRef]
36. Cameron, D.J.A.; Shaver, M.P. Aliphatic polyester polymer stars: Synthesis, properties and applications in biomedicine and nanotechnology. *Chem. Soc. Rev.* **2011**, *40*, 1761–1776. [CrossRef] [PubMed]
37. Yuan, W.; Liu, X.; Zou, H.; Li, J.; Yuan, H.; Ren, J. Synthesis, self-assembly, and properties of homoarm and heteroarm star-shaped inorganic–organic hybrid polymers with a POSS Core. *Macromol. Chem. Phys.* **2013**, *214*, 1580–1589. [CrossRef]
38. Milenin, S.A.; Cherkaev, G.V.; Demchenko, N.V.; Serkova, E.S.; Krasnova, I.Y.; Selezneva, E.V.; Buzin, M.I.; Bakirov, A.V.; Vasil'ev, V.G.; Shifrina, Z.B.; et al. Influence of the growing flexible shell on the molecular behavior of hybrid dendrimers. *Macromolecules* **2020**, *53*, 9706–9716. [CrossRef]
39. Muzafarov, A.M.; Gorbatsevich, O.B.; Rebrov, E.A.; Ignat'eva, G.M.; Chenskaya, T.B.; Myakushev, V.D.; Bulkin, A.F.; Papkov, V.S. Organosilicon dendrimers: Volume-growing polyallylcarbosilanes. *Polym. Sci.* **1993**, *35*, 1575–1580.

molecules

MDPI

Article

Reliable Condensation Curing Silicone Elastomers with Tailorable Properties

Alena Jurásková [1,2], Stefan Møller Olsen [2], Kim Dam-Johansen [3], Michael A. Brook [4] and Anne Ladegaard Skov [1,*]

1 DPC, Department of Chemical and Biochemical Engineering, Technical University of Denmark (DTU), Building 227, 2800 Kgs. Lyngby, Denmark; alejur@kt.dtu.dk
2 Hempel A/S, Lundtoftegårdsvej 91, 2800 Kgs. Lyngby, Denmark; STMO@hempel.com
3 CoaST, Department of Chemical and Biochemical Engineering, Technical University of Denmark (DTU), Building 229, 2800 Kgs. Lyngby, Denmark; kdj@kt.dtu.dk
4 Chemistry and Chemical Biology, McMaster University, 1280 Main St. W, Hamilton, ON L8S 4M1, Canada; mabrook@mcmaster.ca
* Correspondence: al@kt.dtu.dk; Tel.: +45-23-65-21-56; Fax: +45-45-88-22-58

Abstract: The long-term stability of condensation curing silicone elastomers can be affected by many factors such as curing environment, cross-linker type and concentration, and catalyst concentration. Mechanically unstable silicone elastomers may lead to undesirable application failure or reduced lifetime. This study investigates the stability of different condensation curing silicone elastomer compositions. Elastomers are prepared via the reaction of telechelic silanol-terminated polydimethylsiloxane (HO-PDMS-OH) with trimethoxysilane-terminated polysiloxane ($(MeO)_3Si\text{-}PDMS\text{-}Si(OMe)_3$) and ethoxy-terminated octakis(dimethylsiloxy)-T8-silsesquioxane ($(QM^{OEt})_8$), respectively. Two post-curing reactions are found to significantly affect both the stability of mechanical properties over time and final properties of the resulting elastomers: Namely, the condensation of dangling and/or unreacted polymer chains, and the reaction between cross-linker molecules. Findings from the stability study are then used to prepare reliable silicone elastomer coatings. Coating properties are tailored by varying the cross-linker molecular weight, type, and concentration. Finally, it is shown that, by proper choice of all three parameters, a coating with excellent scratch resistance and electrical breakdown strength can be produced even without an addition of fillers.

Keywords: silicone elastomers; stability; network structure; condensation curing; reinforcing cross-linker domains; coatings

Citation: Jurásková, A.; Møller Olsen, S.; Dam-Johansen, K.; Brook, M.A.; Skov, A.L. Reliable Condensation Curing Silicone Elastomers with Tailorable Properties. *Molecules* **2021**, *26*, 82. https://dx.doi.org/10.3390/molecules26010082

Received: 5 December 2020
Accepted: 23 December 2020
Published: 27 December 2020

1. Introduction

Condensation curing silicone elastomers are commonly used as protective coatings and sealants [1–5]. They are prepared via a condensation reaction between a HO-PDMS-OH and a silane cross-linker with hydrolysable groups such as amino, amide, acyloxy, ketoxime, or alkoxy [6]. Alkoxysilanes are preferred for this reaction, as they are non-corrosive, inexpensive, and do not produce toxic by-products during curing [7]. The condensation reaction allows efficient curing in ambient environment, which represents a significant advantage over silicone elastomers prepared by addition, radical, or UV curing. Besides the mild curing conditions, condensation curing silicone elastomer applications also benefit from the superior properties common to silicone elastomers, such as high flexibility, low surface energy, high chemical resistance, and excellent thermal stability [1,8,9]. Nevertheless, one drawback of condensation curing silicone elastomers is poor control over the curing reaction [6,9,10].

In our previous study [10], we showed that an inappropriate choice of the network formulation significantly compromises long-term elastomer stability such that the elastomer's properties change continuously over time (the study was terminated after 6 months, at which time changes were still being observed). Several factors, such as cross-linker volatility and purity, as well as catalyst concentration, were shown to significantly affect network formation and thus also contribute to poor reliability and reproducibility of silicone elastomers. For example, low molecular weight alkoxysilane cross-linkers tend to evaporate from the elastomer mixture and are prone to premature hydrolysis-condensation reactions (Figure 1a,b) [10–14]. In addition, HO-PDMS-OH condenses in the presence of a tin catalyst, which further contributes to network formulation difficulties due to the loss of reactive groups participating in the crosslinking reaction (Figure 1c) [10,15]. The cross-linker volatility and condensation of HO-PDMS-OH affect the true stoichiometric ratio between the functional group of cross-linker and polymer: r = f[cross-linker]/2[HO-PDMS-OH, where [...] denotes the original concentration of the species. Depending on the final *r*, we hypothesized two types of post-curing effect leading to long-term elastomer instability: The reaction between cross-linker molecules in the case of cross-linker excess, and the condensation of dangling and/or unreacted silanol-terminated polymer chains in the case of cross-linker deficit [10].

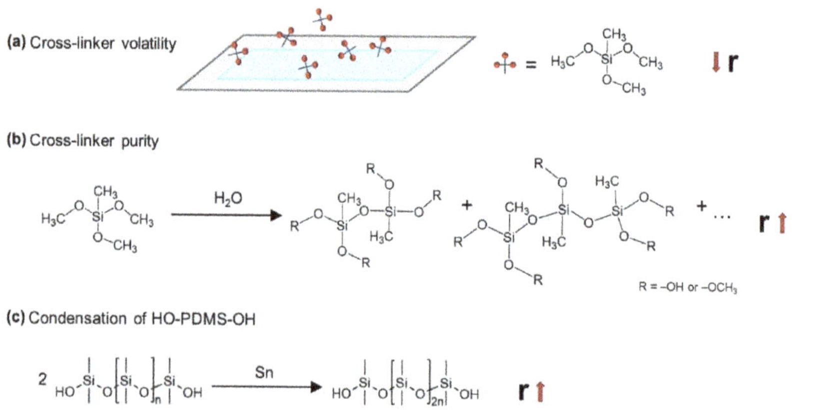

Figure 1. Factors influencing the true stoichiometric ratio, r = f[cross-linker]/2[HO-PDMS-OH], in a condensation curing silicone elastomer formulation [10]. The example shown is a system consisting of HO-PDMS-OH, methyltrimethoxysilane (f = 3), and a tin catalyst: (**a**) Cross-linker volatility, (**b**) hydrolysis-condensation reaction of the cross-linker molecules during storage leading to cross-linker impurity, and (**c**) condensation reaction of HO-PDMS-OH in the presence of a tin catalyst.

It was obvious that in order to develop stable condensation curing silicone elastomers, whose properties do not change significantly over time, potential post-curing effects must be investigated. The first part of this study is therefore focused on optimizing elastomer formulations, paying particular attention to the relationship between the stoichiometric ratio and the type/extent of the post-curing reaction. Trimethoxysilane-terminated polysiloxanes ($(MeO)_3Si$-PDMS-$Si(OMe)_3$) and ethoxy-terminated silsesquioxane ($(QM^{OEt})_8$) are used as cross-linkers. The choice of the cross-linkers is based on previous research [10] in which we showed that low molecular weight alkoxysilane cross-linkers constitute the largest obstacle towards developing stable silicone elastomers with reproducible material performance. This is because of their volatility, which causes the curing to become highly dependent on both sample dimensions and the surrounding environment. In addition, we also showed that methyltrimethoxysilane does not allow curing of thin films ($\leq$200 μm), unless the films are covered during the initial stage of the curing reaction. For thin silicone elastomer films, which need to be cured in open air, higher molecular weight alkoxysiloxane cross-linkers should therefore be used.

After the thorough study of the relation between condensation curing silicone elastomer formulation and elastomer stability, the post-curing effects were well understood. The second part of this work therefore focuses on the performance of stable elastomer films and the possibility to tailor their properties, in particular Young's modulus, elongation at break, electrical breakdown strength, and scratch resistance.

2. Results and Discussion

2.1. Optimization of Condensation Curing Silicone Elastomer Formulations

In our previous study, we hypothesized two post-curing effects leading to condensation curing silicone elastomer instability: The condensation of dangling and/or unreacted silanol-terminated polymer chains, and the reaction between cross-linker molecules [10]. In order to design stable condensation curing silicone elastomers, a better understanding of these post-curing reactions is needed. The work presented in this chapter is therefore focused on the mechanical stability of different condensation curing silicone elastomer formulations. Non-volatile alkoxy-terminated polysiloxanes and silsesquioxane cross-linkers were used to eliminate the negative effects of the volatile alkoxysilane cross-linkers. All elastomers were prepared using the same dibutyltin-dilaurate (Sn_DL) concentration (0.5 wt%) and film thickness (~100 μm). The individual silicone elastomer compositions are summarized in Table A2.

In the ideal case, an optimum network, so-called a model network, will be obtained using a stoichiometry between functional groups of cross-linker and polymer (r = 1) [9]. In such network, all functional groups will be reacted, as illustrated in Figure 2. Thereby, any eventual post-curing reactions causing network instability will be hindered. However, due to the numerous side reactions that take place during condensation curing and the steric hindrance of functional groups, such a model network is never actually formed. Hence, in order to prepare well-characterized and stable silicone elastomers, a stoichiometric ratio optimization study must be conducted for each individual cross-linker.

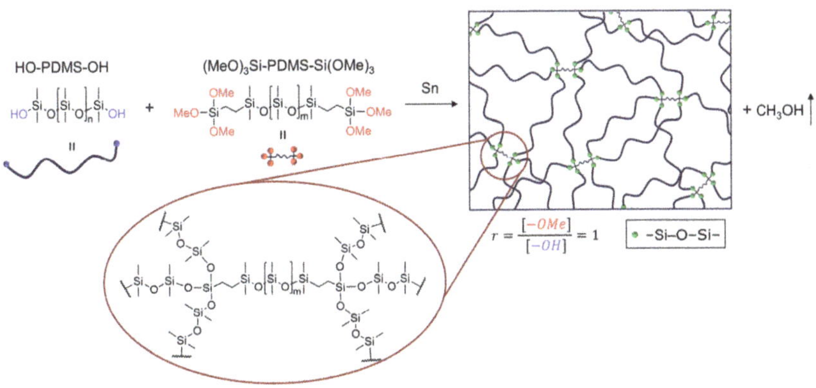

Figure 2. Example of an ideal network structure obtained by the reaction of HO-PDMS-OH with $(MeO)_3Si$-PDMS-$Si(OMe)_3$. The ideal network will be obtained if the components are used in a stoichiometric balance (r = 1), and no side reactions, such as condensation of HO-PDMS-OH and reaction between cross-linker molecules, are present. In addition, all functional groups will possess the same steric accessibility.

2.1.1. Trimethoxysilane-Terminated Polysiloxane Cross-Linkers

The stability of elastomer films (~100 μm) consisting of C2T and Di-10, Di-50, and Di-400, respectively, was evaluated via changes in Young's modulus over time. To investigate the impact of functional group imbalance on post-curing reactions, the stoichiometric ratio of these films was varied from r = 1.5 to r = 20 (Table A2). Figure 3 shows that E_C2T+Di-10 and E_C2T+Di-50 displayed similar trends with respect to alteration of the Young´s modulus over time. In particular, low stoichiometric ratio films (r = 1.5 and 2) showed a three- to four-fold increase in Young´s modulus over the course of four months. High stoichiometric ratio films (r = 15 and 20), on the other hand, exhibited a steep increase in Young´s modulus during the first three weeks, after which it reached a stable value. Finally, films with stoichiometric ratios of $r = 5$ and $r = 10$ displayed the smallest change in Young´s modulus over time. However, E-C2T+Di-400 showed a different trend, in which the Young´s modulus increased during first 30 days independently of the stoichiometric ratio, after which it reached a stable value in films with $r \geq 2$.

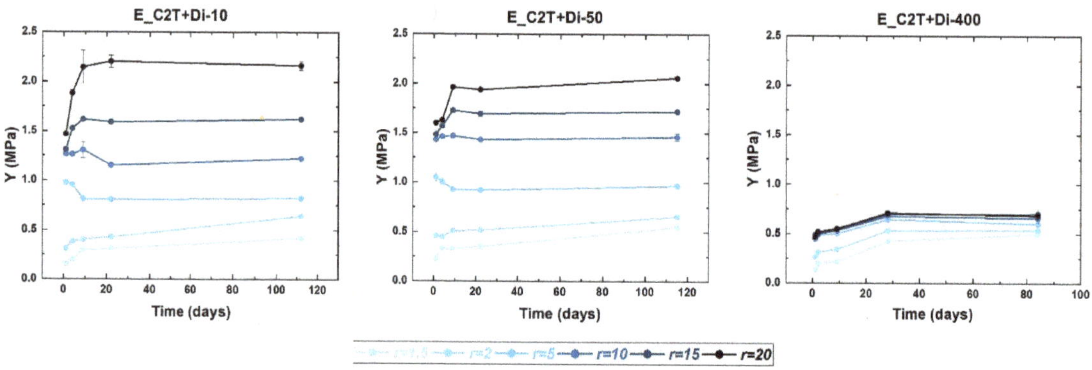

Figure 3. Stability of elastomer films assessed via change in Young´s modulus over time. The elastomers were prepared via the reaction between C2T and Di-10, Di-50, and Di-400, respectively. The stoichiometric ratio was varied from 1.5 to 20. Film thickness was ~100 μm.

To better understand the processes behind the increase in Young´s modulus over time, ^{1}H-NMR and SEC analysis of extracts from the elastomer films were conducted and the results are summarized in Figure 4. The amount of unreacted PDMS over time was calculated from ^{1}H-NMR spectra (Figure A2). Elastomers E_C2T+Di-50 with $r = 5$, 10, and 20 contained between 2 to 4 wt% of extractable PDMS. The corresponding SEC eluograms of the extracts showed a double peak at retention volumes between 17 and 22 mL. This double peak, whose intensity did not decrease over time, was also found in the eluograms of the C2T and Silmer cross-linkers (Figure A1). The extractable PDMS eluting at the retention volume 17–22 mL is therefore assumed to be a non-functional, low molecular weight PDMS originating from the manufacturing of the polymers/cross-linkers. Elastomer E_C2T+Di-50_r1.5 contained a significant amount of extractable PDMS, which decreased over time. The SEC eluograms showed, apart from the double peak at 17–22 mL, a peak at retention volumes between 11 and 17 mL, which corresponds to the retention volume of unreacted HO-PDMS-OH (C2T). The amount of PDMS extractable from E_C2T+Di-400 decreased over the first 30 days regardless of r. The peak at retention volumes between 11 and 17 mL can be then attributed to unreacted C2T and Silmer Di-400. After 27 days, extracts from E_C2T+Di-400 with $r = 5$, 10, and 20 reached stable values of ~5 wt% as a natural result of non-functional PDMS residues from the starting material.

Combining the knowledge gained from the development of Young´s modulus over time (Figure 3) with the extract analysis (Figure 4), several conclusions can be drawn regarding the post-curing reactions. First, the post-curing reaction between unreacted and/or dangling polymer chains is generally a slow process, lasting several months. As expected, this reaction takes place at lower stoichiometric ratios, where more dangling substructures are present. For the formulations studied here, this post-curing reaction occurred for $r < 5$ when Di-10 and Di-50 were used as cross-linkers, and occurred to a minor extent for all tested stoichiometric ratios when Di-400 was used as the cross-linker. This is due to the high molecular weight of Di-400, which is comparable to that of C2T. Second, the post-curing reaction between cross-linker molecules is a comparatively fast process that is completed within approximately 3 weeks, as evidenced by the rapid increase in Young´s modulus over time and the 0 wt% of extractable HO-PDMS-OH in the elastomer network. In the formulations tested here, the reaction between cross-linker molecules become significant at $r > 10$. As demonstrated later on in this work, the right choice of cross-linker leads to a favorable formation of cross-linker domains that provide reinforcement to the elastomer without the addition of fillers. Third, post-curing effects were smallest at $2 < r < 10$ as elastomers prepared within this range contain minimal amounts of both unreacted/dangling polymer chains and cross-linker domains. This is recognized by the combination of the minimum change in Young´s modulus over time and 0 wt% of extractable HO-PDMS-OH. The fact that the stoichiometric ratio, which produces the fewest post-curing reactions, is relatively high—far above the stoichiometric balance of $r = 1$—appears counterintuitive, but may be due to following factors: (1) Steric hindrance of the $-OCH_3$ groups of $(MeO)_3Si$-PDMS-$Si(OMe)_3$ cross-linker may hinder the reaction of all cross-linker functional groups; (2) the cross-linker molecules may undergo a condensation reaction during the main curing process, thereby creating unavoidable cross-linker domains, which decrease the true stoichiometry of the reaction mixture. A schematic illustration of the above-described effects of stoichiometric ratio and elastomer age on elastomer network structure and stability is summarized in Figure 5.

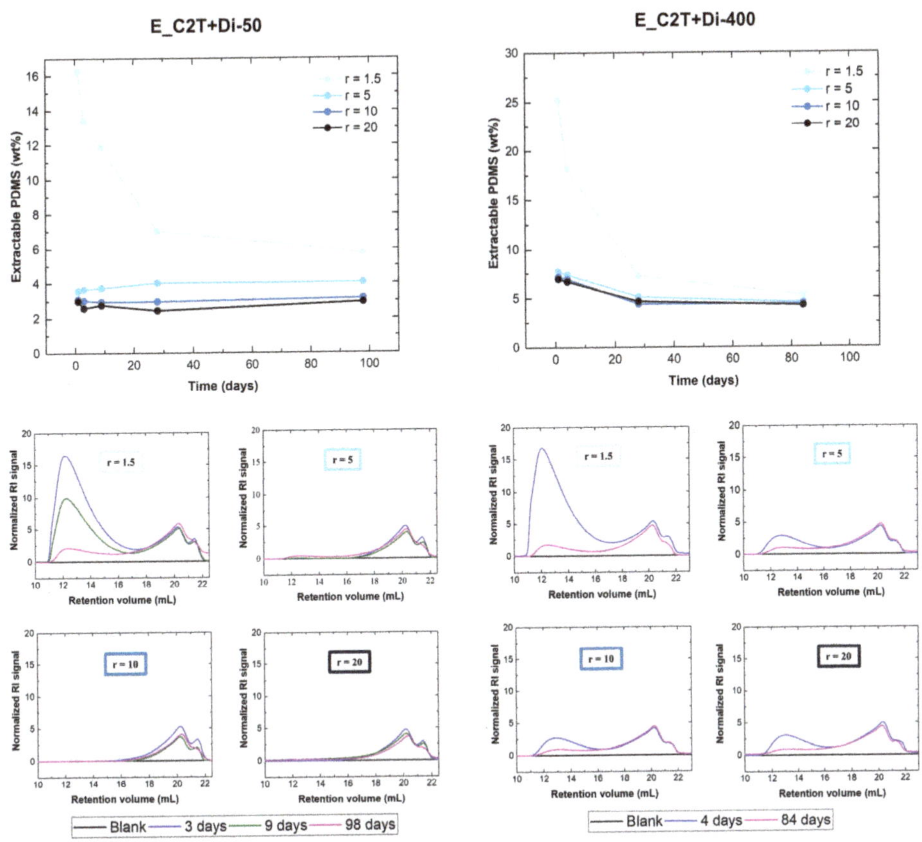

Figure 4. (**Top**) Extractable PDMS (wt%) as a function of time (determined from ^1H NMR analysis). (**Bottom**) SEC eluograms of the extractable PDMS. Analyses were performed on elastomers prepared via the reaction of C2T with Di-50 and Di-400, respectively. The stoichiometric ratio was varied from 1.5 to 20. Film thickness was ~100 μm.

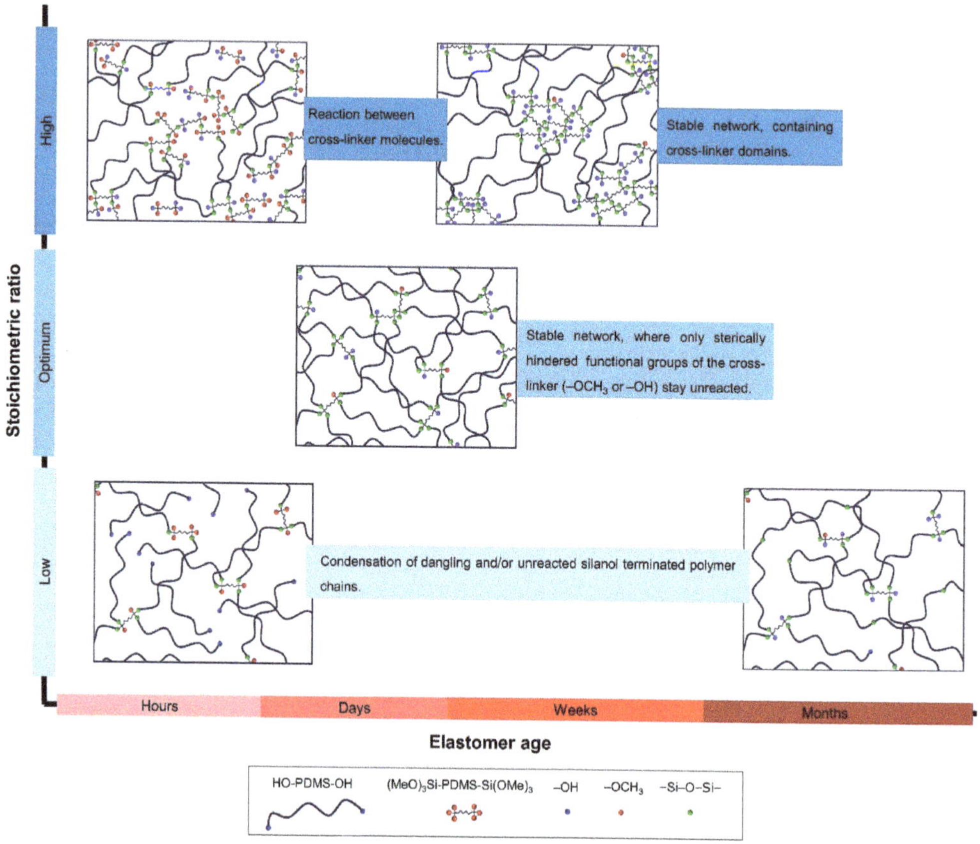

Figure 5. Schematic representation of the effects of stoichiometric ratio and elastomer age on elastomer network structure and stability.

2.1.2. Silsesquioxane Cross-Linker

A formulation optimization study similar to that presented for trimethoxysilane-terminated polysiloxanes cross-linkers was also conducted for $(QM^{OEt})_8$. The silsesquioxane cross-linker benefited from the fact that the $Si(CH_3)_2$ signal of the $(QM^{OEt})_8$ cross-linker is distinguishable from the $Si(CH_3)_2$ signal of the PDMS (C2T) (Figure 6), enabling an even more comprehensive investigation of the condensation curing process.

The amounts of unreacted PDMS and $(QM^{OEt})_8$ over time were calculated from ^{1}H-NMR spectra (Figure A3), and the results are summarized in Figure 7. Silicone elastomers with r = 0.5 and 1 contained 0 wt% of unreacted $(QM^{OEt})_8$ and a significant amount of extractable PDMS, which decreased over time due to the slow condensation of HO-PDMS-OH. Silicone elastomers with r = 3 and 5 contained close to 0 wt% of unreacted $(QM^{OEt})_8$ and only ~4 wt% of extractable, non-functional low molecular weight PDMS originating from the manufacturing of the C2T polymer (Figure A1). Elastomers with a stoichiometric ratio in the interval $3 \leq r \leq 5$ are thus the most stable elastomers, experiencing the fewest post-curing effects. Again, this ratio is far above the expected optimal stoichiometric balance (r = 1), suggesting the formation of cross-linker domains during the main curing reaction, as in the previously studied formulations. Silicone elastomers with r = 8 and 15 contained 3 and 6.5 wt% of unreacted $(QM^{OEt})_8$ molecules, respectively. The concentration of the unreacted cross-linker decreased over time, reaching 0 wt% after 17 days as additional cross-linker domains were created. Since silsesquioxanes are generally known to be self-reinforcing cross-linkers [16–21], $(QM^{OEt})_8$ cross-linker domains are expected to have a positive effect on elastomer film strength.

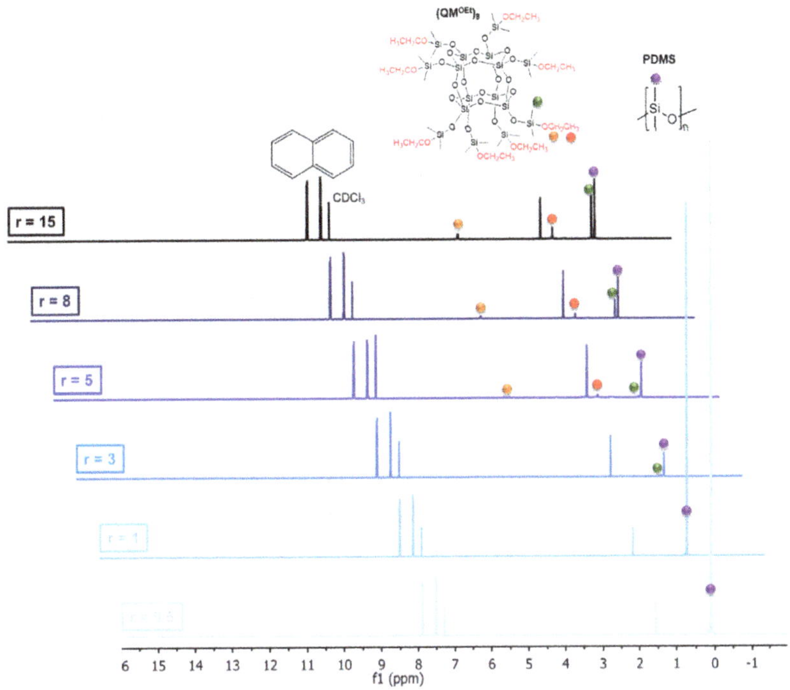

Figure 6. ^{1}H-NMR spectra of extracts from 1-day old elastomer films prepared via the reaction between C2T and $(QM^{OEt})_8$. The stoichiometric ratio was varied from 0.5 to 15.

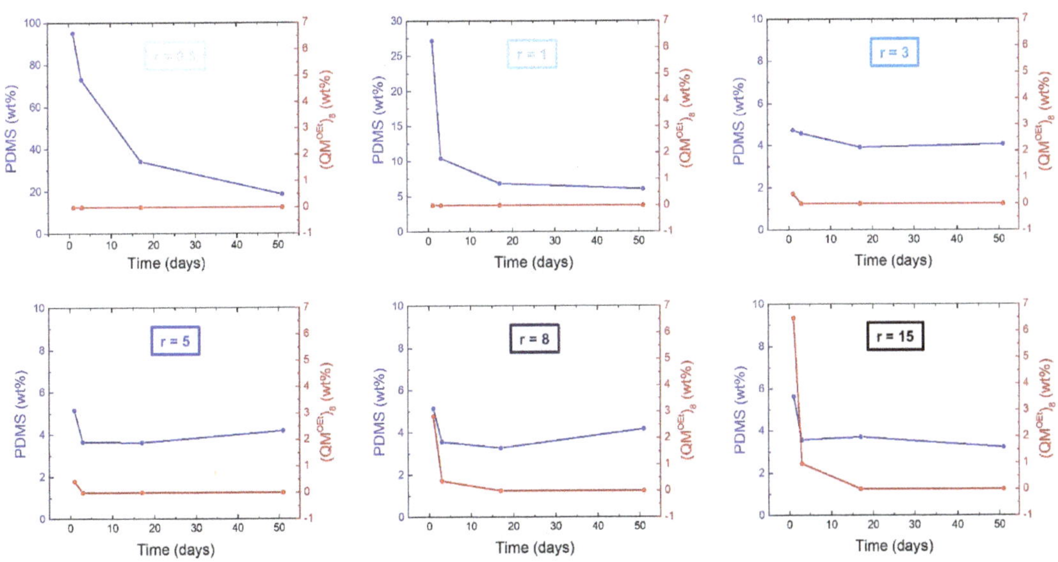

Figure 7. The amount of extractable PDMS and $(QM^{OEt})_8$ as a function of time. Elastomers were prepared via the reaction between C2T and $(QM^{OEt})_8$ using stoichiometric ratios of 0.5, 1, 3, 5, 8, and 15, respectively.

2.2. Mechanically Stable Silicone Elastomer Films and Their Properties

Findings from the formulation optimization study presented above were used to design stable condensation curing silicone elastomer films. With both possible post-curing reactions in mind, different network structures were prepared by changing the cross-linker type (polysiloxane or silsesquioxane), cross-linker chain length (Di-10, 50, or 200), or stochiometric ratio (r = 3, 5 or 15). The individual silicone elastomer compositions are summarized in Table A3. While all films were cured within a few hours, the measurements presented in this chapter were conducted after storage in a climate chamber for 27 days, the time required to achieve complete cross-linker domains creation (see Section 2.1). After 27 days, all films, except for commercial coating E_Ref, contained low amounts of sol fraction ranging from 3 to 5 wt% (Table A4), which can be assigned to the non-reactive PDMS created during polymer/cross-linker synthesis (Figures 4 and A1). The sol fraction of E_Ref was found to be 6–7 wt%. Apart from the double peak at retention volumes between 17 and 22 mL, SEC analysis also showed a peak at retention volumes between 11 and 17 mL (Figure A4). This second peak can be assigned to a silicone oil, either added as a plasticizer or originating from non-reacted HO-PDMS-OH.

Network elasticity and rigidity were tailored via cross-linker domain concentration and density. Figures 8 and 9 show that, as expected, Young´s modulus was increasing, and elongation at break was decreasing with increasing r. These changes were even more pronounced when using Di-10 and $(QM^{OEt})_8$, as the cross-linker domains become more dense. In contrast, the smallest change in both Young´s modulus and elongation at break with increasing r was found in films cross-linked by Di-400 due to its high molecular weight, which is comparable to that of C2T (Table 1). It should be noted that, despite the relatively low Young´s modulus of elastomer films E_C2T+Di-400_r5, E_C2T+Di-400_r15, and E_Di-400 (0.61, 0.67 MPa, and 0.7 MPa, respectively), the sol fraction of each remained below 5 wt% (Table A4). Additionally, the elastomer E_C2T+$(QM^{OEt})_8$_r15 exhibited opacity when elongated. For better understanding of this behavior, which was not observed for any of the other elastomers, the elastomer E_C2T+$(QM^{OEt})_8$_r15 was investigated using SEM in both unstretched and stretched state. The sample preparation procedure can be found in Figure A5. As it can be seen in Figure 10, the elastomer undergoes a significant surface change when stretched, which explains the observed opacity and is believed to be a result of $(QM^{OEt})_8$ domains aggregation and crystallization.

Electrical breakdown (EBD) strength is an indicator of film homogeneity, since film imperfections lead to inhomogeneous fields and thus premature film failure [22]. Increased Young's modulus leads to increased electrical breakdown strength, but only if the film is homogenous [23,24]. The presence of cross-linker domains was found to positively affect electrical breakdown strength. The highest electric breakdown strength of 130 μm/V was obtained for elastomer E_C2T+$(QM^{OEt})_8$_r15 (Figure 11). This is approximately 30% higher than the electrical breakdown strength of the reference coating (E_Ref), which is a commercial condensation curing silicone coating containing reinforcing fillers. It is also significantly higher than values reported for addition curing silicone elastomers, such as Sylgard, Ecoflex, and Elastosil [25].

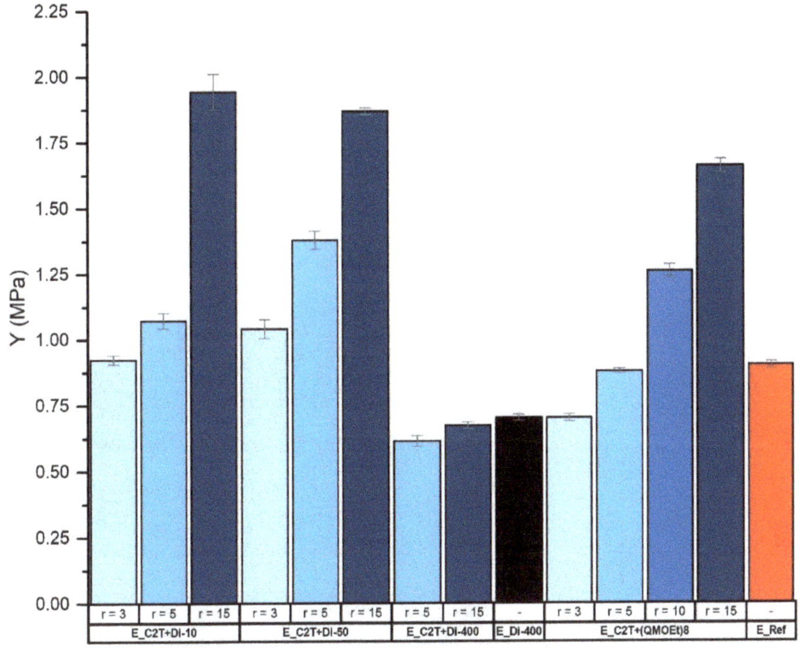

Figure 8. Young´s moduli (MPa) of commercial coating E_Ref and elastomers prepared via the reaction between C2T and Di-10, Di-50, Di-200, and (QMOEt)8, respectively. Film thickness was ~100 µm.

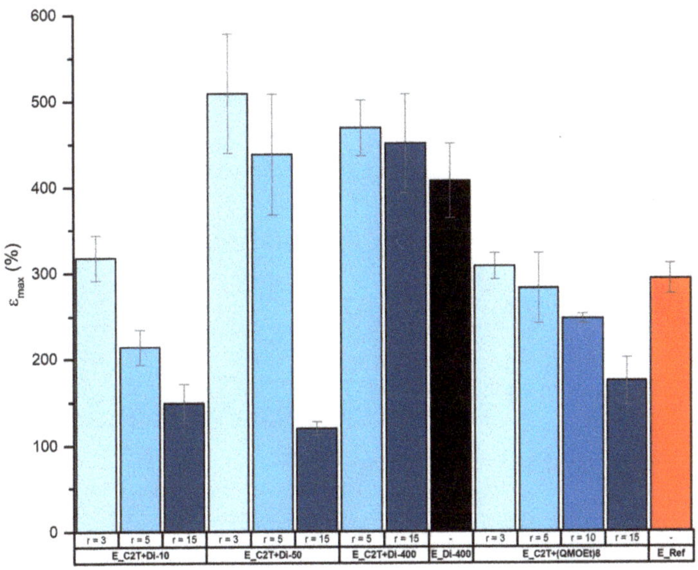

Figure 9. Elongation at break (%) of commercial coating E_Ref and elastomers prepared via the reaction between C2T and Di-10, Di-50, Di-200, and $(QM^{OEt})_8$, respectively. Film thickness was ~100 µm.

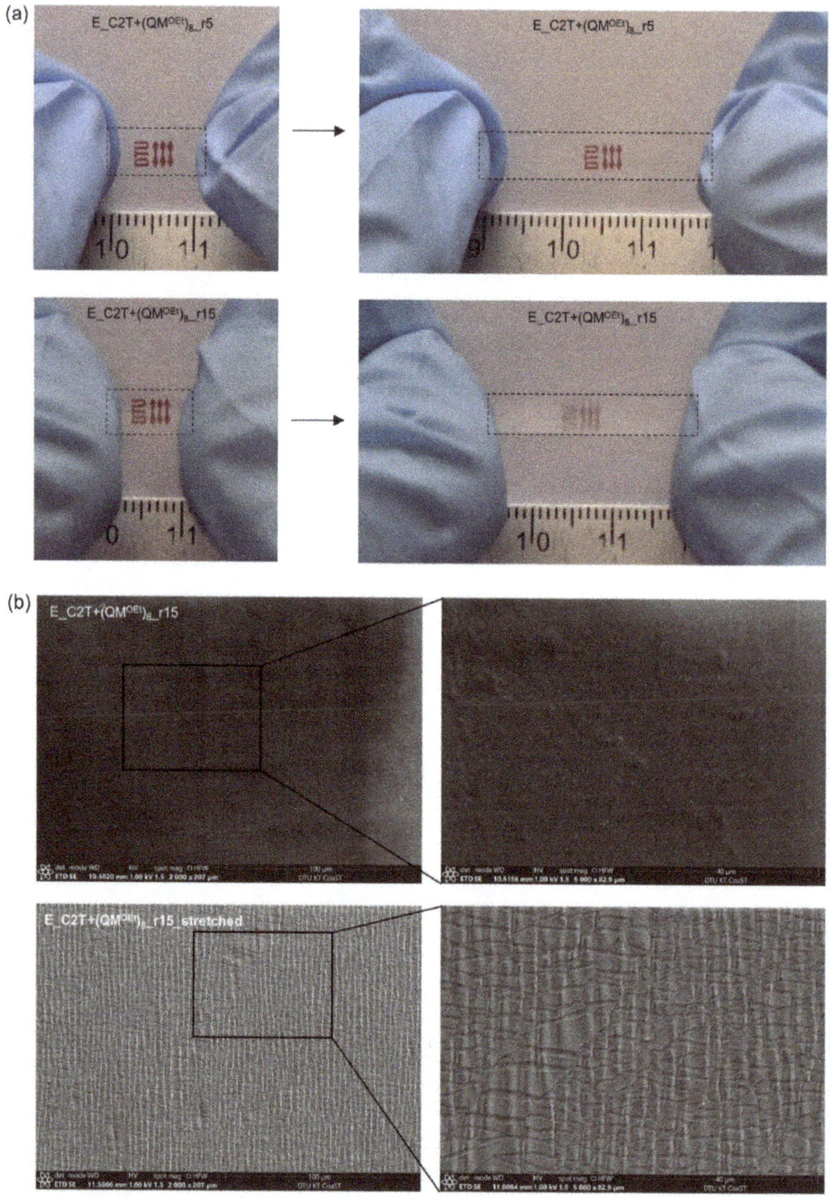

Figure 10. (**a**) Elastomers E_C2T+(QM^{OEt})$_8$_r5 and E_C2T+(QM^{OEt})$_8$_r15 before and after elongation. Elastomer samples are marked with a dashed line to improve visibility. The DTU logo is printed on the paper below the sample. (**b**) SEM images of E_C2T+(QM^{OEt})$_8$_r15 in unstretched and stretched state, respectively.

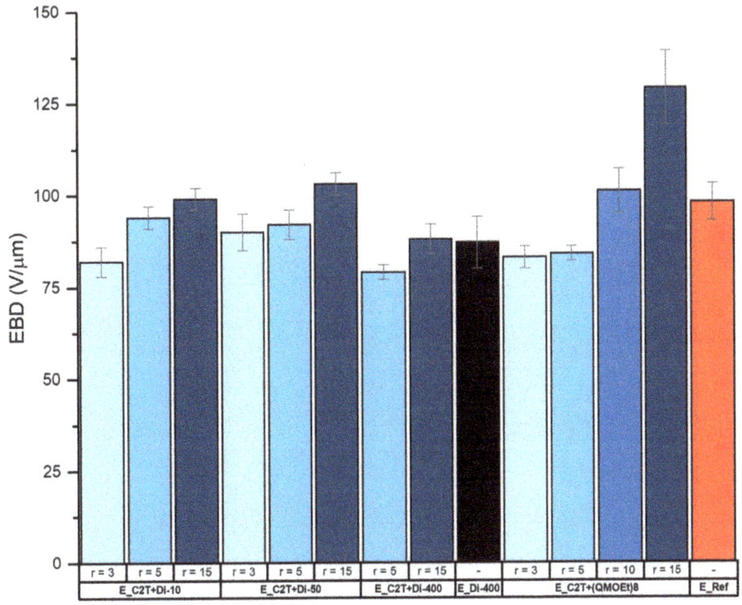

Figure 11. Electrical breakdown (EBD) strength (V/μm) of commercial coating E_Ref and elastomer films prepared via the reaction between C2T and Di-10, Di-50, Di-200, and $(QM^{OEt})_8$, respectively. Film thickness was ~100 μm.

Scratch resistance, together with coating/substrate adhesion, is one of the most important parameters for materials used as protective coatings. A poor scratch resistance and/or adhesion to the substrate can lead to reduction of coating performance, reliability, and lifetime [1,26–28]. In this study, all the tested elastomers, independently of their composition, showed initial cohesive failure before failing adhesively when scratched, signifying a good adhesion to their substrate, namely Hempel's Nexus II 27400 [26,27]. The scratch resistance was then found to improve significantly with increasing *r* for elastomers cross-linked by Di-10 and $(QM^{OEt})_8$ (Figure 12). Elastomers E_C2T+Di-10_r15 and E_C2T+$(QM^{OEt})_8$_r15 displayed a scratch resistance comparable to that of the reference coating, E_Ref, which, unlike the elastomers investigated here, contains reinforcing fillers. While this finding once again shows the positive effect of Di-10 and $(QM^{OEt})_8$ cross-linker domains, a negative correlation between increasing *r* and scratch resistance was observed for elastomers cross-linked by Di-50. This can be explained by the high weight percentage of Di-50 cross-linker in the elastomer (Table A3), which caused the elastomer E_C2T+Di-50_r15 (~58 wt% of the Di-50) to lose its elasticity. Zero or negligible difference in scratch resistance with increasing *r* was reported for E_C2T+Di-400 and E_Di-400, as the high molecular weight of the Di-400 cross-linker does not allow the formation of strongly reinforcing domains due to the long distance between cross-links. Noticeably, even though E_C2T+$(QM^{OEt})_8$_15 showed excellent "single" scratch resistance, its "multiple" scratch resistance was significantly lower. On the other hand, E_C2T+Di-10_r15 displayed excellent "single" and "multiple" scratch resistance. Both elastomers (E_C2T+$(QM^{OEt})_8$_15 and E_C2T+Di-10_r15) contain similar weight percentages of cross-linker (20 and 23 wt%, respectively), suggesting that the reduced "multiple" scratch resistance displayed by E_C2T+$(QM^{OEt})_8$_15 can be attributed to the rigidity of the $(QM^{OEt})_8$ domains. The more elastic Di-10 domains performed well in both "single" and "multiple" scratch tests, indicating that not only cross-linker domains size, but also rigidity/elasticity, is important for scratch resistance. For better understanding of the difference in the network structure containing Di-10, Di-50, Di-400, and $(QM^{OEt})_8$ domains, respectively, Figure 13 introduces a simplified cartoon of network structures prepared at high *r*.

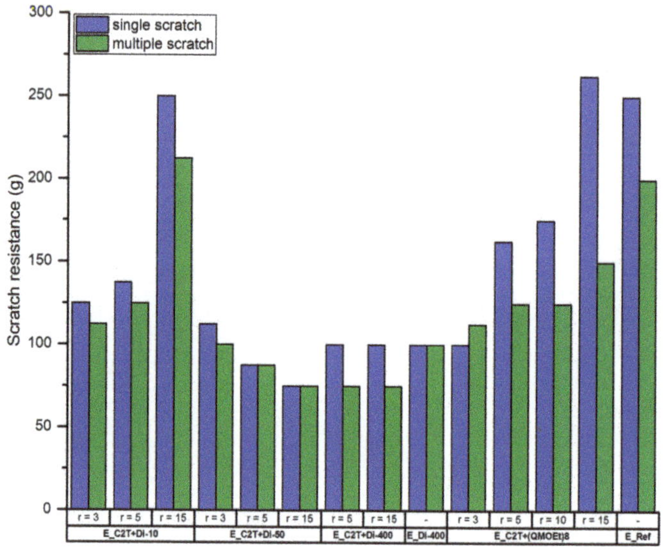

Figure 12. Scratch resistance of commercial coating E_Ref and elastomer films prepared via the reaction between C2T and Di-10, Di-50, Di-200, and $(QM^{OEt})_8$, respectively. Film thickness was ~100 μm.

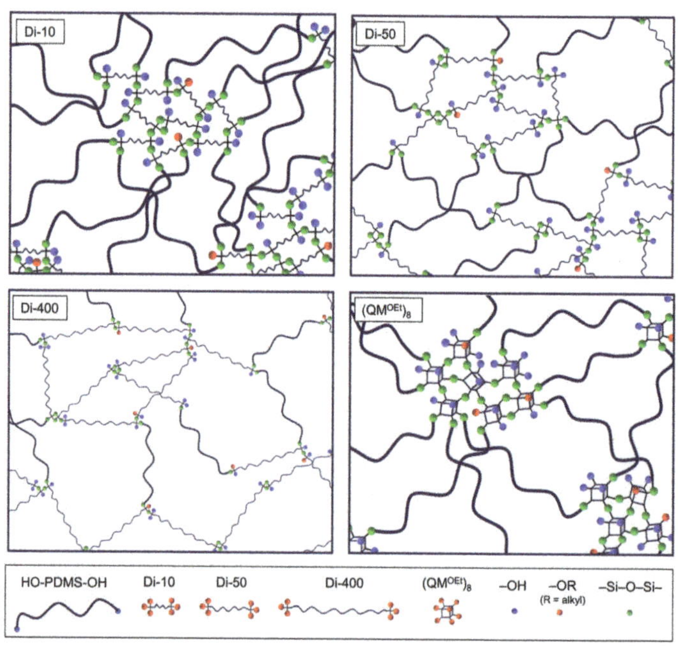

Figure 13. Simplified illustration of network structures prepared at high *r* using Di-10, Di-50, Di-400, and $(QM^{OEt})_8$ cross-linker, respectively.

The results above demonstrate that introducing cross-linker domains with specific structural properties can significantly improve silicone elastomer performance. For example, silicone elastomers E_C2T+$(QM^{OEt})_8$_15 and E_C2T+Di-10_r15 were shown to match, or even outperform, commercial coating E_Ref containing fillers and other additives in both electrical breakdown strength and scratch resistance. The addition of fillers is commonly used to improve the mechanical properties of silicone elastomers. However, it often leads to an undesirable Mullins effect, a loss of coating transparency, and an increase in Young´s modulus [29]. Creating cross-linker domains in the elastomer network represents an alternative method for improving elastomers' mechanical properties without introducing the Mullins effect or compromising elastomer transparency. Although the Young´s modulus of elastomers E_C2T+$(QM^{OEt})_8$_15 and E_C2T+Di-10_r15 is significantly higher than that of commercial coating E_Ref, it should be noted that the latter contains a silicone oil in the sol fraction, which lowers its Young´s modulus (Figure A4).

3. Materials and Methods

3.1. Materials

Dibutyltin-dilaurate catalyst (Sn_DL)—Tib kat 218 (produced by TIB Chemicals) and Hempel´s Nexus II 27400 was kindly provided by Hempel. Linear, trimethoxysilane-terminated polysiloxanes ($(MeO)_3$Si-PDMS-Si$(OMe)_3$)—Silmer TMS Di-10, Silmer TMS Di-50, and Silmer TMS Di-400), were purchased from Siltech. The molecular weights of $(MeO)_3$Si-PDMS-Si$(OMe)_3$ were determined by SEC and ^{1}H NMR (Tables 1 and A1). The names of the polysiloxane cross-linkers are shortened to Di-10, Di-50, and Di-400, respectively, throughout the manuscript. Silanol-terminated polydimethylsiloxane (HO-PDMS-OH; C2T) was obtained from Wacker Chemie. The molecular weight of C2T was determined by SEC (Table 1). Ethoxy-terminated octakis(dimethylsiloxy)-T8-silsesquioxane ($(QM^{OEt})_8$) was synthetized via ethanolysis of octakis(dimethylsiloxy)-T8-silsesquioxane ($(QM^H)_8$) in the presence of Karstedt´s catalyst [30]. Hostaphan RN 190/190 µm (polyester plastic carrier) was supplied by Mitsubishi Polyester Film. Syringe filter (diameter: 25 mm, pore size: 0.45, PTFE membrane) was purchased from Fisher Scientific.

Table 1. Summary of compounds used in the elastomer formulations.

Compound	Abbreviation	Molecular Weight (g/mol)	Chemical Structure
HO-PDMS-OH	C2T	20,200 *	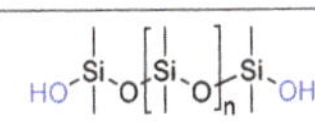
$(MeO)_3$Si-PDMS-Si$(OMe)_3$	Di-10 Di-50 Di-400	1200 * 5500 * 20,300 *	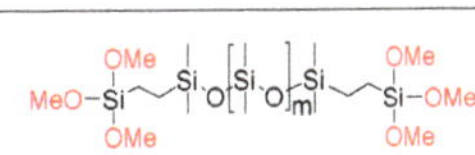
$(QM^{OEt})_8$	$(QM^{OEt})_8$	1370.4 **	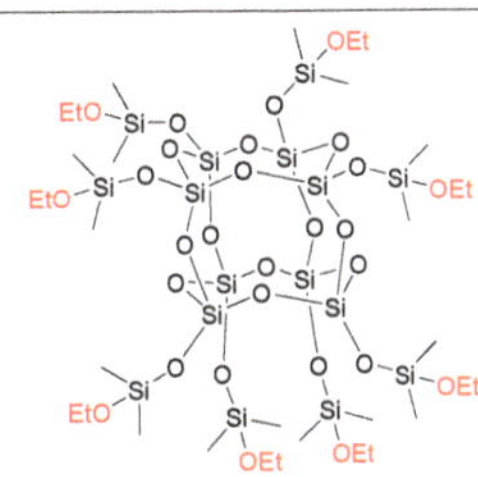

* Number average molecular weight (M_n) determined by SEC; ** determined from molecular structure.

3.2. Elastomer Film Preparation

Example of the preparation procedure for elastomer E_C2T+Di-10_r3: Silanol-terminated polydimethylsiloxane—C2T and trimethoxysilane-terminated polysiloxane—Di-10 were mixed in a mixing container using a FlackTech speed mixer at 2700 rpm for 3 min. The stoichiometric ratio between cross-linker and polymer functional groups (r = 6[$(MeO)_3$Si-PDMS-Si$(OMe)_3$]/2[HO-PDSM-OH]) was r = 3. Subsequently, 0.5 wt% of Sn_DL was added before mixing in a FlackTech speed mixer at 2700 rpm for another 3 min. After a homogeneous mixture was obtained, the sample was coated into a thin film by a coating knife with a nominal coating height of 200 μm onto a polyester carrier. For the scratch resistance measurements, the mixture was coated onto a tie-coat—Hempel´s Nexus II 27400, which was previously coated onto a polyester carrier using a coating knife with a nominal coating height of 200 μm. The films were subsequently cured in a climate chamber at 25 °C and 80% relative humidity.

The elastomer compositions for the long-term stability study can be found in Table A2.

The elastomer compositions for tensile, electrical breakdown strength, and scratch resistance measurements can be found in Table A3.

The abbreviation, molecular weight, and chemical structure of the compounds used in the elastomer formulations are summarized in Table 1.

3.3. *Evaluation of the Elastomer Film Stability*

The silicone elastomer films, for which composition details can be found in Table A2, were used to study the effect of elastomer composition on network structure and stability. The following methods were used to characterize film stability: **Mechanical stability** (Figure 14a) was evaluated by tracking changes in Young´s modulus over time. The Young´s modulus was determined from the tangent of the linear region of the stress–strain curves at low strains (up to approximately 20% strain). Standard deviations were calculated from five tensile measurements for each sample composition. The tensile measurements were performed on an ARES-G2 rheometer using SER geometry according to a previously reported procedure [10]. **Chemical composition of the extract and wt% of extractable PDMS/cross-linker** (Figure 14b) was determined using Proton Nuclear Magnetic Resonance (^{1}H NMR). A specimen 25 mm in length and 6 mm in width was cut from the silicone elastomer film (~100 μm), weighed, and extracted in 1 mL of chloroform-*d* for 24 h. After the extraction, a known amount of naphthalene was added, and the solution was transported into a NMR tube. ^{1}H NMR characterization was performed on a 7 Tesla Spectrospin-Bruker AC 300MHz spectrometer at room temperature. ^{1}H NMR spectra were analyzed using MestReNova. Detailed calculation of the wt% of extractable PDMS can be found in Figures A2 and A3. **Molecular weight of extractable PDMS** (Figure 14c) was obtained using Size-Exclusion Chromatography (SEC). Two specimens, 25 mm in length and 6 mm in width, were cut from the silicone elastomer film (~100 μm), weighed, and extracted in 1.2 mL of toluene for 24 h. The extract was filtered through a syringe filter. SEC was performed on a TOSOH EcoSEC HLC-8320GPC system equipped with an EcoSEC RI detector. This system was fitted with two SDV LINEAR S 5 μm 8 × 300 mm columns in series, protected by a GUARD column (SDV 5 μm 8 × 50 mm), all supplied by PSS. Samples were run in toluene at 35 °C. Molar mass characteristics were calculated using PSS WinGPC Unity, Build 9350 software, and linear PDMS standards acquired from PSS.

3.4. *Evaluation of the Silicone Elastomer Films Performance*

The silicone elastomer film compositions can be found in Table A3. The following properties were tested in order to evaluate elastomer performance: **Sol fraction** was determined using both ^{1}H NMR (sol fraction consisting purely from the extractable PDMS species) and sample weight lost (sol (%) = $(m_i - m_d)/m_i \times 100$, where m_i is the original sample weight and m_d is the sample weight after the extraction and drying). For more details on the extraction, see the procedure in Section 3.3. The 24-h extraction time was found to be sufficient, as no additional weight loss was observed after a second extraction of the same specimen. **Tensile properties** were performed on an ARES-G2 rheometer using SER geometry according to a previously described procedure [10]. **Scratch resistance** was measured using a Motorized Clemen Scratch Tester equipped with a Ø 1mm ball tool. Film destruction was observed visually. Two different parameters were evaluated—"single" scratch resistance and "multiple" scratch resistance. The "single" scratch resistance was determined as the maximum load at which the coating continues to resist penetration.

The "multiple" scratch resistance was determined as the maximum load at which the coating remained unpenetrated after three consecutive scratches at the same place. Both "single" and "multiple" scratch procedures were repeated three times for each coating composition. If the coating was penetrated during one or more of the three repetitions, the load was lowered and the whole procedure was repeated. **Electrical breakdown strength** was measured using an instrument built in-house following the international standards (IEC 60243-1 (1998) and IEC 60243-2 (2001)) [31,32]. A silicone elastomer film approximately 100 μm thick was placed on a plastic frame and positioned inside the breakdown instrument between two semi-spherical stainless-steel electrodes. A voltage ramp of 100 V/s was then applied until the elastomer short-circuited. The electrical breakdown strength was calculated as the voltage at breakdown divided by sample thickness. The standard deviation was calculated from 10 breakdown measurements for each elastomer film. **Scanning electron microscope (SEM)**—Inspect S, FEI company—was used to evaluate the changes in the elastomer E_C2T+$(QM^{OEt})_8$_15 before and after stretch. Images were taken in a low vacuum mode. The sample preparation can be found in Figure A5.

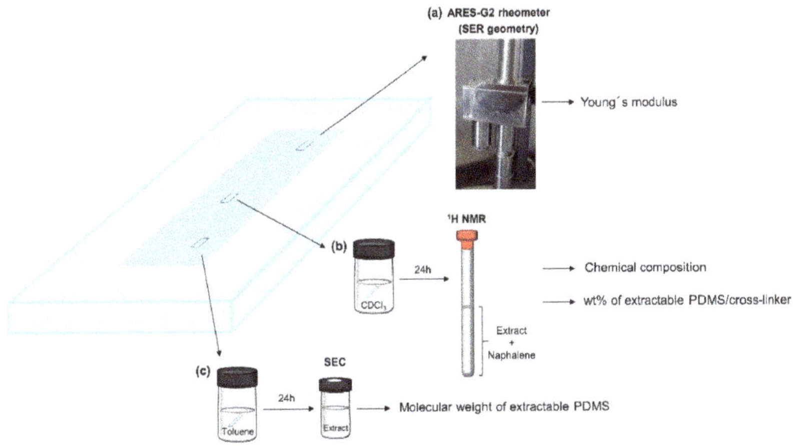

Figure 14. Methods used to evaluate the stability and network structure of elastomer films: (**a**) ARES-G2 rheometer equipped with SER geometry was used to track changes in Young´s modulus over time;(**b**) ^{1}H NMR was used to analyze the chemical composition of the extract and to determine the wt% of extractable PDMS/cross-linker; (**c**) SEC was used to measure the molecular weight of extractable PDMS.

The exact sample thickness for the tensile, scratch resistance, and electrical breakdown strength measurements was determined using an optical microscope according to a previously described procedure [10].

4. Conclusions

In this study, post-curing reactions occurring in condensation curing silicone elastomers were thoroughly investigated using rheology, SEC, and ^{1}H NMR, findings from which were correlated to the elastomers' network structure. Depending on the stoichiometric ratio, two types of post-curing reaction were shown to occur: Reactions between unreacted and/or dangling polymer chains, and reactions between cross-linker molecules. The exact stoichiometry, *r*, required to initiate these post-curing reactions depends on many factors, such as the type and molecular weight of the cross-linker used, as well as the concentration of the catalyst. Thus, in order to provide a guideline on how to develop stable and reliable silicone elastomers via condensation curing, a formulation optimization study was conducted. For hereby studied elastomer formulations, a stoichiometric ratio between 3 and 5 was found to produce the fewest post-curing reactions. Lower *r* values produced elastomers that become unstable over time due to the reaction between unreacted and/or dangling polymer chains. At higher *r* values, so-called cross-linker domains were formed via the reaction between redundant cross-linker molecules. In contrast to the condensation of unreacted and/or dangling polymer chains, this reaction occurred relatively quickly—within approximately 3 weeks—and therefore did not contribute to long-term elastomer instability.

The long-term stability study served as a tool for preparation of stable condensation curing silicone elastomers. The elastomer properties were then tailored by cross-linker domains density and concentration. The cross-linker domains density was altered by changing either cross-linker chain length (Di-10, Di-50, and Di-400) or cross-linker type ($(MeO)_3Si$-PDMS-$Si(OMe)_3$ and $(QM^{OEt})_8$). The cross-linker domains concentration was altered by varying *r*. As expected, the Young´s modulus of the elastomers was increasing and elongation at break was decreasing with increasing *r*. The smallest change of these properties was observed when using the highest molecular weight cross-linker (Di-400), as the resulting cross-linker domains were less dense due to the long distance between cross-links. The presence of cross-linker domains was found to positively affect electrical breakdown strength, with the highest value obtained for E_C2T+$(QM^{OEt})_8$_r15 (130 μm/V). Scratch resistance was found to be highly dependent on both the size and rigidity/elasticity of the cross-linker domains. Due to their high weight percentage in the elastomer network, cross-linker domains Di-400 and Di-50 were found to have negligible and negative effects, respectively, on scratch resistance. On the other hand, scratch resistance was significantly improved by the presence of both Di-10 and $(QM^{OEt})_8$ cross-linker domains. Scratch resistance comparable to that of the reference coating (E_Ref), which contained reinforcing fillers, was obtained for both E_C2T+Di-10_r15 and E_C2T+$(QM^{OEt})_8$_15. While E_C2T+Di-10_r15 performed well in both "single" and "multiple" scratch resistance tests, "multiple" scratch resistance was significantly lower for E_C2T+$(QM^{OEt})_8$_15, indicating the importance of not only cross-linker domains size, but also rigidity/elasticity.

Author Contributions: Conceptualization, A.J., A.L.S., and S.M.O.; formal analysis, A.J.; methodology, A.J.; investigation, A.J.; resources, S.M.O. and M.A.B.; data curation, A.J.; writing—original draft preparation, A.J.; writing—review and editing, A.J., A.L.S., S.M.O., M.A.B., and K.D.-J.; supervision, A.L.S. and S.M.O.; project administration, A.L.S., K.D.-J., and S.M.O.; funding acquisition, K.D.-J. All authors have read and agreed to the published version of the manuscript.

Funding: This research was funded by the financial support from the Hempel Foundation to CoaST.

Data Availability Statement: The data can be obtained from the corresponding author.

Acknowledgments: Narayanan Rajagopalan is acknowledged for help with the SEM images.

Conflicts of Interest: The authors declare no conflict of interest.

Sample Availability: Samples of the compounds are not available from the authors.

Appendix A

Table A1. Number average molecular weights (M_n) and polydispersity indices (PDI) of Di-10, Di-50, Di-400, and C2T. The data were obtained by SEC and ^{1}H NMR. The molecular weights obtained by SEC were used for the further calculation of the silicone elastomer network compositions.

Compound	SEC		^{1}H NMR
	M_n (g/mol)	PDI	M_n (g/mol)
Di-10	1200	1.49	1300
Di-50	5500	4.06	4600
Di-400	20,300	2.25	23,000
C2T	20,200	2.18	-

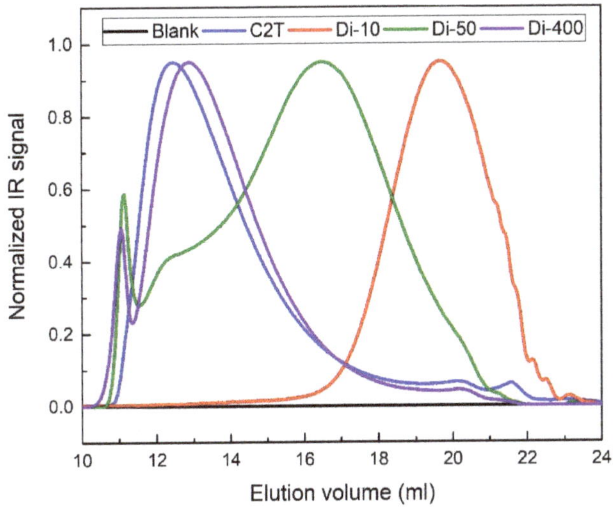

Figure A1. SEC eluograms of HO-PDMS-OH (C2T) and $(MeO)_3$Si-PDMS-Si$(OMe)_3$ cross-linkers (Di-10, Di-50, and Di-400).

Table A2. Compositions of silicone elastomers for long-term stability tests.

C2T (~20,200 g/mol) + Di-10 (~1200 g/mol) *							
Name	*r*	**C2T (g)**	**Di-10**		**Sn_DL**		
			wt%	**g**	**wt%**	**g**	**μL**
E_C2T+Di-10_r1.5	1.5	3.88	3	0.13	0.5	0.02	19
E_C2T+Di-10_r2	2	3.85	4	0.15			
E_C2T+Di-10_r5	5	3.64	9	0.36			
E_C2T+Di-10_r10	10	3.34	17	0.66			
E_C2T+Di-10_r15	15	3.08	23	0.92			
E_C2T+Di-10_r20	20	2.87	28	1.13			
C2T (~20,200 g/mol) + Di-50 (~5500 g/mol) *							
Name	*r*	**C2T (g)**	**Di-50**		**Sn_DL**		
			wt%	**g**	**wt%**	**g**	**μL**
E_C2T+Di-50_r1.5	1.5	3.52	12	0.48	0.5	0.02	19
E_C2T+Di-50_r2	2	3.39	15	0.61			
E_C2T+Di-50_r5	5	2.75	31	1.25			
E_C2T+Di-50_r10	10	2.10	48	1.90			
E_C2T+Di-50_r15	15	1.69	58	2.31			
E_C2T+Di-50_r20	20	1.42	65	2.58			
C2T (~20,200 g/mol) + Di-400 (~20,300 g/mol) *							
Name	*r*	**C2T (g)**	**Di-400**		**Sn_DL**		
			wt%	**g**	**wt%**	**g**	**μL**
E_C2T+Di-400_r1.5	1.5	2.66	34	1.34	0.5	0.02	19
E_C2T+Di-400_r2	2	2.40	40	1.6			
E_C2T+Di-400_r5	5	1.50	63	2.50			
E_C2T+Di-400_r10	10	0.92	77	3.08			
E_C2T+Di-400_r15	15	0.66	84	3.34			
E_C2T+Di-400_r20	20	0.52	87	3.48			

Table A2. *Cont.*

C2T (~20,200 g/mol) + $(QM^{OEt})_8$ (1370.404 g/mol) **							
Name	*r*	C2T (g)	$(QM^{OEt})_8$		Sn_DL		
			wt%	μL ***	wt%	g	μL
E_C2T+$(QM^{OEt})_8$_r0.5	0.5	0.5	1	5	0.5	0.02	19
E_C2T+$(QM^{OEt})_8$_r1	1	0.49	2	10			
E_C2T+$(QM^{OEt})_8$_r3	3	0.48	5	30			
E_C2T+$(QM^{OEt})_8$_r5	5	0.46	8	48			
E_C2T+$(QM^{OEt})_8$_r8	8	0.44	12	73			
E_C2T+$(QM^{OEt})_8$_r15	15	0.40	20	125			

* Total mass 4 g, ** Total mass 0.5 g, *** Solution in toluene (0.81 g/mL).

Table A3. Compositions of silicone elastomers for tensile, electrical breakdown strength, and scratch resistance measurements.

C2T (~20,200 g/mol) + Di-10 (~1200 g/mol) *							
Name	*r*	C2T (g)	Di-10		Sn_DL		
			wt%	g	wt%	g	μL
E_C2T+Di-10_r3	3	5.66	6	0.34	0.5	0.03	28
E_C2T+Di-10_r5	5	5.46	9	0.54			
E_C2T+Di-10_r15	15	4.63	23	1.37			
C2T (~20,200 g/mol) + Di-50 (~5500 g/mol) *							
Name	*r*	C2T (g)	Di-50		Sn_DL		
			wt%	g	wt%	g	μL
E_C2T+Di-50_r3	3	4.72	21	1.28	0.5	0.03	28
E_C2T+Di-50_r5	5	4.13	31	1.87			
E_C2T+Di-50_r15	15	2.54	58	3.46			
C2T (~20,200 g/mol) + Di-400 (~20,300 g/mol) **							
Name	*r*	C2T (g)	Di-400		Sn_DL		
			wt%	g	wt%	g	μL
E_C2T+Di-400_r5	5	1.50	63	2.50	0.5	0.03	28
E_C2T+Di-400_r15	15	0.66	83	3.34			
E_Di-400	-	-	100	4			
C2T (~20,200 g/mol) + $(QM^{OEt})_8$ (1370.404 g/mol) **							
Name	*r*	C2T (g)	$(QM^{OEt})_8$		Sn_DL		
			wt%	μL ***	wt%	g	μL
E_C2T+$(QM^{OEt})_8$_r3	3	3.81	5	255	0.5	0.02	19
E_C2T+$(QM^{OEt})_8$_r5	5	3.69	8	411			
E_C2T+$(QM^{OEt})_8$_r10	10	3.42	15	763			
E_C2T+$(QM^{OEt})_8$_r15	15	3.19	20	1067			
E_Ref	Reference coating = commercially used silicone elastomer coating prepared via condensation curing chemistry						

* Total mass 6 g, ** Total mass 4 g, *** Solution in toluene (0.76 g/mL).

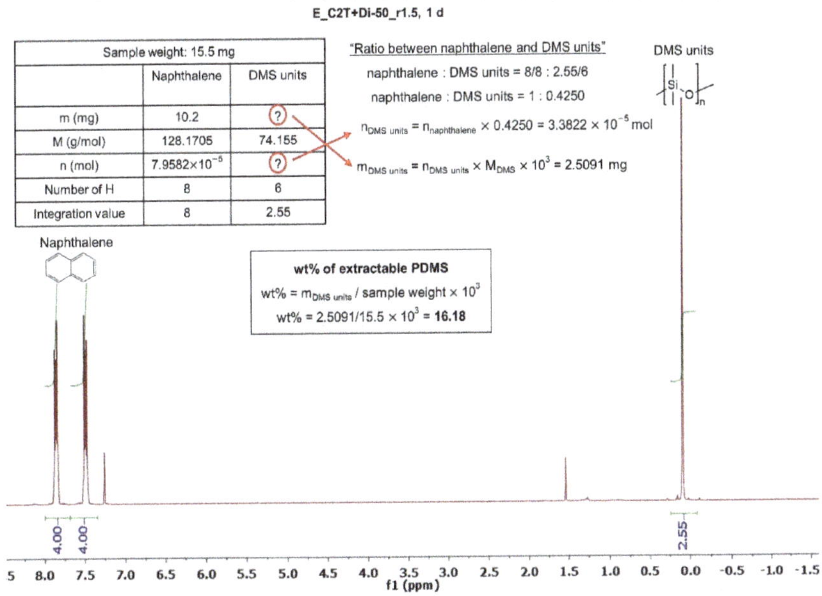

Figure A2. An example of calculation of the extractable PDMS from E_C2T+Di-50_r1.5 after 1 day in a climate chamber.

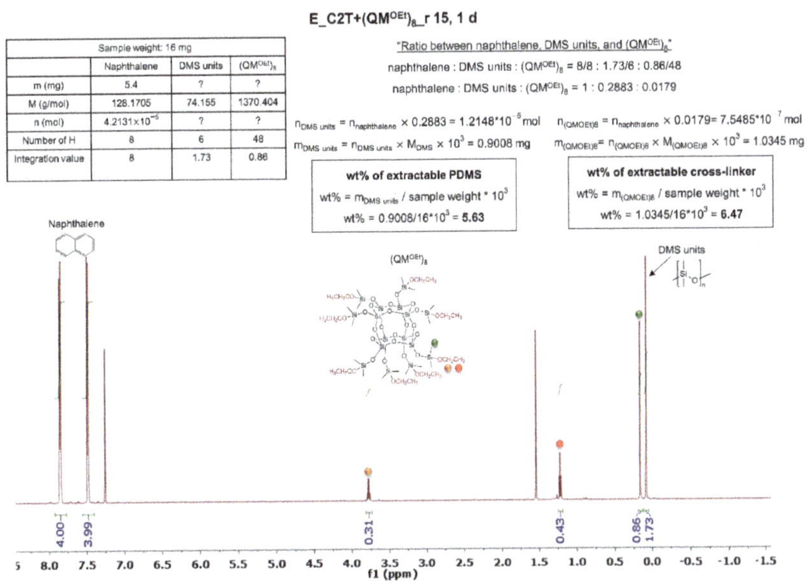

Figure A3. An example of calculation of the extractable PDMS from E_C2T+$(QM^H)_8$_r15 after 1 day in a climate chamber.

Table A4. Amounts of extractable PDMS (determined by ^{1}H NMR) and sol fraction (determined from the weight loss) of the stable elastomer films.

Elastomer	Extractable PDMS (wt%)	Sol Fraction (wt%)
E_C2T+Di-10_r3	4.1	4.2
E_C2T+Di-10_r5	4.1	3.8
E_C2T+Di-10_r15	3.8	3.1
E_C2T+Di-50_r3	4.2	3.3
E_C2T+Di-50_r5	3.5	3.5
E_C2T+Di-50_r15	3.5	3.6
E_C2T+Di-400_r5	4.6	4.7
E_C2T+Di-400_r15	4.6	3.8
E_Di-400	4.3	2.5
E_C2T+$(QM^{OEt})_8$_r3	4.3	5.0
E_C2T+$(QM^{OEt})_8$_r5	4.2	3.7
E_C2T+$(QM^{OEt})_8$_r10	3.9	2.7
E_C2T+$(QM^{OEt})_8$_r15	3.6	5.0
E_Ref	7.0	6.3

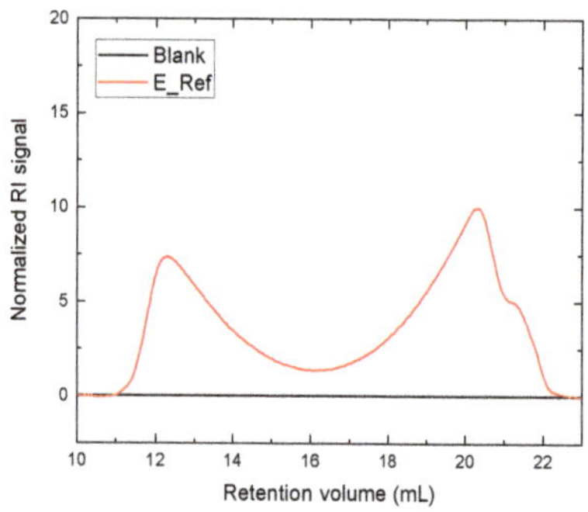

Figure A4. SEC eluogram of the sol fraction of a commercial coating (E_Ref). The elastomer age was 27 days.

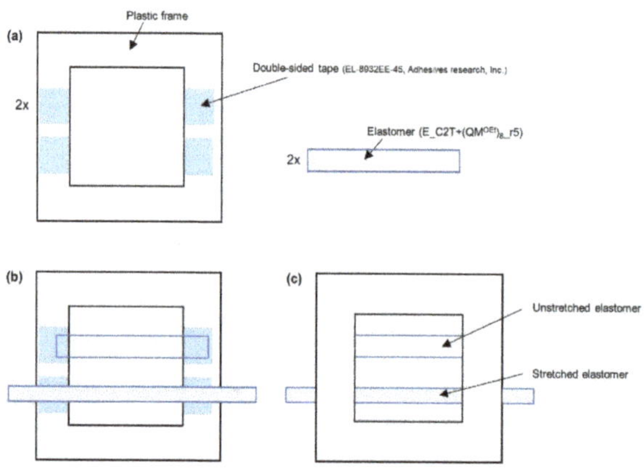

Figure A5. A schematic illustration of SEM sample preparation: (**a**) A double-sided tape was attached onto two plastic frames made of hard plastic. Two specimens 25 mm in length and 6 mm in width were cut from the elastomer film E_C2T+$(QM^{OEt})_8$_r15. (**b**) One of the two specimens was attached to the frame in unstretched state. The other specimen was stretched by ~100% (a sufficient strain to observe the film opacity) before attaching to the frame. (**c**) To ensure a proper specimen attachment, the second plastic frame was placed facing down by the double side tape on the top of the setup.

References

1. Eduok, U.; Faye, O.; Szpunar, J. Recent developments and applications of protective silicone coatings: A review of PDMS functional materials. *Prog. Org. Coat.* **2017**, *111*, 124–163. [CrossRef]
2. De Buyl, F. Silicone sealants and structural adhesives. *Int. J. Adhes. Adhes.* **2001**, *21*, 411–422. [CrossRef]
3. Gutowski, W.S.; Wolf, A.; Deng, P.; Zhang, Y.; Yin, Y. Application of polydimethylsiloxane (PDMS) polymers as structural adhesives, sealants, and high-performance functional coatings. *J. Mach. Constr. Maint.* **2019**, *3*, 7–17.
4. Beers, M.D.; Thompson, J.E. Oil Resistant Low Modulus Silicone Sealant Composition. U.S. Patent 4,514,529, 30 April 1985.
5. Stein, J.; Burnell, T.B. Sprayable, Condensation Curable Silicone Foul Release Coatings and Articles Coated Therewith. U.S. Patent 5,904,988, 18 May 1999.
6. Kimura, T.; Ikeno, M. Room Temperature Fast Curable Compositions. U.S. Patent 6,306,998 B1, 23 October 2001.
7. Lucas, G.M. Process for Producing Alkoxy-Terminated Polysiloxanes. U.S. Patent 4,599,394, 8 July 1986.
8. Ogliani, E.; Yu, L.; Skov, A.L.; Skov, A.L. Designing reliable silicone elastomers for high-temperature applications. *Polym. Degrad. Stab.* **2018**, *157*, 175–180. [CrossRef]
9. Skov, A.L.; Vudayagiri, S.; Skov, A.L. How to tailor flexible silicone elastomers with mechanical integrity: A tutorial review. *Chem. Soc. Rev.* **2019**, *48*, 1448–1464. [CrossRef]
10. Jurásková, A.; Dam-Johansen, K.; Olsen, S.M.; Skov, A.L. Factors influencing mechanical long-term stability of condensation curing silicone elastomers. *J. Polym. Res.* **2020**, *27*, 1–14. [CrossRef]
11. Tang, M.Y.; Mark, J.E. Effect of composition and cross-link functionality on the elastomeric properties of bimodal networks. *Macromolecules* **1984**, *17*, 2616–2619. [CrossRef]
12. Sprung, M.M.; Guenther, F.O. The Partial Hydrolysis of Methyltrimethoxysilane. *J. Am. Chem. Soc.* **2005**, *77*, 4173–4175. [CrossRef]
13. Smith, K.A. Polycondensation of methyltrimethoxysilane. *Macromolecules* **1987**, *20*, 2514–2520. [CrossRef]
14. Issa, A.A.; Luyt, A.S. Kinetics of Alkoxysilanes and Organoalkoxysilanes Polymerization: A Review. *Polymers* **2019**, *11*, 537. [CrossRef]
15. Clarson, S.J.; Wang, Z.; Mark, J.E. Effect of stannous 2-ethylhexanoate on the network formation and chain extension reactions of α,ω-dihydroxy terminated poly(dimethylsiloxane). *Eur. Polym. J.* **1990**, *26*, 621–622. [CrossRef]
16. Yuan, Z.; Wang, J. Preparation and characterization of incompletely condensed POSS and its application in RTV composites. *J. Elastomers Plast.* **2016**, *49*, 157–172. [CrossRef]
17. Ji, J.-Y.; Ge, X.; Pang, X.; Liu, R.; Wen, S.; Sun, J.; Liang, W.; Ge, J.; Chen, X. Synthesis and Characterization of Room Temperature Vulcanized Silicone Rubber Using Methoxyl-Capped MQ Silicone Resin as Self-Reinforced Cross-Linker. *Polymers* **2019**, *11*, 1142. [CrossRef] [PubMed]
18. Shi, Y.; Gao, X.; Zhang, D.; Liu, Y.; Huang, G. Synthesis and thermal properties of modified room temperature vulcanized (RTV) silicone rubber using polyhedral oligomeric silsesquioxane (POSS) as a cross linking agent. *RSC Adv.* **2014**, *4*, 41453–41460. [CrossRef]
19. Sun, J.; Kong, J.; He, C. Liquid polyoctahedral silsesquioxanes as an effective and facile reinforcement for liquid silicone rubber. *J. Appl. Polym. Sci.* **2019**, *136*, 46996. [CrossRef]
20. Zhang, D.; Huang, G.; Shi, Y.; Zhang, G.; Liu, Y. Polyhedral oligomeric silsesquioxane/silica/polydimethylsiloxane rubber composites with enhanced mechanical and thermal properties. *J. Appl. Polym. Sci.* **2015**, *132*, 1–8. [CrossRef]
21. Mohammad, S.A.; Wee, A.G.; Rumsey, D.J.; Schricker, S.R. Maxillofacial Materials Reinforced with Various Concentrations of Polyhedral Silsesquioxanes. *J. Dent. Biomech.* **2010**, *1*, 701845. [CrossRef]
22. Silau, H.; Stabell, N.B.; Petersen, F.R.; Pham, M.; Yu, L.; Skov, A.L. Weibull Analysis of Electrical Breakdown Strength as an Effective Means of Evaluating Elastomer Thin Film Quality. *Adv. Eng. Mater.* **2018**, *20*, 1–8. [CrossRef]
23. Dissado, L.A.; Fothergill, J.C. Electrical Degradation and Breakdown in Polymers. *Polymer* **1993**, *34*, 3966. [CrossRef]
24. Yu, L.; Skov, A.L. Molecular Strategies for Improved Dielectric Elastomer Electrical Breakdown Strengths. *Macromol. Rapid Commun.* **2018**, *39*, 1–6. [CrossRef]
25. Vaicekauskaite, J.; Skov, A.L.; Vudayagiri, S.; Skov, A.L. Mapping the mechanical and electrical properties of commercial silicone elastomer formulations for stretchable transducers. *J. Mater. Chem. C* **2020**, *8*, 1273–1279. [CrossRef]
26. Chalker, P.; Bull, S.; Rickerby, D. A review of the methods for the evaluation of coating-substrate adhesion. *Mater. Sci. Eng. A* **1991**, *140*, 583–592. [CrossRef]
27. Li, J.; Beres, W. Scratch Test for Coating/Substrate Systems—A Literature Review. *Can. Met. Q.* **2007**, *46*, 155–173. [CrossRef]
28. Sangermano, M.; Messori, M. Scratch Resistance Enhancement of Polymer Coatings. *Macromol. Mater. Eng.* **2010**, *295*, 603–612. [CrossRef]
29. Madsen, F.B.; Daugaard, A.E.; Hvilsted, S.; Skov, A.L. The Current State of Silicone-Based Dielectric Elastomer Transducers. *Macromol. Rapid Commun.* **2016**, *37*, 378–413. [CrossRef] [PubMed]
30. Jurásková, A.; Skov, A.L.; Brook, M.A. A Mild Route to Convert SiH Compounds to Their Alkoxy Analogues. *Ind. Eng. Chem. Res.* **2020**, *59*, 18412–18418. [CrossRef]
31. Zakaria, S.; Madsen, F.B.; Skov, A.L. Post Curing as an Effective Means of Ensuring the Long-term Reliability of PDMS Thin Films for Dielectric Elastomer Applications. *Polym. Technol. Eng.* **2017**, *56*, 83–95. [CrossRef]
32. Gale, C.B.; Brook, M.A.; Skov, A.L. Compatibilization of porphyrins for use as high permittivity fillers in low voltage actuating silicone dielectric elastomers. *RSC Adv.* **2020**, *10*, 18477–18486. [CrossRef]

molecules

Article

New Antiadhesive Hydrophobic Polysiloxanes

Maria Nowacka [1], Anna Rygała [2], Dorota Kręgiel [2] and Anna Kowalewska [1,*]

[1] Centre of Molecular and Macromolecular Studies, Polish Academy of Sciences, Sienkiewicza 112, 90-363 Łódź, Poland; mnowacka@cbmm.lodz.pl

[2] Department of Environmental Biotechnology, Faculty of Biotechnology and Food Sciences, Lodz University of Technology, Wólczańska 171/173, 90-924 Łódź, Poland; anna.rygala@p.lodz.pl (A.R.); dorota.kregiel@p.lodz.pl (D.K.)

* Correspondence: anko@cbmm.lodz.pl; Tel.: +48-42-6803-350

Abstract: Intrinsic hydrophobicity is the reason for efficient bacterial settlement and biofilm growth on silicone materials. Those unwelcomed phenomena may play an important role in pathogen transmission. We have proposed an approach towards the development of new anti-biofilm strategies that resulted in novel antimicrobial hydrophobic silicones. Those functionalized polysiloxanes grafted with side 2-(carboxymethylthioethyl)-, 2-(*n*-propylamidomethylthioethyl)- and 2-(mercaptoethylamidomethylthioethyl)- groups showed a wide range of antimicrobial properties towards selected strains of bacteria (reference strains *Staphylococcus aureus*, *Escherichia coli* and water-borne isolates *Agrobacterium tumefaciens*, *Aeromonas hydrophila*), fungi (*Aureobasidium pullulans*) and algae (*Chlorella vulgaris*), which makes them valuable antibacterial and antibiofilm agents. Tested microorganisms showed various levels of biofilm formation, but particularly effective antibiofilm activity was demonstrated for bacterial isolate *A. hydrophila* with high adhesion abilities. In the case of modified surfaces, the relative coefficient of adhesion for this strain was 18 times lower in comparison to the control glass sample.

Keywords: functionalized polysiloxanes; hydrophobic; antimicrobial properties

check for updates

Citation: Nowacka, M.; Rygała, A.; Kręgiel, D.; Kowalewska, A. New Antiadhesive Hydrophobic Polysiloxanes. *Molecules* **2021**, *26*, 814. https://doi.org/10.3390/molecules26040814

Academic Editors: Bruce P. Lee and Matthias Schnabelrauch

Received: 20 December 2020
Accepted: 2 February 2021
Published: 4 February 2021

1. Introduction

Polysiloxanes (silicones) are an important class of inorganic polymers of high temperature and oxidative stability, excellent low temperature flexibility, high resistance to weathering and many chemicals, that has found many industrial and household uses [1–3]. Silicone materials exhibit also good biocompatibility, tolerance to sterilization, bio-durability and hemocompatibility [4,5]. That is the reason for their widespread application for medical purposes and in healthcare. Silicone-coated surfaces operate in both single-use and reusable applications, e.g., as artificial implants, catheters and drains or contact lenses. They are hypoallergenic and have a user-friendly appearance and feel. Due to the low rotation barriers around polar siloxane bonds, most poly(alkylsiloxanes) are very flexible [1,6]. Nonpolar side groups reduce the critical surface tension of polysiloxanes and make them capable of wetting most surfaces.

Despite those obvious advantages, polysiloxanes also have limitations and one of them is their intrinsic hydrophobicity [7] which, unfortunately, makes them attractive supports for bacterial settlement and biofilm growth [8–10]. Adhered bacterial cells in mature biofilms produce extracellular polymeric substances (EPSs) which promote the development of biofilm structures. Moreover, they usually express starvation phenotypes and defense mechanisms (antibiotic resistance or multidrug resistance). That is why traditional disinfectants may not be efficient enough to destroy microbial membranes. Moreover, positive selection of the most resistant cells may even enhance the biofilm adhesion.

Various polymeric materials (including silicones) may be modified to become biofilm-resistant and enhance their antibacterial properties [11–14]. Initial tests that were directed towards killing the microorganisms were mainly based on the use of toxic quaternary ammonium-based materials [15]. Moreover, sediments of dead cells that may become the source of secondary infections are an additional problem. Environmental concerns led to the development of new methods that provided long-term antimicrobial activity, hampering the growth of microorganisms while not adversely affecting the properties and stability of synthetic polymers. Strategies that would improve the resistance of silicones towards biofilm formation and make them bacteria-repellent while preserving their biocompatibility became the main concern in studies on antimicrobial silicones of the new generation. For example, antimicrobial and antifouling activity was noted for polysiloxanes grafted with uncharged hydrophilic poly(ethylene glycol) (PEG) [16,17] and polyacrylate [18] chains as well as hydrolysable block polysiloxane-polymethacrylate copolymers [19]. Such amphiphilic polymers were designed as non-antibiotic antimicrobials of minimal toxicity to mammalian cells, and their antiadhesive activity can be linked to their hydrophobic/hydrophilic balance. Hydrophilic units improve the biocompatibility of macromolecules and their binding to polyanionic cell membranes while hydrophobic segments insert into the lipid bilayer and cause its reorganization and damage. More recent solutions for controlling biofilms by synthetic biology methods are based on the protein engineering of biofilm-related enzymes and secretion of signaling molecules for cell–cell communication (quorum sensing) [20]. The idea of grafting polysiloxanes with chemical entities that would mime components of natural quorum-sensing molecules seems to be an attractive alternative. The library of antimicrobial and bacteriostatic polysiloxanes was thus expanded by us and others for systems containing enzymes [21] as well as residues of *N*-acetylcysteine [22,23] and guanidine [24]. Those polymeric materials were hydrophilic and some of them were even fairly soluble in water.

Looking for species that would exhibit targeted anti-biofilm activity, not changing the characteristic hydrophobicity of silicones, we have tested new polysiloxanes grafted with 2-(carboxymethylthioethyl)-(P-1), 2-(mercaptoethylamidomethylthioethyl)-(P-2) and 2-(*n*-propylamidomethylthioethyl)-(P-3) side groups (Scheme 1). The premise for this approach was the fact that carboxylic acids, including ibuprofen [25] and acetic acid [26,27], are regarded as non-antibiotic drugs and are well-known for their antimicrobial behaviour. Polysiloxanes with amide linkers in their side chains resemble, to some extent, antibacterial silicones modified with peptides [28,29]. The functionalization of polysiloxanes with mercaptoethyl residues was inspired by the well-known ability of thiols to take part in the disruption of proteins through oxidation into S–S bonds and/or SH-induced S–S interchange reactions [30,31]. Surface energy measurements proved that all the prepared polymers are hydrophobic. The materials were tested against a wide range of microorganisms (gram-positive and gram-negative bacteria, fungi and algae) that belong to different systematic groups (prokaryotes, eukaryotes) with different cell organizations and cell wall structures. Those microorganisms commonly occur in water environments. Most of them exhibit high adhesion abilities and are also responsible for nosocomial infections. Tests with selected bacterial strains (*Staphylococcus aureus, Escherichia coli, Agrobacterium tumefaciens, Aeromonas hydrophila*), fungi (*Aureobasidium pullulans*) and algae (*Chlorella vulgaris*) indicated good anti-microbial properties of the functionalized polysiloxanes. A considerable decrease in adhesion was observed in the amount of *A. hydrophila* cells, depending on the kind of functional groups that were grafted at polysiloxane chains. The obtained results show that certain active organic groups may be used for the valuable modification of hydrophobic silicones and improvement of their antimicrobial properties.

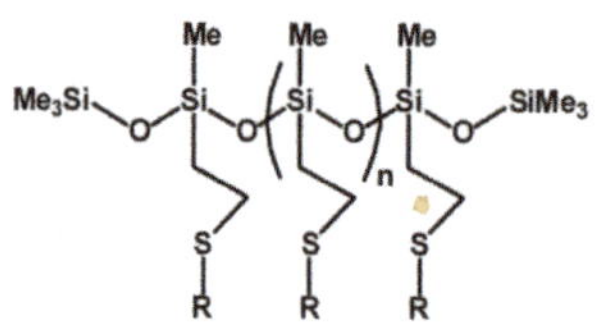

R^1 = -CH_2COOH P-1

R^2 = -$CH_2C(O)NH$-$(CH_2)_2SH$ P-2

R^3 = -$CH_2C(O)NH$-$(CH_2)_2CH_3$ P-3

Scheme 1. Structure of functionalized hydrophobic polysiloxanes of antimicrobial properties.

2. Results and Discussion

2.1. Synthesis and Properties of the Functionalized Polysiloxanes

Polymethylsiloxane homopolymers grafted with side 2-(carboxymethylthioethyl) groups (P-1) were obtained by the photoinitiated addition of thioglycolic acid to poly(vinylmethylsiloxanes), adapting the procedure described earlier for the preparation of poly(2-(carboxymethylthioethyl)silsesquioxanes) [32]. The quantitative reaction resulted exclusively in β-addition product, which was subsequently functionalized with 2-(mercaptoethylamido-methylthioethyl)-(P-2) and 2-(*n*-propylamidomethylthioethyl)methyl- (P-3) moieties through direct amidation of the carboxylic groups with, respectively, cysteamine and *n*-propylamine, using SiO_2 as a heterogeneous, reusable catalyst. The environmentally benign silica catalyst effectively catalyzed amide synthesis from acids and amines without production of toxic by-products [33,34]. The degree of conversion was estimated following the relative increase in new signals corresponding to -CH_2- in the products (3.5 ppm for P-2 and 1.7 ppm for P-3) with respect to the resonances of Si-Me groups in ^{1}H-NMR spectra. The structure of products was characterized using ^{13}C and ^{29}Si-NMR and FTIR spectroscopies. Characteristic vibration modes proved the formation of the secondary amide bonds (amide I band at 1650 cm^{-1} (ν C = O) and amide II band at 1578 cm^{-1} (δ NH in plane + ν C-N)) as well as the presence of S-H groups (ν S-H at 2460 cm^{-1}) in P-2. The conversion of COOH groups into amide bonds was not quantitative due to the steric congestion that becomes an obstacle once about 70 and 60% of substitution was reached in the case of P-2 and P-3, respectively. The reaction yield was 52–56%. The products were purified by three-fold precipitation in nonpolar solvents. Their properties were studied in comparison to those of the precursor VMS-T11.

Calorimetric analysis of the obtained products showed (Figure 1a) that the characteristic temperature of devitrification recorded for all the studied polymers is well below 0 °C while being significantly higher than the characteristic glass transition temperature for poly(methylvinylsiloxanes) [35]. The effect may be correlated with the presence of hydrogen bonds between carboxylic groups or amide linkages in the side chains. The polymers are thermally stable up to 100 °C in the nitrogen atmosphere (Figure 1b, Table 1) which is significantly different to the rapid depolymerization of VMS-T11 (the first decomposition step with T_{d1} at 117 °C, at V_{d1} of 8.41%·min/°C, followed by the slower degradation of the thermally crosslinked polysiloxane). Thermolysis of the siloxane precursor proceeds in a way typical of poly(alkylsiloxanes) (chain scission and backbiting) [36,37], although the ceramic residue (10 wt.%) suggests that the sample must have partly crosslinked at T > 350 °C through the radical polymerization of vinyl groups [38,39].

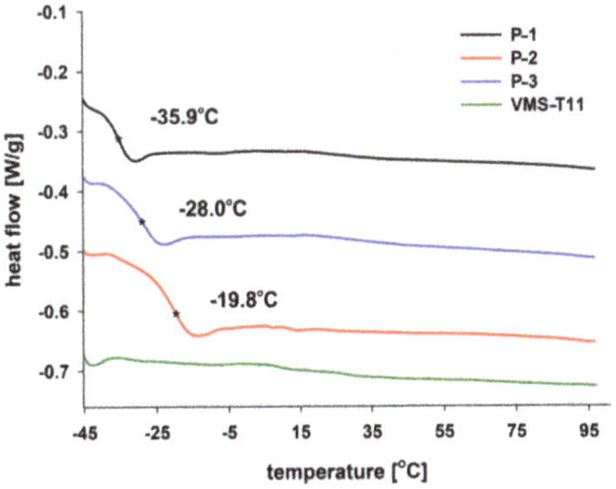

(a)

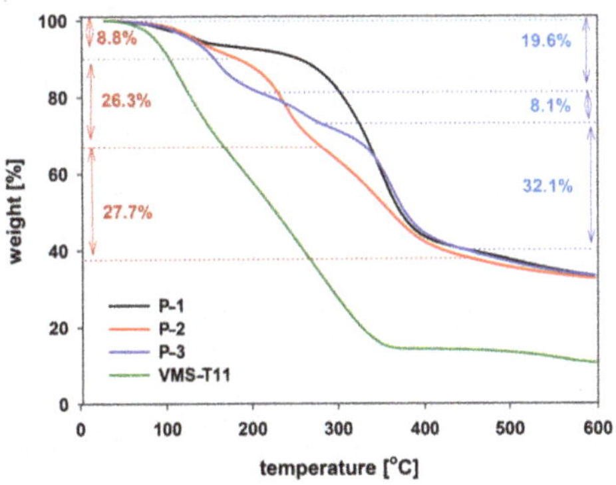

(b)

Figure 1. Thermal characteristics of the functionalized silicones: (**a**) Differential Scanning Calorimetry (DSC), 3rd heating run; (**b**) Thermogravimetric Analysis (TGA) in N_2 10 °C/min.

Table 1. Thermal degradation characteristics (the prepared materials and their precursor).

Sample	$T_{5\%}$ (°C)	T_{d1} (°C)	T_{d2} (°C)	T_{d3} (°C)	Rate (%·min/°C)			Residue (%)
					V_{d1}	V_{d2}	V_{d3}	
VMS-T11	87.6	117.2	272.0	554.1	8.41	4.59	0.38	10.1
P-1	137.6	135.6	350.3	506.5	0.30	5.77	0.34	33.4
P-2	135.4	134.1	238.2	357.4	0.81	6.10	2.77	32.7
P-3	129.7	159.5	256.7	361.9	1.73	0.71	5.80	33.3

T_d—peak temperature of weight loss derivative; V—rate of weight loss at major decomposition step; $T_{5\%}$—5% weight loss temperature; Residue—char residue at 600 °C.

Thermal degradation of the studied products P-1, P-2 and P-3 starts at T > 120 °C. Within 120–150 °C, traces of moisture and volatiles may desorb from the material, as previously shown for ladder [2-(carboxymethylthio)ethyl]polysilsesquioxanes [40] and octahedral silsesquioxanes grafted with *N*-ethylhexanamide groups [41]. However, for P-3, the initial weight decrease overlaps with another process with a maximum at 159 °C (19.6 wt.% decrease at 1.7%·min/°C). For P-1, the main degradation step (60% weight loss, T_{d2} = 350.3 °C, V_{d2} = 5.77%·min/°C) occurred within 250–400 °C and may be correlated to the decarboxylation of side -COOH groups [42,43]. Thermal decomposition of P-2 and P-3 is more complicated due to the more complex composition of the side groups. For both P-2 and P-3, two additional phenomena may be distinguished: a weight loss within 200–300 °C and another one at 300–400 °C. Degradation of P-2 within 200–300 °C (T_{d2} = 238.2 °C, V_{d2} = 6.10%·min/°C) may be linked to partial siloxane chain scission/chain transfer as a result of interactions of nucleophilic thiols and electrophilic silicon atoms in the main chain, analogously to the degradation pattern observed for silicones functionalized with γ-chloropropyl groups [44,45]. The next process observed for P-2 was also severe and quite rapid (T_{d3} = 357.4 °C, V_{d3} = 2.77%·min/°C), and we have attributed it to amide bonds degradation (analogously to those reported in the literature [34,41,43,46]). The residual carboxylic moieties (~30% of the siloxane units) may also participate in the decomposition of amide bonds (acidolysis of amides [47,48]). The decomposition steps observed for P-3 during the thermal treatment overlap largely with those observed for P-2. However, the rate of decomposition at T_{d3} = 361.9 °C (V_{d3} = 5.80%·min/°C) is larger than that observed for P-2, which may be attributed to the lack of reactive moieties that could induce crosslinking of the polysiloxane matrix. The ceramic residue at 600 °C was similar for all the studied functionalized silicones (33–34 wt.%).

Wettability of the studied polymers was estimated using the sessile drop technique for samples coated onto glass slides. Contact angles of the test liquids (water and glycerol) were measured at room temperature and free surface energy was determined (Figure 2). Despite the presence of hydrophilic units in the functional groups, the hydrophobicity of the studied samples was similar to that of poly(vinylmethylsiloxanes) [32,49]. Thin films of P-1, P-2 and P-3 reduced the surface free energy of both clean and 3-aminopropyltriethoxysilane (APTES)-modified glass supports with respect to the effect of VMS-T11 which indicates that, analogously to poly(vinyl-co-(2-hydroxyethylthioethyl))methylsiloxanes [49], the functional macromolecules adapted their conformation in order to form the most energetically favoured interface with hydrophobic air. The effect was found to be even more pronounced for P-1 coated on APTES-glass (the free surface energy of VMS-T11, P-2 and P-3 did not change upon changing the type of support). In this case, -COOH functional groups may be involved in the formation of hydrogen bonds and ionic structures with aminopropyl groups grafted on the surface of slides. As a result, the polymer chain adopts the most suitable conformation for those interactions, leaving Me groups more exposed to air, and thus causing the relative decrease in surface energy. The observed changes in free surface energy illustrate the ability of the functionalized polysiloxanes to adapt to the environmental conditions, including specific components of biofilms.

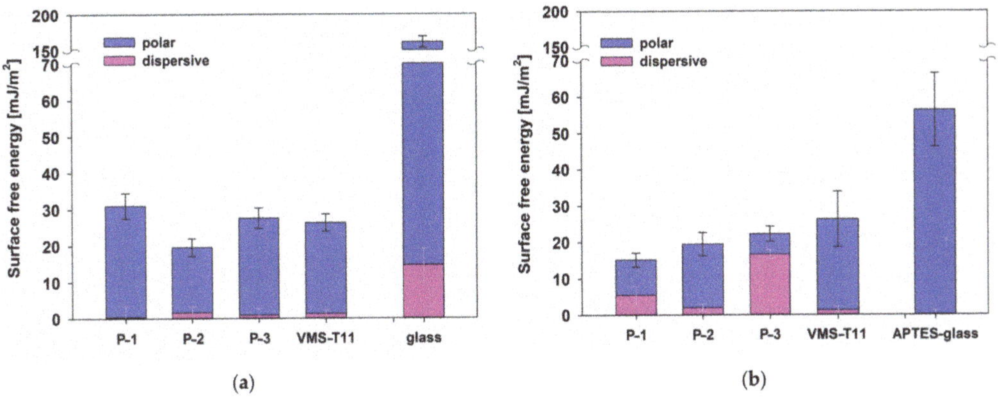

Figure 2. Surface free energy of functionalized polysiloxanes in comparison to their precursor VMS-T11 coated on (**a**) bare glass and (**b**) APTES-glass; (data presented as the mean ± standard deviation (SD, $n = 4$)).

2.2. Antiadhesive Properties of the Functionalized Silicones

The functionalized polysiloxane materials were tested against a wide range of prokaryotic and eukaryotic microorganisms: selected bacterial strains (*S. aureus, E. coli, A. tumefaciens, A. hydrophila*), fungi (*A. pullulans*) and algae (*C. vulgaris*) with different cell organizations and cell wall structures. In our studies on cell adhesion, we decided to use a rather long, 6-day contact of the polymer with the tested microorganism. This strategy is not only most suitable for testing the anti-adhesive activity of a polymer surface, it also allows for evaluation of the reaction of microbial cells to the presence of functional groups grafted on the polysiloxane materials. A similar methodology was used in our previous research works on bioactive polymers [22,23]. The activity tests for the modified polysiloxanes were carried out in comparison to glass (control) and hydrophobic poly(vinylsilsesquioxane) (LPSQ) [50], which was used as a general model of hydrophobic silicone materials (instead of liquid VMS-T11) due to the higher durability of coatings formed on glass supports. The rapid and simple luminometric method that was chosen for this study allowed the estimation of biological material on the test surfaces. This method is based on bacterial ATP quantification and can be used not only to indirectly evaluate the total number of adhered cells, but also the cell viability, including cells that are unable to grow. This is worth noting, especially in the case of modified surfaces with potential antimicrobial activity.

Contrary to the hydrophobic LPSQ, all the functionalized polysiloxanes showed strong antimicrobial activities; however, their action depended on both the type of active group and the type of microorganism. Figure 3 presents the microbial abilities of cell adhesion after 6 days of incubation, expressed as the relative adhesion coefficient A. This parameter allows for estimation of the tendency of microbial cells to adhere, in relation to planktonic cells in the tested sample.

Significant differences were found by one-way ANOVA test in terms of adhesion results for the very adhesive bacterial strain of *A. hydrophila*. This water-borne isolate shows strong adhesive abilities, probably due to the presence of many peritrichous lateral flagella and non-pillar polysaccharide adhesins that play an important role in cell adhesion [51]. For this strain, the A coefficient in the presence of functionalized silicones was 18 times lower in comparison to the control glass sample. There is a striking difference in the adhesion of microorganisms on the surfaces of P-1, P-2 and P-3 in comparison to the model of a hydrophobic silicone analogue (LPSQ). All the tested functionalized polysiloxane materials also exhibited antimicrobial properties against gram-positive and gram-negative bacteria, fungi and algae that showed weaker adhesive abilities in comparison to *A. hydrophila* strain. The relative adhesion coefficient values for these microorganisms and glass carriers were lower and equal from 1 to 1.14. However, for strains with much weaker adhesion abilities, the differences between the results for the control glass sample and its modifications were not so significant. In general, the reference bacterial strains *E. coli* and *S. aureus* were found to be less adhesive than the wild-type strains *A. hydrophila* and *A. tumefaciens*. However, a significant difference can be observed comparing the behavior of *E. coli* and *S. aureus* on surfaces covered with P-1, P-2 and P-3 in comparison to the sample of hydrophobic LPSQ. P-1 seems to be the most effective, although the difference between P-1, P-2 and P-3 is not high. The qualitative effect of the modified polysiloxanes on *C. vulgaris* was similar to that on *A. tumefaciens*.

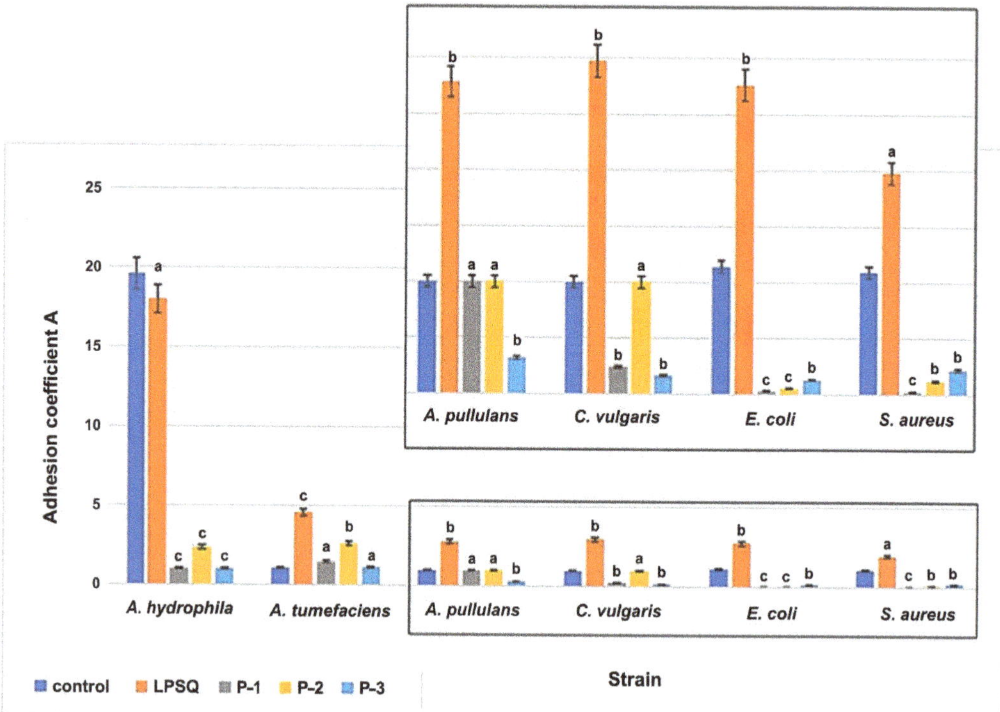

Figure 3. Adhesion coefficient (A) for microbial strains and polysiloxanes functionalized with reactive side groups and a model hydrophobic silicone (LPSQ). The results were compared to those for the control glass carriers. Values with letters show statistically significant differences: a, $p \geq 0.05$; b, $0.005 < p < 0.05$; c, $p < 0.005$.

Our observations suggest that presence of bioactive groups: $-CH_2CH_2SCH_2COOH$, $-CH_2CH_2SCH_2C(O)NHCH_2CH_2SH$ or $-CH_2CH_2SCH_2C(O)NHCH_2CH_2CH_3$ grafted to silicon atoms of polysiloxane chains may be responsible for the antimicrobial activity of these systems. The structural segments of those active groups are widely known for their antimicrobial activity. They may act alone or as a synergistic group against a wide range of microorganisms. [52]. Köllnberger and co-workers documented recently that carboxylic-acid-modified silicone materials exhibited a specific antimicrobial activity against clinically relevant pathogens [53]. The presence of thiol and/or methyl groups also markedly increases the potency of different antimicrobials [54,55].

Microbial adhesion was also detected by microscopic observations (Figure 4). On modified surfaces, in the case of the *A. hydrophila* strain, groups of loosely packed cells were noticed. Therefore, this observation confirmed the results of luminometric studies. However, in the case of other tested microorganisms, the final effect varied and depended on the kind of microorganism.

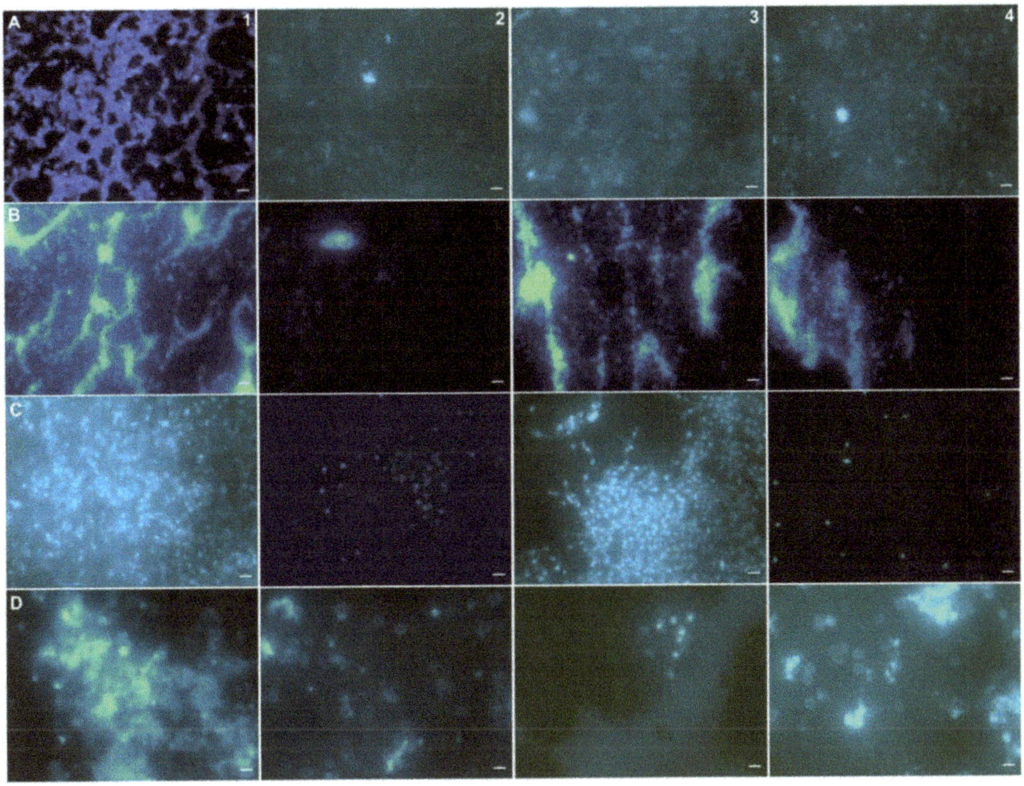

Figure 4. Microbial attachment observed by fluorescence microscopy: (**A**) *A. hydrophila* biofilms; (**B**) *A. tumefaciens* biofilms; (**C**) *A. pullulans* biofilms; (**D**) *C. vulgaris* biofilms; (**1**) control; (**2**) P-1; (**3**) P-2; (**4**) P-3 (scaling bar: 10 μm).

A. tumefaciens rods are often isolated from water environments; they show the ability to create various exopolysaccharides that play a major role in biofilm formation. These compounds, namely cyclic-β-(1,2)-glucan, cellulose, curdlan, succinoglycan, and other unipolar polysaccharides, protect bacteria against environmental stresses [56]. The cell wall of microalgae and fungi also consists of a polysaccharide and glycoprotein matrix providing the cells with a formidable defense against different environmental stress conditions.

Special attention was paid to *A. pullulans* cells. This yeast-like fungus can be found in different environments: soil, water or air. *A. pullulans* cells show high adhesion abilities. This yeast-like fungus forms chlamydoconidia and blastoconidia with high environmental resistance, colonizes different surfaces and extensively produces extracellular polysaccharide pullulan (poly-α-1,6-maltotriose) [57]. Its biofilm and spore structures play an important role in protecting cells against adverse environmental conditions. For example, *A. pullulans* strains were able to inactivate some antimicrobials, namely antibiotics (virginiamycin) and disinfectants (phenols) [58,59]. However, in the case of *A. pullulans,* we observed altered morphology, especially after contact with both P-2 and P-3 surfaces (Figure 5). The cells showed abnormal elongated shapes in comparison to the control sample.

The results of cell-cycle disturbance of those dimorphic fungi in the presence of polymers are very interesting. Planktonic cell and mycelium morphological transitions are not clear. It was documented that *A. pullulans* morphology may be influenced by numerous environmental factors. The effects of some toxic substances, namely alcohols, on the morphology of *A. pullulans* were studied by Sevilla and co-workers [60]. These antimicrobials induced the transition from yeast-like cells to mycelium forms. A similar phenomenon was observed in our studies. Interestingly, only swollen cells and chlamydospores, not mycelial cells, are responsible for pullulan formation [61]. This fact is important because pullulan enhances cell adhesion abilities, and it may explain the poor cell aggregation and biofilm formation by hyphae forms on the P-2 and P-3 surfaces.

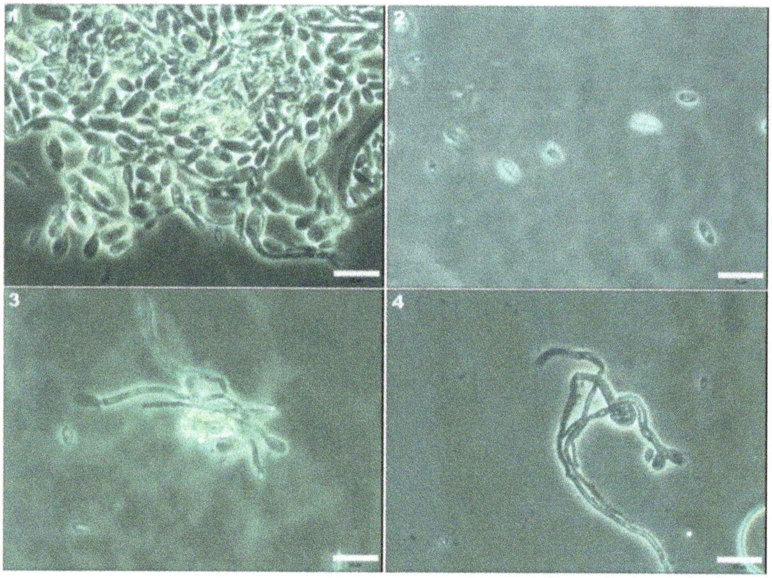

Figure 5. Morphology of *A. pullulans* cells observed by light microscopy: (**1**) control; (**2**) P-1; (**3**) P-2; (**4**) P-3 (scaling bar: 20 μm).

In vitro evaluating antimicrobial activity was based on the agar well diffusion method. Because of the fact that P-2 and P-3 are too viscous and, as such, could not be precisely dosed into the agar plates, we decided to test the activity of P-1 and a range of hydrophobic organosilicon compounds as well as substances used for modification of P-1 in order to obtain P-2 (cysteine) and P-3 (*n*-propylamine). Additional tests with $Me_3SiCH_2CH_2SCH_2COOH$ and neat thioglycolic acid (a substrate in the synthesis of P-1 and $Me_3SiCH_2CH_2SCH_2COOH$) were also carried out.

The obtained results showed that almost all tested substances, except LPSQ, Me_3SiVi, VMS-T11 (not grafted with bioactive groups), inhibited the growth of tested microorganisms (Table 2). Correlation of the activity of thioglycolic acid, $Me_3SiCH_2CH_2SCH_2COOH$ and P-1 with respect to Me_3SiVi and VMS-T11 indicate that the grafting of a bioactive molecule on organosilicon compounds indeed enhances the antimicrobial properties of the latter. On the other hand, the polymeric nature of materials such as P-1 makes more feasible practical use of antimicrobial species of low molecular weight.

Table 2. Growth inhibition zones (mm) determined by agar well diffusion method.

Strain	1	2	3	4	5	6	7	8
A. hydrophila	0	0	0	35	38	40	18	16
A. tumefaciens	0	0	0	30	33	30	12	14
E. coli	0	0	0	10	23	29	11	10
S. aureus	0	0	0	28	27	27	8	10
A. pullulans	0	0	0	33	32	36	7	15
C. vulgaris	0	0	0	28	17	13	8	10

(1) LPSQ; (2) Me_3SiVi; (3) VMS-T11; (4) thioglycolic acid; (5) cysteamine; (6) *n*-propylamine; (7) $Me_3SiCH_2CH_2SCH_2COOH$; (8) P-1.

The strain *A. hydrophila* showed a high sensitivity to the compounds used. This feature could have influenced the best anti-adhesive effect of the functional polymers for this strain (Figure 4). In the case of the *A. pullulans* strain, the zones of growth inhibition were surrounded by a white mycelium area without dark pigmentation (Figure 6). This confirms that the tested antimicrobials markedly alter the growth cycle of these yeast-like fungi.

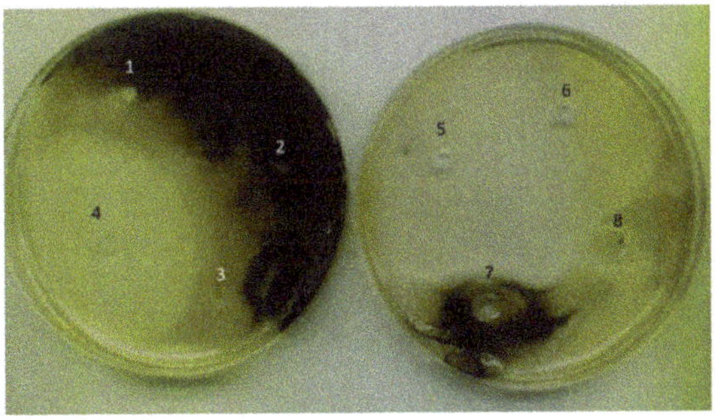

Figure 6. In vitro evaluation antimicrobial activity of selected chemicals on *A. pullulans* strain. (**1**) LPSQ; (**2**) Me_3SiVi; (**3**) VMS-T11; (**4**) thioglycolic acid; (**5**) cysteamine; (**6**) *n*-propylamine; (**7**) $Me_3SiCH_2CH_2SCH_2COOH$; (**8**) P-1.

Therefore, our research into cell adhesion conducted during the 6-day incubation of yeast-like cells with bioactive compounds grafted as side chains on polysiloxane backbone as well as on the agar plates (agar well diffusion method) confirmed the antimicrobial and antiadhesive activities of tested functional polymers. Modified polymer surfaces are becoming more widely investigated for possible use in various settings including clinics, industry, and even the home. These results are promising but require further research on the mechanisms of possible cell destruction and biofouling inhibition by the new hydrophobic polysiloxanes. The future studies of the antimicrobial activity of polymers, either in solution or immobilized on surface, will provide interesting data and possibly new insight into the mechanisms of their antimicrobial action. It is worth paying attention to the long-term use of antimicrobial polymers and the potential risk of increasing microbial resistance.

3. Materials and Methods

3.1. Synthesis of Functionalized Polysiloxanes

The polymeric siloxane precursor—poly(vinylmethylsiloxane) terminated with trimethylsilyl groups (VMS-T11; M_{nGPC} = 1800 g/mol, M_{nNMR} = 1000 g/mol, PDI = 1.3)—was purchased from Gelest (Morrisville, PA, USA). Other reagents: thioglyclic acid (≥99%); cysteamine (95%); *n*-propylamine ((≥99%); 2,2-dimethoxy-2-phenylacetophenone (DMPA, 99%) and silica (high-purity grade, pore size 60 Å, 70–230 mesh) were purchased from Sigma Aldrich (St. Louis, MO, USA) and used as received. Solvents (toluene (Chempur, Karlsruhe, Germany, pure p.a.) and tetrahydrofuran (THF, Chempur, pure p.a.)), were purified according to the literature procedures [62].

Trimethylvinylsilane (Me_3SiVi, 97%), used for the in vitro tests of antimicrobial activity and for the synthesis of $Me_3SiCH_2CH_2SCH_2COOH$, was purchased from Sigma Aldrich and used as received. Poly(vinylsilsesquioxane) terminated with trimethylsilyl groups (LPSQ) was obtained as previously reported [50].

3.1.1. Synthesis of poly[2-(carboxymethylthioethyl)methylsiloxane] (P-1)

The functionalized polysiloxane was prepared with 90% yield by thiol-ene addition of thioglycolic acid to vinyl groups of poly(vinylmethylsiloxane), adapting the earlier published procedure [32].

NMR (THF-d_8) δ (ppm):

^{1}H: 0.1 (-$OSiMe_3$), 0.2 (-SiMe), 0.9 (-$SiCH_2$-), 2.7 (-CH_2S-), 3.2 (-SCH_2-)

^{13}C: −1.3 (-SiMe), 1.0 (-OSiMe$_3$), 17.6 (-SiCH$_2$-), 26.9 (-CH$_2$S), 33.3 (-SCH$_2$), 173.2 (-COOH)
^{29}Si: −23.6, −21.3 (-CH$_2$(Me)SiO$_{2/2}$), 8.7 (-OSiMe$_3$)

FTIR (cm^{-1}): 3500–3000 (ν O-H), 3000–2560 (ν C-H), 1710 (ν C=O), 1418 (δ C-OH), 1381 (δ C-H), 1275 (ν C-O), 1261 (δ Si-C), 1172 (δ Si-CH$_2$), 1087 (ν Si-O-Si), 798 (ρ Si-CH$_3$)

Me$_3$SiCH$_2$CH$_2$SCH$_2$COOH was prepared following exactly the same procedure (except the ratio [-CH = CH$_2$]$_0$/[S-H]$_0$ = 1) using trimethylvinylsilane as the substrate. The volatiles were removed under reduced pressure to leave liquid product (Y = 98%). The product was used without purification.

NMR (THF-d$_8$) δ (ppm):
^{1}H: 0.0 (-SiMe$_3$), 0.85 (-SiCH$_2$-), 2.7 (-CH$_2$S-), 3.2 (-SCH$_2$-)

3.1.2. A General Procedure for the Synthesis of poly[2-(mercaptoethylamidomethylthioethyl)methylsiloxane] (P-2) and poly[2-(*n*-propylamidomethylthioethyl)methylsiloxane] (P-3)

P-2 and P-3 were obtained using P-1 as the substrate. A procedure of amidation carried out with the use of a heterogeneous catalyst (SiO$_2$) was applied [33,34]. P-1 was charged under argon into a Schlenk flask and dissolved in freshly distilled THF (Table 3). Cysteamine dissolved in DMF (P-2) or *n*-propylamine (P-3, no additional solvent) was added to this solution, followed by the addition of toluene, and the mixture was stirred for 15 min at room temperature. SiO$_2$ was added and the mixture was heated at 100 °C for one week. The reaction progress was monitored through FTIR analysis. When no changes in the intensity of the bands characteristic of C(O)OH (decrease in ν(C=O) IR band at around 1700 cm^{-1}) and C(O)NH (increase in amide I band at 1650 cm^{-1} (ν C=O) and amide II band at 1578 cm^{-1} (δ NH in plane + ν C-N)) were observed, the reaction mixture was heated for another 24 h. FTIR spectrum showed no further changes and the reaction was terminated. Once the maximum possible amidation of carboxylic groups of P-1 was reached, the reaction mixture was cooled down and filtered. The filtrate was concentrated under reduced pressure and the product still containing a small amount of THF was purified by precipitation into pentane (THF/pentane). After separation, the product was reprecipitated from THF/diethyl ether and then MeOH/pentane. All products were very viscous liquids (semi-waxes). The described conversion of the carboxyl groups (approx. 70%) was the maximum value achieved under the described reaction conditions. The observed presence of free carboxyl groups resulted from their incomplete substitution.

Table 3. Composition of reaction mixtures during the synthesis of P-2 and P-3.

Sample	P-2	P-3
P-1 (g)	3	3
cysteamine (mol)	0.0168	-
n-propylamine (mol)	-	0.0168
[COOH]$_0$/[NH$_2$]$_0$	1	1
THF (mL)	10	10
toluene (mL)	90	80
DMF (mL)	45	-
SiO$_2$ (g)	0.4	0.4
X (%)	72	63
Y (%)	56	52

Y—reaction yield; X—degree of conversion of -COOH groups into (-C(O)NH-).

P-2

NMR (THF-d$_8$) δ (ppm):
^{1}H: 0.2 (s. -SiMe + -OSiMe$_3$), 0.9 (m. -SiCH$_2$-), 2.7 (m. -CH$_2$S-), 3.3 (m. -SCH$_2$-), 3.5 (NH-CH$_2$- + -CH$_2$-SH), 3.43 (-SH), 8.2 (-NH-)
^{13}C: −0.9 (-SiMe + -OSiMe$_3$), 17.3 (-SiCH$_2$-), 27.2 (-CH$_2$SH), 33.2–35.0 (-SCH$_2$-), 37.0–39.2 (-CH$_2$S-), 40.6 (-NHCH$_2$-), 170.2 (-C(O)NH-), 173.2 (COOH) (residue)
^{29}Si: –23.4, –21.3 (-CH$_2$(Me)SiO$_{2/2}$), 10.0 (-OSiMe$_3$)

FTIR (cm^{-1}): 3500–3000 (ν O-H), 3287 (ν N-H), 3000–2940 (ν C-H), 2460 (ν S-H), 1700 (ν C=O) (carboxylic groups), 1660 amide I band (ν C=O), 1536 amide II band (δ NH in plane + ν C-N), 1381 (δ C-H), 1261 (δ Si-C), 1169 (δ Si-CH_2), 1076 (ν Si-O-Si), 798 (ρ Si-CH_3)

P-3

NMR (THF-d_8) δ (ppm):

^{1}H: 0.2 (-SiMe + -OSiMe_3), 0.9–1.0 (-SiCH_2- + -CH_3), 1.7 (-CH_2-CH_3), 2.7 (-CH_2S-), 3.2 (-SCH_2- + -NHCH_2-), 8.3 (-NH-)

^{13}C: 0.8 (-SiMe + -OSiMe_3), 10.6 (-CH_2-CH_3), 17.6 (-SiCH_2-), 20.8 (-CH_2-CH_3), 26.5 (-SCH_2-), 35.4 (-CH_2S-), 41.1 (-NHCH2-), 170.8 (-C(O)NH-), 174.6 (-COOH) (residue)

^{29}Si: −23.5, −21.2 (-CH_2(Me)$SiO_{2/2}$), 8.6 (-OSiMe_3)

FTIR (cm^{-1}): 3500–3000 (ν O-H), 3250 (ν N-H), 3000–2500 (ν C-H), 1700 (ν C=O) (carboxylic groups), 1650 amide I band (ν C=O), 1578 amide II band (δ NH in plane + ν C-N), 1379 (δ C-H), 1264 (δ Si-C), 1164 (δ Si-CH_2), 1080 (ν Si-O-Si), 798 (ρ Si-CH_3)

3.2. *Analytic Methods*

3.2.1. Nuclear Magnetic Resonance Spectroscopy (NMR)

Liquid state ^{1}H-, ^{13}C- and ^{29}Si-NMR spectra for the functionalized polysiloxanes were recorded in THF-d_8 as a solvent on a Bruker DRX-500 MHz spectrometer, with TMS as the reference (Billerica, MA, USA).

3.2.2. Fourier Transform Infra-Red Spectroscopy (FT-IR)

The absorption spectra were recorded for thin films of samples cast on crystal KBr windows using a Nicolet 380 FTIR spectrometer (Thermo Scientific, Waltham, MA, USA). The spectra were obtained by adding 32 scans at a resolution of 1 cm^{-1}. The position of characteristic vibration modes was correlated with the appropriate literature data [34,63].

3.2.3. Thermogravimetric Analysis (TGA)

The thermal stability of the studied polymeric materials was analyzed with a Hi-Res TGA 2950 Thermogravimetric Analyzer (TA Instruments, New Castle, DE, USA) in nitrogen atmosphere (heating rate 10 °C/min, resolution 3, sensitivity 3).

3.2.4. Differential Scanning Calorimetry (DSC)

The samples were studied using a DSC 2920 Modulated apparatus (TA Instruments, New Castle, DE, USA). Thermograms were taken for samples (sealed in aluminum pans) heated in N_2 atmosphere at the rate of 10 °C/min from room temperature to 100 °C, ten cooled down at the same rate to −50 °C and heated again to 100 °C. The cooling/heating cycle was repeated and the temperatures of characteristic phase transitions were taken from the third measurement.

3.2.5. Surface Energy Measurements

Thin polymer films were spread on commercially available clean glass slides (clean-ready to use, Lab Glass) and 3-aminopropyltriethoxysilane pre-treated glass supports (APTES-glass, silane-prep slides, Sigma Aldrich) using a slit coating applicator (film thickness of 150 μm). Surface free energy was estimated by contact-angle measurements (sessile drop technique) at the film–air interface, as described earlier [22], using deionized water and glycerol (Chempur, pure p.a., anhydrous) as the reference liquids. Surface energies (including their polar and dispersive components) were estimated by the Owens–Wendt method [64].

3.3. *Biological Studies*

3.3.1. Biological Material

Reference strains *E. coli* ATCC 8738, and *S. aureus* ATCC 6538, as well as water-borne bacterial isolates *A. hydrophila* KC756842 and *A. tumefaciens* KJ719245 deposited at LOCK

Culture Collection [65], were subcultured on Triptic Soy Agar slants (Merck Millipore, Burlington, MA, USA). In turn, mold *A. pullulans* LOCK0461 was cultivated on Malt Extract Agar slants (Merck Millipore). Tested microorganisms were cultivated at 28–30 °C for 48 h, and then maintained at 4 °C. For *C. vulgaris* incubation in liquid Bold Basal medium [66] at 25 °C for 7 days, the light was provided by a cool white LED (T5 15W 6400 K, 80 $\mu mol \cdot m^{-2} \cdot s^{-1}$) with continuous illumination within the experimental period.

3.3.2. Abiotic Surfaces

Bacterial adhesion was evaluated under laboratory conditions. Samples of the studied polymers (P-1, P-2, P-3 and LPSQ) were cast on clean glass slides (clean, ready to use, Lab Glass) using a slit coating applicator (film thickness of 150 µm). Experiments were conducted using white glass as the control carrier (Star Frost 76 mm × 26 mm, Waldemar Knittel Glasbearbeitungs GmbH, Bielefeld, Germany). Glass carriers were sterilized by autoclaving at 121 °C, and modified carriers were sterilized using UV light (265 nm, 2 h per each side).

3.3.3. Assessment of Bacterial Adhesion

Sterile carriers (10 mm × 10 mm) were placed in sterile 25 mL Erlenmeyer flasks with 20 mL of culture medium (Table 4).

Table 4. Culture media.

Kind of Microorganisms	Culture Medium	Reference
Bacteria Fungi	50-fold diluted buffered peptone water (Merck KGaA, Darmstadt, Germany)	[65]
Algae	Bold basal medium	[66]

The amount of cell inoculum (0.1 mL) was standardized densitometrically (1°McF). The samples were incubated at 25 °C with agitation on a laboratory shaker (135 rpm) for 6 days. For *C. vulgaris* incubation, the cool white LED (T5 15 W 6400 K, 80 $\mu mol \cdot m^{-2} \cdot s^{-1}$) with continuous illumination was used. Cell adhesion to the carriers was evaluated using both fluorescence microscopy and luminometry using ATP-free sampling pens (Merck KGaA, Darmstadt, Germany). Luminometric measurements were expressed in Relative Light Units (RLU) using a HY-LiTE2 luminometer (Merck KGaA, Darmstadt, Germany). The relative adhesion coefficient (A) was then calculated: the RLU result for adhered cells was divided by the RLU results for culture suspension in the given sample [67]. Adhered bacterial cells were observed after DAPI staining and using the fluorescence microscope NICON type BX41 fitted with a 50× lens and with top illumination of the tested surfaces by an external lamp. Images were captured with a digital camera.

3.3.4. Determination of Antimicrobial Activity of Working Solutions Used to Create Functional Polymers and Other Model Compounds

The routine antimicrobial susceptibility testing was based on the agar well diffusion method [68]. The agar plate surface was inoculated by spreading 200 µL of the microbial inoculum over the entire agar surface (TSA (Merck KGaA, Darmstadt, Germany) for bacteria, MEB (Merck KGaA, Darmstadt, Germany) for fungi, bold basal agar for algae). Then, a hole with a diameter of 5 mm was punched aseptically with a sterile cork borer, and a volume of 10 µL of the tested substance at the desired concentration was introduced into the well. Then, agar plates were incubated at 25 °C. The antimicrobial agent diffused in the agar medium and inhibited the growth of the microbial strain tested. After incubation, the antimicrobial activity of the tested molecules was detected by the appearance of the inhibition zone (mm) around the well.

3.3.5. Statistical Methods

The results of microbial adhesion were calculated as the means and standard deviations in the data from three independent tests. Analysis of variance (ANOVA) was used to examine the differences between group means, representing the adhesion results (OriginLab Corporation, Northampton, MA, USA). The results were compared to those for the control samples (glass carriers). Values with letters show statistically significant differences: a, $p \geq 0.05$; b, $0.005 < p < 0.05$; c, $p < 0.005$.

4. Conclusions

We have shown that specific functionalization of polysiloxanes may increase their antiadhesive properties against a range of prokaryotic and eukaryotic microorganisms of different cell organization and cell wall structures, while not decreasing the inherent hydrophobic nature of silicone materials. Polysiloxanes bearing 2-(carboxymethylthioethyl)-, 2-(*n*-propylamidomethylthioethyl)- and 2-(mercaptoethylamidomethylthioethyl)- side groups have shown good antimicrobial and antibiofilm activity towards selected strains of bacteria (*A. hydrophila*, *S. aureus*, *E. coli*), fungi (*A. pullulans*) and algae (*C. vulgaris*). The results obtained for model thin-coating samples are promising for application of the studied polymers incorporated as structural units in more complicated macromolecular systems, e.g., crosslinked silicone rubber or gels.

This research is in line with current trends and developments for antimicrobial materials. The antimicrobial and antiadhesive polymers not only play a critical role in cell biology, but they are a powerful tool in many applications, where the control of microbial adhesion is necessary. However, it is worth paying attention to the mechanisms of antimicrobial action of modified compounds, effects of their lasting action on microbial cells and whether the long-term use of such antimicrobial materials is not associated with the risk of increasing cell resistance to the biocides used.

Author Contributions: Conceptualization, A.K.; formal analysis, M.N.; investigation, M.N., A.R. and D.K.; writing—original draft preparation, A.K., D.K.; writing—review and editing, A.K., M.N.; supervision, A.K. All authors have read and agreed to the published version of the manuscript.

Funding: This research received no external funding

Institutional Review Board Statement: Not applicable.

Informed Consent Statement: Not applicable.

Data Availability Statement: Data presented in this study are available on request.

Acknowledgments: The studies were carried out within statutory funds of Centre of Molecular and Macromolecular Studies, Polish Academy of Sciences and Department of Environmental Biotechnology, Faculty of Biotechnology and Food Sciences, Lodz University of Technology.

Conflicts of Interest: The authors declare no conflict of interest.

Sample Availability: Samples of the compounds are available from the authors only for scientific collaboration.

References

1. Jones, R.G.; Ando, W.; Chojnowski, J. (Eds.) *Silicon-Containing Polymers. The Science and Technology of Their Synthesis and Applications*; Springer Science+Business Media: Berlin, Germany, 2000; Section 1; pp. 1–244. [CrossRef]
2. Abe, Y.; Gunji, T. Oligo- and polysiloxanes. *Progr. Polym. Sci.* **2004**, *29*, 149–182. [CrossRef]
3. Shit, S.C.; Shah, P. A Review on Silicone Rubber. *Natl. Acad. Sci. Lett.* **2013**, *36*, 355–365. [CrossRef]
4. Curtis, J.; Colas, A. Chapter II. 5.18—Medical applications of silicones A2—Ratner, Buddy, D. In *Biomaterials Science*, 3rd ed.; Hoffman, A.S., Schoen, F.J., Lemons, J.E., Eds.; Academic Press: Cambridge, MA, USA, 2013; pp. 1106–1116. ISBN 9780080877808.
5. Mojsiewicz-Pieńkowska, K. *Safety and Toxicity Aspects of Polysiloxanes (Silicones) Applications. Chapter 16 in Concise Encylopedia of High Performance Silicones*; Tiwari, A., Soucek, M.D., Eds.; Scrivener Publishing LLC.: Beverly, MA, USA, 2014; pp. 243–252; ISBN 9781118938478. [CrossRef]
6. Mazurek, M.H. 3.12 Silicones. In *Comprehensive Organometallic Chemistry III*; Michael, D., Mingos, P., Crabtree, R.H., Eds.; Elsevier: Oxford, UK, 2007; pp. 651–697. ISBN 978-0-08-045047-6.

7. Owen, M.J. Silicone hydrophobicity and oleophilicity. *Silicon* **2014**, *9*, 651–655. [CrossRef]
8. Krishnan, S.; Weinman, C.J.; Ober, C.K. Advances in polymers for anti-biofouling surfaces. *J. Mater. Chem.* **2008**, *18*, 3405–3413. [CrossRef]
9. Yuan, Y.; Hays, M.P.; Hardwidge, P.R.; Kim, J. Surface characteristics influencing bacterial adhesion to polymeric substrates. *RSC Adv.* **2017**, *7*, 14254–14261. [CrossRef]
10. Lam, M.; Migonney, V.; Falentin-Daudre, C. Review of silicone surface modification techniques and coatings for antibacterial/antimicrobial applications to improve breast implant surfaces. *Acta Biomater.* **2021**, *119*, 42–56. [CrossRef]
11. Chen, A.; Peng, H.; Blakey, I.; Whittaker, A.K. Biocidal Polymers: A Mechanistic Overview. *Polym. Rev.* **2007**, *57*, 276–310. [CrossRef]
12. Muñoz-Bonilla, A.; Fernandez-Garcia, M. Polymeric materials with antimicrobial activity. *Progr. Polym. Sci.* **2012**, *37*, 281–339. [CrossRef]
13. Kottmann, A.; Mejía, E.; Hémery, T.; Klein, J.; Kragl, U. Recent Developments in the Preparation of Silicones with Antimicrobial Properties. *Chem. Asian J.* **2017**, *12*, 1168–1179. [CrossRef]
14. Ganewatta, M.S.; Chuanbing Tang, C. Controlling macromolecular structures towards effective antimicrobial polymers. *Polymer* **2015**, *63*, A1–A29. [CrossRef]
15. Jiao, Y.; Niu, L.; Ma, S.; Li, J.; Tay, F.R.; Chen, J. Quaternary ammonium-based biomedical materials: State-of-the-art, toxicological aspects and antimicrobial resistance. *Progr. Polym. Sci.* **2017**, *71*, 53–90. [CrossRef] [PubMed]
16. Hawkins, M.L.; Faÿ, F.; Réhel, K.; Linossier, I.; Grunlan, M.A. Bacteria and diatom resistance of silicones modified with PEO-silane amphiphiles. *Biofouling* **2014**, *30*, 247–258. [CrossRef] [PubMed]
17. Hawkins, M.L.; Schott, S.S.; Grigoryan, B.; Rufin, M.A.; Ngo, B.K.D.; Vanderwal, L.; Stafslien, S.J.; Grunlan, M.A. Anti-protein and anti-bacterial behavior of amphiphilic silicones. *Polym. Chem.* **2017**, *8*, 5239–5251. [CrossRef]
18. Kuliasha, C.A.; Finlay, J.A.; Franco, S.C.; Clare, A.S.; Stafslien, S.J.; Brennan, A.B. Marine anti-biofouling efficacy of amphiphilic poly(coacrylate) grafted PDMSe: Effect of graft molecular weight. *Biofouling* **2017**, *33*, 252–267. [CrossRef]
19. Guazzelli, E.; Galli, G.; Martinelli, E.; Margaillan, A.; Bressy, C. Amphiphilic hydrolyzable polydimethylsiloxane-*b*-poly(ethyleneglycol methacrylate-*co*-trialkylsilyl methacrylate) block copolymers for marine coatings. I. Synthesis, hydrolysis and surface wettability. *Polymer* **2020**, *186*, 121954. [CrossRef]
20. Fang, K.; Park, O.-J.; Hong, S.H. Controlling biofilms using synthetic biology approaches. *Biotechnol. Adv.* **2020**, *40*, 107518. [CrossRef]
21. Kim, Y.D.; Dordick, J.S.; Clark, D.S. Siloxane-Based Biocatalytic Films and Paints for Use as Reactive Coatings. *Biotechnol. Bioeng.* **2001**, *72*, 475–482. [CrossRef]
22. Nowacka, M.; Rygała, A.; Kręgiel, D.; Kowalewska, A. Poly(silsesquioxanes) and poly(siloxanes) grafted with N-acetylcysteine for eradicating mature bacterial biofilms in water environment. *Colloids Surf. B Biointerfaces* **2018**, *172*, 627–634. [CrossRef] [PubMed]
23. Kregiel, D.; Rygala, A.; Kolesinska, B.; Nowacka, M.; Herc, A.S.; Kowalewska, A. Antimicrobial and antibiofilm n-acetyl-l-cysteine grafted siloxane polymers with potential for use in water systems. *Int. J. Mol. Sci.* **2019**, *20*, 2011. [CrossRef] [PubMed]
24. Drozdov, F.V.; Tarasenkov, A.N.; Parshina, M.S.; Cherkaev, G.V.; Strukova, E.N.; Muzafarov, A.M. Synthesis of guanidinopropyl triethoxysilane and its homopolymer as a new class of organosilicon antibacterial agents. *J. Organomet. Chem.* **2020**, *918*, 121243. [CrossRef]
25. Obad, J.; Šušković, J.; Kos, B. Antimicrobial activity of ibuprofen: New perspectives on an "Old" non-antibiotic drug. *Eur. J. Pharm. Sci.* **2015**, *71*, 93–98. [CrossRef] [PubMed]
26. Fraise, A.P.; Wilkinson, M.A.C.; Bradley, C.R.; Oppenheim, B.; Moiemen, N. The antibacterial activity and stability of acetic acid. *J. Hosp. Infect.* **2013**, *84*, 329–331. [CrossRef]
27. Halstead, F.D.; Rauf, M.; Moiemen, N.S.; Bamford, A.; Wearn, C.M.; Fraise, A.P.; Lund, P.A.; Oppenheim, B.A.; Webber, M.A. The Antibacterial Activity of Acetic Acid against Biofilm-Producing Pathogens of Relevance to Burns Patients. *PLoS ONE* **2015**, *10*, e0136190. [CrossRef]
28. Jebors, S.; Pinese, C.; Nottelet, B.; Parra, K.; Amblard, M.; Mehdi, A.; Martinez, J.; Subra, G. Turning peptides in comb silicone polymers. *J. Pept. Sci.* **2015**, *21*, 243–247. [CrossRef] [PubMed]
29. Martin, J.; Wehbi, M.; Echalier, C.; Hunger, S.; Bethry, A.; Garric, X.; Pinese, C.; Martinez, J.; Vezenkov, L.; Subra, G.; et al. Direct Synthesis of Peptide-Containing Silicones: A New Way to Bioactive Materials. *Chem. Eur. J.* **2020**, *26*, 12839–12845. [CrossRef]
30. Trivedi, M.V.; Laurence, J.S.; Siahaan, T.J. The role of thiols and disulfides in protein chemical and physical stability. *Curr. Protein Pept. Sci.* **2009**, *10*, 614–625. [CrossRef]
31. Costa, F.; Sousa, D.M.; Parreira, P.; Lamghari, M.; Gomes, P.; Martins, M.C.L. N-acetylcysteine-functionalized coating avoids bacterial adhesion and biofilm formation. *Sci. Rep.* **2017**, *7*, 17374. [CrossRef]
32. Kowalewska, A.; Nowacka, M.; Tracz, A.; Makowski, T. Supramolecular self-assembly of linear oligosilsesquioxanes on mica-AFM surface imaging and hydrophilicity studies. *Soft Matter* **2015**, *11*, 4818–4829. [CrossRef]
33. Comerford, J.W.; Clark, J.H.; Macquarrie, D.J.; Breeden, S.W. Clean, reusable and low cost heterogeneous catalyst for amide synthesis. *Chem. Commun.* **2009**, 2562–2564. [CrossRef] [PubMed]
34. Nowacka, M.; Makowski, T.; Kowalewska, A. Hybrid Fluorescent Poly(silsesquioxanes) with Amide- and Triazole-Containing Side Groups for Light Harvesting and Cation Sensing. *Materials* **2020**, *13*, 4491. [CrossRef]

35. Dünki, S.J.; Cuervo-Reyes, E.; Opris, D.M. A facile synthetic strategy to polysiloxanes containing sulfonyl side groups with high dielectric permittivity. *Polym. Chem.* **2017**, *8*, 715–724. [CrossRef]
36. Thomas, T.H.; Kendrick, T.C. Thermal analysis of polydimethylsiloxanes. I. Thermal degradation in controlled atmospheres. *J. Polym. Sci. Part A2 Polym. Phys.* **1969**, *7*, 537–549. [CrossRef]
37. Camino, G.; Lomakin, S.M.; Lageard, M. Thermal polydimethylsiloxane degradation. Part 2. The degradation mechanisms. *Polymer* **2002**, *43*, 2011–2015. [CrossRef]
38. Fina, A.; Tabuani, D.; Carniato, F.; Frache, A.; Boccaleri, E.; Camino, G. Polyhedral oligomeric silsesquioxanes (POSS) thermal degradation. *Thermochim. Acta* **2006**, *440*, 36–42. [CrossRef]
39. Nowacka, M.; Fischer, C.; Kowalewska, A.; Hebda, M.; Hodor, K. Thermally induced phenomena leading to degradation of poly(silsesquioxane) materials. *Eur. Polym. J.* **2017**, *86*, 17–28. [CrossRef]
40. Kowalewska, A.; Nowacka, M.; Makowski, T.; Michalski, A. Thermal stability of self-assembled surfaces and micropatterns made of ladder polysilsesquioxanes. *Polymer* **2016**, *90*, 147–155. [CrossRef]
41. Männle, F.; Rosquist Tofteberg, T.; Skaugen, M.; Bu, H.; Peters, T.; Dietzel, P.D.C.; Pilz, M. Polymer nanocomposite coatings based on polyhedral oligosilsesquioxanes: Route for industrial manufacturing and barrier properties. *J. Nanopart. Res.* **2011**, *13*, 4691–4701. [CrossRef]
42. Søndergaard, R.R.; Norrman, K.; Krebs, F.C. Low-temperature side-chain cleavage and decarboxylation of polythiophene esters by acid catalysis. *J. Polym. Sci. Part A Polym. Chem.* **2012**, *50*, 1127–1132. [CrossRef]
43. Wu, W.H.; Thomas, P.; Hume, P.; Jin, J. Effective Conversion of Amide to Carboxylic Acid on Polymers of Intrinsic Microporosity (PIM-1) with Nitrous Acid. *Membranes* **2018**, *8*, 20. [CrossRef]
44. Voronkov, M.G. Reactions of α-elimination of silanones as a path for formation and destruction of siloxane structures. *Russ. Chem. Bull.* **1998**, *47*, 795–806. [CrossRef]
45. Dong, F.; Tang, X.; Ma, L.; Tan, X.; Feng, S. Thermal degradation kinetics of functional polysiloxane with pendent γ-chloropropyl groups. *Polym. Bull.* **2020**, in press. [CrossRef]
46. Zhang, K.; Ishida, H. Smart Synthesis of High-Performance Thermosets Based on ortho-Amide–Imide Functional Benzoxazines. *Front. Mater.* **2015**, *2*, 5. [CrossRef]
47. Kotliar, A.M.J. Interchange reactions involving condensation polymers. *Polym. Sci. Macromol. Rev.* **1981**, *16*, 367–395. [CrossRef]
48. Van Bennekom, A.C.M.; Willemsen, P.A.A.T.; Gaymans, R.J. Amide-modified poly(butylene terepthalate): Thermal stability. *Polymer* **1996**, *37*, 5447–5459. [CrossRef]
49. Nowacka, M.; Herc, A.S.; Kowalewska, A. Thiol-ene addition of mercaptoalcohols to poly(vinylsiloxanes) under visible light photocatalysis—An approach towards cross-linkable hydrophilic silicones. *Polyhedron* **2020**, *185*, 114588. [CrossRef]
50. Kowalewska, A.; Nowacka, M. Synthesis of Ladder Silsesquioxanes by in situ Polycondensation of Cyclic Tetravinylsiloxanetetraols. *Silicon* **2015**, *7*, 133–146. [CrossRef]
51. Kręgiel, D.; Berlowska, J.; Mizerska, U.; Fortuniak, W.; Chojnowski, J.; Ambroziak, W. Chemical modification of polyvinyl chloride and silicone elastomer in inhibiting adhesion of *Aeromonas hydrophila*. *World J. Microbiol. Biotechnol.* **2013**, *29*, 1197–1206. [CrossRef]
52. Zi, Y.; Zhu, M.; Li, X.; Xu, Y.; Wei, H.; Li, D.; Mu, C. Effects of carboxyl and aldehyde groups on the antibacterial activity of oxidized amylose. *Carbohydr. Polym.* **2018**, *192*, 118–125. [CrossRef] [PubMed]
53. Köllnberger, A.; Schrader, R.; Briehn, C.A. Carboxylic acid mediated antimicrobial activity of silicone elastomers. *Mater. Sci. Eng. C* **2020**, *113*, 111001. [CrossRef]
54. Peterson, L.R. Quinolone molecular structure-activity relationships: What we have learned about improving antimicrobial activity. *Clin. Infect. Dis.* **2001**, *33* (Suppl. 3), S180–S186. [CrossRef]
55. Wiradharma, N.; Khan, M.; Yong, L.-K.; Hauser, C.A.E.; Seow, S.V.; Zhang, S.; Yang, Y.-Y. The effect of thiol functional group incorporation into cationic helical peptides on antimicrobial activities and spectra. *Biomaterials* **2011**, *32*, 9100–9108. [CrossRef] [PubMed]
56. Matthysse, A.G. Exopolysaccharides of *Agrobacterium tumefaciens*. In *Agrobacterium Biology*; Springer: Berlin/Heidelberg, Germany, 2018; pp. 111–141.
57. Sabev, H.A.; Robson, G.D.; Handley, P.S. Influence of starvation, surface attachment and biofilm growth on the biocide susceptibility of the biodeteriogenic yeast *Aureobasidium pullulans*. *J. Appl. Microbiol.* **2006**, *101*, 319–330. [CrossRef]
58. Dos Santos, V.L.; Monteiro Ade, S.; Braga, D.T.; Santoro, M.M. Phenol degradation by *Aureobasidium pullulans* FE13 isolated from industrial effluents. *J. Hazard. Mater.* **2009**, *161*, 1413–1420. [CrossRef] [PubMed]
59. Leathers, T.D.; Rich, J.O.; Nunnally, M.S.; Anderson, A.M. Inactivation of virginiamycin by *Aureobasidium pullulans*. *Biotechnol. Lett.* **2018**, *40*, 157–163. [CrossRef]
60. Sevilla, M.J.; Landajuela, L.; Uruburu, F. The effect of alcohols on the morphology of *Aureobasidium pullulans*. *Curr. Microbiol.* **1983**, *9*, 169–171. [CrossRef]
61. Campbell, B.S.; Siddique, A.B.M.; McDougall, B.M.; Seviour, R.J. Which morphological forms of the fungus *Aureobasidium pullulans* are responsible for pullulan production? *FEMS Microbiol. Lett.* **2004**, *232*, 225–228. [CrossRef]
62. Armarego, W.L.F.; Chai, C.L.L. *Purification of Laboratory Chemicals*, 5th ed.; Elsevier Science: London, UK, 2003.
63. Kowalewska, A.; Fortuniak, W.; Rózga-Wijas, K.; Handke, B. Thermolysis of new hybrid silsesquioxane-carbosilane materials. *Thermochim. Acta* **2009**, *494*, 45–53. [CrossRef]

64. Owens, D.K.; Wendt, R.C. Estimation of the surface free energy of polymers. *J. Appl. Polym. Sci.* **1969**, *13*, 1741–1747. [CrossRef]
65. Rygala, A.; Berlowska, D.; Kregiel, D. Heterotrophic plate count for bottled water safety management. *Processes* **2020**, *8*, 739. [CrossRef]
66. Wong, Y.K.; Ho, Y.H.; Ho, K.C.; Leung, H.M.; Yung, K.K.L. Growth medium screening for *Chlorella vulgaris* growth and lipid production. *J. Aquac. Mar. Biol.* **2017**, *6*, 00143. [CrossRef]
67. Kregiel, D. Attachment of *Asaia lannensis* to materials commonly used in beverage industry. *Food Control* **2013**, *32*, 537–542. [CrossRef]
68. Balouiri, M.; Sadiki, M.; Ibnsouda, S.K. Methods for in vitro evaluating antimicrobial activity: A review. *J. Pharm. Anal.* **2016**, *6*, 71–79. [CrossRef] [PubMed]

molecules

MDPI

Article

Chlorine-Functional Silsesquioxanes (POSS-Cl) as Effective Flame Retardants and Reinforcing Additives for Rigid Polyurethane Foams

Anna Strąkowska *, Sylwia Członka, Karolina Miedzińska and Krzysztof Strzelec

Institute of Polymer and Dye Technology, Lodz University of Technology, 90-537 Lodz, Poland; sylwia.czlonka@dokt.p.lodz.pl (S.C.); karolina.miedzinska@dokt.p.lodz.pl (K.M.); krzysztof.strzelec@p.lodz.pl (K.S.)

* Correspondence: anna.strakowska@p.lodz.pl

Abstract: The subject of the research was the production of silsesquioxane modified rigid polyurethane (PUR) foams (POSS-Cl) with chlorine functional groups (chlorobenzyl, chloropropyl, chlorobenzylethyl) characterized by reduced flammability. The foams were prepared in a one-step additive polymerization reaction of isocyanates with polyols, and the POSS modifier was added to the reaction system in an amount of 2 wt.% polyol. The influence of POSS was analyzed by performing a series of tests, such as determination of the kinetics of foam growth, determination of apparent density, and structure analysis. Compressive strength, three-point bending strength, hardness, and shape stability at reduced and elevated temperatures were tested, and the hydrophobicity of the surface was determined. The most important measurement was the determination of the thermal stability (TGA) and the flammability of the modified systems using a cone calorimeter. The obtained results, after comparing with the results for unmodified foam, showed a large influence of POSS modifiers on the functional properties, especially thermal and fire-retardant, of the obtained PUR-POSS-Cl systems.

Keywords: rigid polyurethane foams; POSS-Cl; flame retardants; porous composite; mechanical properties; hydrophobicity

Citation: Strąkowska, A.; Członka, S.; Miedzińska, K.; Strzelec, K. Chlorine-Functional Silsesquioxanes (POSS-Cl) as Effective Flame Retardants and Reinforcing Additives for Rigid Polyurethane Foams. *Molecules* **2021**, *26*, 3979. https://doi.org/10.3390/molecules26133979

Academic Editors: Pradip K. Bhowmik, Sławomir Rubinsztajn, Marek Cypryk and Wlodzimierz Stanczyk

Received: 5 May 2021
Accepted: 26 June 2021
Published: 29 June 2021

1. Introduction

Foams are the main group of polyurethane materials that dominate the market with over 65% of the world's polyurethane production [1,2]. They are characterized by a porous structure, low apparent density, and high strength. Their properties can be modified by the appropriate selection of components and their mutual ratios [3–5]. Polyurethane porous materials fall into two main groups: flexible polyurethane foams and rigid polyurethane foams. One of the most important functional features of foams is the low thermal conductivity coefficient determined by the cell structure of porous materials. The content of open cells in the structure increases the heat transfer capacity [2]. Rigid polyurethane foams are strongly cross-linked materials characterized by a closed-cell structure which affects a high compressive strength and a low thermal insulation coefficient [6,7]. As for the application, due to their good thermal insulation, mechanical and physical properties, and lightweight polyurethane foams are generally used as a low-cost material widely used in building insulation, transportation, electronics, packaging, furniture, and others [5,8]. In addition, flexible polyurethane foams show excellent sound-absorbing properties and the ability to damp vibrations, which allows them to be used as acoustic materials [9].

The main drawback of PUFs is their easy ignition and high flame spreadability, which makes their use limited in many engineering applications [10]. It is known that the degradation of the urethane bonds of rigid polyurethane foams begins at 200 °C [11]. However, it is possible to reduce the combustibility of polyurethane foams by applying various flame retardants [12]. There are various types of flame retardants, including: phosphorous compounds [13], bromine compounds [14], melamine compounds [15], expandable graphite [14,16], and inorganic salts [17]. Moreover, inorganic metal oxides and hydroxides including compounds of magnesium, silicon, and aluminum also play an important role in reducing combustibility [18–21]. Halogen compounds have also been used previously as flame retardants. However, due to the increasing environmental requirements in recent times, their application is limited because of the release of toxic and harmful gases from their combustion [5,22].

Recently, there has been an increasing amount of research into a multifunctional group of flame retardants, polyhedral oligomeric silsesquioxanes (POSSs), that combine characteristics of organic and inorganic materials [2,23]. Their synthesis dates back to the 1950s, but in recent years they have gained growing interest as hybrid organic/inorganic precursors [24]. They are reactive nanofillers with inorganic nanocage ((RSiO1.5)n) surrounded by at most eight organic groups [25]. Polyhedral oligomeric silsesquioxanes can exist in various forms, depending on the spatial isomerism and the type of substituents (alkyl, aryl, or halogenated groups). Scientists discovered that the fireproof effect of these nanofillers is based on the ability to form a ceramic protective layer that reduces the amount of heat released during the pyrolysis and combustion of the material [18,26–28]. However, the organic functional groups in the POSS cage are of great importance for the fireproofing properties, as they break the Si-C bond at a temperature of about 300–350 °C in the air. Immediately after this process, POSS cages are fused to form a thermally insulating and oxidation-resistant layer of carbonized silicon carbide [29,30].

When considering the effect of POSS on flame retardancy, two main types of these nanofillers are distinguished: containing eight identical R groups or seven R groups and one functional group R′ which can be an alcohol, ester, epoxy, isocyanate, amine, or silane. The wide range of these substituents enables their selective selection to achieve compatibility of dispersed nanofillers with the polymer matrix. As with the use of other fillers, the incorporation of POSSs into a polymer matrix can affect both the melt viscosity, mechanical strength, and thermomechanical and electrical properties of the polymers [31]. Moreover, polyhedral oligomeric silsesquioxanes can reduce the amount of heat, smoke and CO released during combustion, thereby increasing thermal stability and mitigating fire hazards [18,24,32,33]. It can be concluded that due to their properties POSS based compounds are commonly used in multi-functional applications. Furthermore, thanks to the presence of both reactive and non-reactive groups, POSSs can be readily attached to the polymer matrix by mixing, grafting, and copolymerization [3,34].

2. Results and Discussion

2.1. The Impact of POSS-Cl on Polyol Viscosity and Processing Parameters of Polyol Premixes

Polyol premixes are typical non-Newtonian shear-thinning fluids for which the viscosity decreases with increasing shear rate. The viscosity of the polyol premixes is a parameter that determines the rate of foam growth. Therefore, it is important that the viscosity is not too high due to the addition of solid excipients, as this would drastically limit the foam growth process, resulting in a composite with a very small pore size and high density.

As expected, the addition of POSS particles increased the viscosity of the polyol (Table 1), with the POSS-Ar-Et-Cl (Table 5) containing system having the highest viscosity. In all premixes the dependence of viscosity decrease as a function of shear rate was visible. The incorporation of solid particles can not only disrupt the porous structure as a result of an increase in viscosity but also, depending on the reactivity of the surface, can create a risk of aggregates and agglomerates forming in the polymer matrix. This phenomenon is very undesirable as it can have a negative effect on the mechanical properties or the insulating properties. Therefore, it is important to choose the modifier so that it disperses well in the reaction mixture and that it does not have a tendency to aggregate.

Table 1. The impact of POSS-Cl on viscosity and growth kinetic of polyol premixes.

Sample Code	Dynamic Viscosity η [mPa·s]			T_{Max} [°C]	Processing Times [s]		
	0.5 RPM	50 RPM	100 RPM		Cream Time	Growth Time	Tack-Free Time
PUR-0	840 ± 9	430 ± 7	320 ± 8	118.2 ± 4	40 ± 2	270 ± 9	365 ± 9
PUR-POSS-Ar-Cl	1120 ± 10	930 ± 9	440 ± 10	135.7 ± 5	47 ± 1	260 ± 11	325 ± 9
PUR-POSS-Pr-Cl	1090 ± 10	870 ± 10	420 ± 11	138.2 ± 4	45 ± 2	215 ± 10	298 ± 12
PUR-POSS-Ar-Et-Cl	1210 ± 10	960 ± 11	470 ± 12	141.2 ± 6	48 ± 2	230 ± 9	314 ± 8

The polymerization reaction of PUR foams is a strongly exothermic reaction with the release of a large amount of heat. For the reference foam, the maximum temperature was 118.2 °C. On the other hand, the introduction of POSS-Cl significantly increased the reactivity of the system, which was manifested by an increase in the maximum temperature during foam synthesis. The largest increase in Tmax was recorded for the PUR-POSS-Ar-Et-Cl system (141.2 °C). In addition, the individual times of foam synthesis have also changed. Due to the higher viscosity value, the initial foaming was slower, which was manifested in the longer creaming time of the composition with the addition of POSS. Compared to PUR-0, creaming time increases from 40 s to more than 45 s. However, well-dispersed solids can act as an additional nucleation center, leading to the formation of more vesicle cells during the nucleation process, which could be manifested by accelerating foam expansion, where the rising time shortened from 270 s for PUR-0 to even 215 s for PUR-POSS-Pr-Cl. The introduction of the modifier also leads to a shortening of the tack-free time. This means that the POSS-Cl particles can act as a hardening accelerator during the foaming process.

2.2. The Impact of POSS-Cl on Morphology and Apparent Density PUR Foams

The cellular structure is an important parameter of porous materials, determining their mechanical and thermal properties. Figure 1 shows that PUR-0 has the most regular pore size and distribution. The shape of the PUR foam cells was typically a closed polyhedron ranging in size from 410 to 475 μm, and the number of open or damaged cells is small. Microscopic photos of the foams with the addition of POSS show that their cell sizes are less uniform than in the case of the reference foam. The number of open or damaged cells has increased and a larger dispersion of pore size can be observed. Presumably, POSS particles can act as nucleation centers, leading to the formation of smaller cells [35]. As shown in previous studies, particulate matter changes the rheology of the system and reduces the nucleation energy. The lowered nucleation barrier facilitates the extensive formation of smaller cells which later fuse into larger cells [36].

Figure 2 shows the SEM pictures taken at a higher magnification, where the filler particles dispersed in the pores of modified foams can be observed. The filler particles can exist both in the cell struts as well as inside the pores. It is visible that for modified foams particles are attached to the cell wall and cell struts which is visible for foam PUR-POSS-Ar-Et-Cl and PUR-POSS-Pr-Cl. In contrast, in the case of foam PUR-POSS-Ar-Cl, filler particles are observed in the void cell space.

One of the most important parameters determining the use of PUR foams as a thermal insulation material is the apparent density, which indirectly affects the mechanical properties, such as compressive strength and hardness. Changes in the structure as a result of the introduction of POSS are manifested by an increase in the apparent density of foams (Table 2). The reason for this slight increase may be the higher initial viscosity of the polyol premix due to the implementation of POSS, which also affects the heterogeneous nucleation process and the formation of more small cells. However, taking into account that the density of polyurethane foam is very sensitive to slight changes in environmental conditions, such as humidity, temperature, or slight changes in mixing time, it can be concluded that the differences in density are not too large [37].

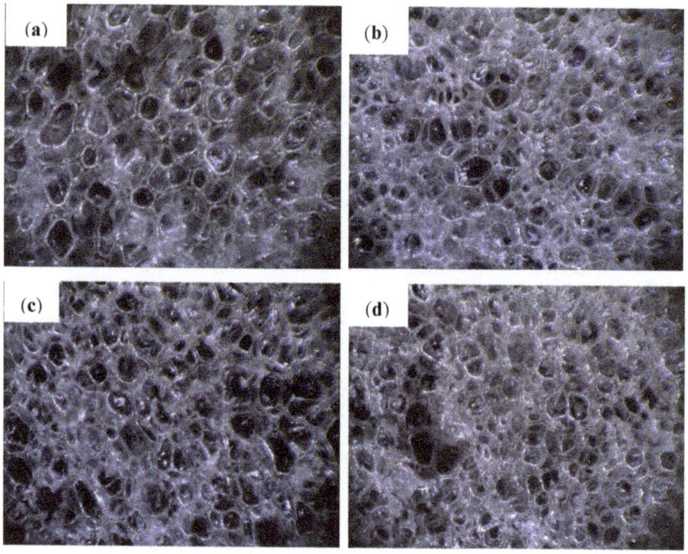

Figure 1. Cellular morphology of (**a**) PUR-0, (**b**) PUR-POSS-Ar-Cl, (**c**) PUR-POSS-Pr-Cl and (**d**) PUR-POSS-Ar-Et-Cl.

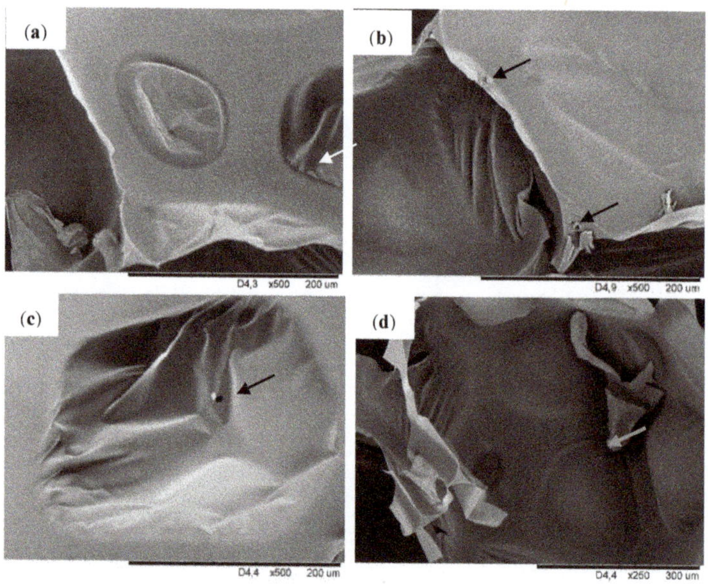

Figure 2. SEM images of (**a**) PUR-0, (**b**) PUR-POSS-Ar-Cl, (**c**) PUR-POSS-Pr-Cl and (**d**) PUR-POSS-Ar-Et-Cl.

Table 2. Parameters of the structure of PUR composite foams.

Sample Code	Cell Size [μm]	Wall Thickness [μm]	Apparent Density [kg m^{-3}]
PUR-0	475 ± 10	63 ± 4	39 ± 1
PUR-POSS-Ar-Cl	442 ± 8	68 ± 5	42 ± 2
PUR-POSS-Pr-Cl	421 ± 9	67 ± 2	43 ± 2
PUR-POSS-Ar-Et-Cl	413 ± 12	69 ± 3	41 ± 2

2.3. Mechanical Properties of PUR Foams

To determine the mechanical properties of the PUR foam, the compressive strength at 10% deformation was tested. Since in rigid PUR foams there is anisotropy of cells manifested by cell elongation in the direction of foam growth, the samples were tested in two directions: parallel and perpendicular to the direction of foam growth (Figure 3). By analyzing the data from the diagram, it can be seen that the addition of POSS-Cl has a significant effect on increasing the compressive strength of PUR. The PUR-0 foam showed a compressive strength parallel to the direction of the foam growth of 207 kPa, while for modified foams there is a tendency to increase the mechanical strength to even 262 kPa for PUR-POSS-Pr-Cl. This 25% increase in strength was due to the higher foam density resulting from the introduction of the POSS cage structure. For other modified foams, a positive effect on strength parameters is also visible. Much lower strengths were obtained in the case of measurements carried out in the direction perpendicular to the direction of foam growth. But also in this case, the introduction of POSS-Cl improved the compressive strength. Only a slight deterioration was noted for PUR-POSS-Ar-Et-Cl, which could be due to the fact that this foam was characterized by the most heterogeneous structure.

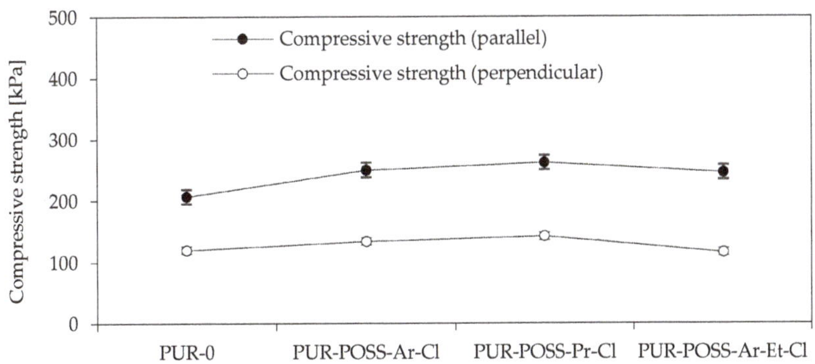

Figure 3. Compressive strength in the direction parallel and perpendicular to the growth direction of PUR foams.

The obtained results of the highest bending stress transmitted by the sample (δfm), called the flexural strength, are shown in Figure 4. Based on the results, it can be seen that the addition of POSS-Cl also significantly changed the flexural strength of the foams obtained. For all modified foams, an increase in the value of the maximum bending stress transferred by the sample was observed. The PUR-POSS-Pr-Cl series is characterized by the highest bending resistance, for which δfm is equal to 442 kPa. This result is higher by almost 15% compared to PUR-0. A very similar tendency can be seen in the case of foam hardness results. The increase in hardness was also caused by the higher density of the samples, although the standard deviation for these results was quite large, which resulted from the heterogeneity of the structure and the large dispersion of the obtained results. Therefore, it should be remembered that the addition of modifiers disrupts the cellular structure, which may cause deterioration of mechanical properties; therefore, it is important to properly select the amount of the introduced modifier.

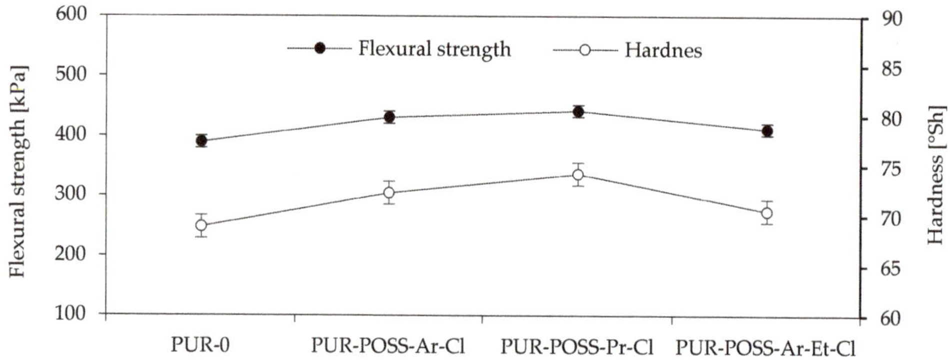

Figure 4. Flexural strength and Hardness of PU foams.

2.4. *Hydrophobic Properties of PUR Foams*

The presence of Cl in the POSS particle resulted in its low surface energy, which could translate into a reduction in the hydrophobicity of the entire system [38]. Therefore, the water absorption and contact angle of the foams' surfaces were measured for the samples (Figure 5). These parameters are important from the operational point of view because it is important that insulation materials made of PUR do not show a tendency to absorb water or moisture from the environment, which could change their general performance parameters and lead to faster degradation.

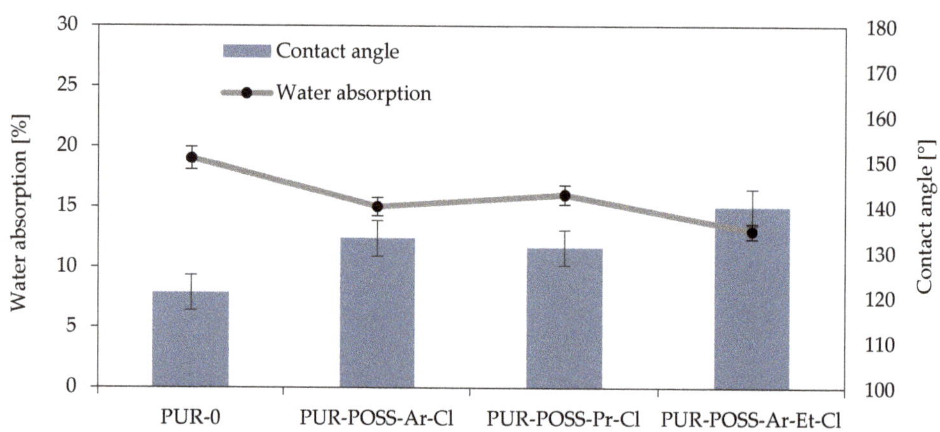

Figure 5. Effect of contact angle on water absorption of RPUFs modified with POSS-Cl.

The measurement of short-term water absorption showed a positive effect on the reduction of water absorption by foam. In all modified foams, it was possible to reduce the water absorption, where the use of POSS-Ar-Et-Cl allowed for a 30% reduction in this parameter. This effect was caused not only by the nature of the halide POSS themselves but also by the more irregular structure which also has a strong influence on the hydrophobicity of the surface.

The reduction in affinity for water is also well illustrated by the results of the measurements of the contact angle of the foam surfaces, which show that the foams with the addition of POSS-Cl are characterized by much lower wettability by water compared to PUR-0. This test also confirmed the most hydrophilic nature of the sample with the addition of POSS-Ar-Et-Cl, for which a contact angle of 140° was obtained (for the PUR-0 sample, the contact angle was 121°).

2.5. Thermal Stability and Fire Behavior of PUR Foams

To evaluate the influence of the applied flame retardants on thermal stability, TGA measurements were performed. Figure 6a,b shows the thermogravimetric (TGA) and derivative thermogravimetric (DTG) curves of analyzed foams. Thermal degradation of RPUFs takes place in three stages. The first stage of decomposition occurs from about 150 to 250 °C and corresponds to about 10% weight loss. It is associated with the dissociation of the urethane bonds corresponding to the degradation of the hard segments [8,11]. The second stage of degradation, manifested by about 50% loss of initial mass starts between 300 and 350 °C and it corresponds to the thermal decomposition of the soft polyol segments of PUR [39]. The last, third degradation stage, in which the weight loss reaches about 70%, occurs between 500 and 600 °C and it is related to the decomposition of the fragments formed in the previous stage into volatile products [40].

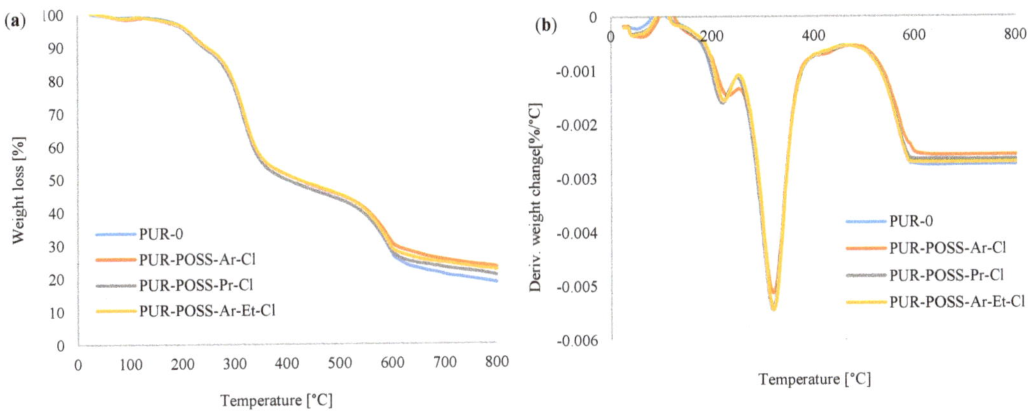

Figure 6. (**a**) TGA and (**b**) DTG curves for PUR foams modified with POSS-Cl.

Table 3 shows the characteristic temperature values corresponding to the successive stages of decomposition. T_5 is the temperature corresponding to a 5% weight loss, similarly T_{10}, T_{50} and T_{70} corresponding to 10, 50 and 70% weight loss to the initial mass of samples. Comparing the foams containing the POSS additives with the reference PUR-0 foam, it has been noticed that modified PUR foams required a higher temperature to pass between the degradation steps. The results summarized in the table show that the PUR-POSS-Pr-Cl foam reached temperatures slightly higher than the reference sample during the subsequent stages of degradation. On the other hand, PUR-POSS-Ar-Cl and PUR-POSS-Ar-Et-Cl samples show increased thermal stability, which is especially noticeable in the analysis of weight loss over 50%.

Table 3. Results of thermogravimetric analysis of PUR foams modified with POSS-Cl.

Sample Code	T_5 [°C]	T_{10} [°C]	T_{50} [°C]	T_{70} [°C]	Char Residue [%]
PUR-0	209	241	397	589	18.6
PUR-POSS-Ar-Cl	211	247	419	605	23.2
PUR-POSS-Pr-Cl	207	241	399	593	20.8
PUR-POSS-Ar-Et-Cl	211	245	417	597	22.4

The determination of char residue by TGA is a relative method of assessing the fire resistance of a polymer [41]. The foams containing the addition of modifiers showed a higher percentage of carbonation residues than the reference foam, where at 800 °C PUR-POSS-Ar-Cl, PUR-POSS-Pr-Cl and PUR-POSS-Ar-Et-Cl the carbon content increased, respectively, up to 23.2, 20.8 and 22.4% compared to 18.6% for the PUR-0 reference sample.

It can be concluded that POSS-Cl can act as a charring agent, which positively influences not only the thermal stability, but also the fire resistance of PUR foams. All these observations may indicate the effect of POSS additives on increasing thermal stability.

The fire resistance properties of PUR based on combustion in a cone calorimeter are represented by the ignition time (TTI), the peak heat release rate (pHRR), and total smoke release (TSR), total heat release (THR), the maximum average rate of heat emission (MAHRE) as well as the total smoke release (TSR) and char residue.

Analyzing the data in Table 4, characterizing the combustion process of PUR foams, a positive effect of POSS-Cl on the reduction of flammability of modified foams can be noticed. TTI for PUR-POSS-Cl is longer than for pure PUR-0. In addition, the parameter informing about the ignition spark of the combusted material was reduced, i.e., the heat release rate—determined by the pHRR. Namely, for the samples PUR-POSS-Ar-Cl, PUR-POSS-Pr-Cl and PUR-POSS-Ar-Et-Cl, the pHRR value was 172, 211, 189 kW/m^2, respectively, compared to 268 kW/m^2 for the reference sample. A similar trend was observed for MARHE. Such an improvement in the fire resistance of PUR composite foams may be related to the formation of a protective carbon layer in the initial decomposition phase, which is an insulating barrier for the material beneath it and prevents the transfer of necessary heat and gases for further combustion. The carbon layer can also reduce the amount of smoke and harmful gases released during combustion. It is also believed that POSS may migrate to the surface during initial ignition and be degraded by homolytic breakage of Si-C bonds. The resulting ceramic layer consisting of stable Si-O bonds creates a reinforced carbon layer, which protects the material against further combustion [42]. In addition, the combustion reactions are radical, and the chlorine atoms formed during the decomposition of POSS act as radical scavengers, additionally inhibiting the combustion of the composite.

Among all series of PUR composite foams, the greatest improvement was observed for PUR composite foams containing POSS with aromatic structures (Figure 7a,b)—the pHRR value decreased by about 30%. Such improvement is a result of the stabilization of the aromatic system by the delocalization of the pi-electrons on the ring [43]. In addition, for all modified PUR-POSS-Cl foams lower TSR values were recorded, which additionally increases safety during a fire, as smoke poisoning during a fire is as dangerous as contact with fire or high temperature. Moreover, the introduction of POSS-Cl slightly lowers the value of total heat release (THR).

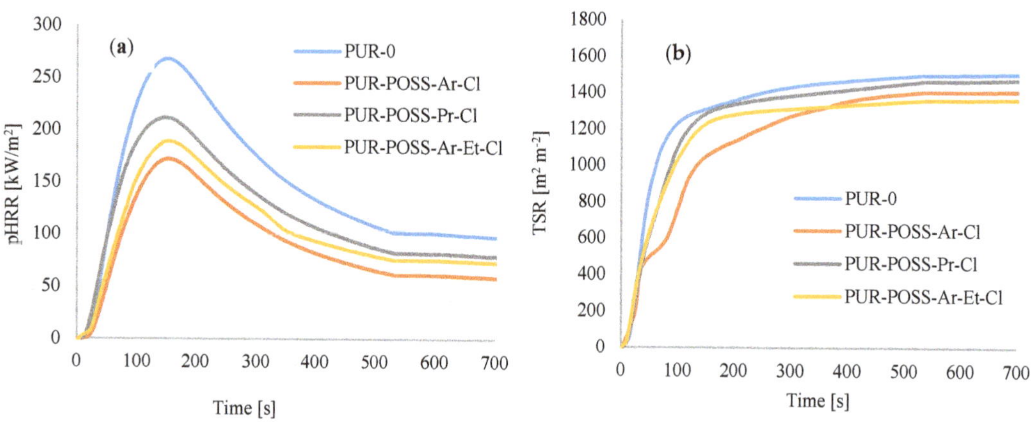

Figure 7. The results of the cone calorimeter test—(**a**) the peak rate of heat release (pHRR), (**b**) the total smoke release (TSR).

Table 4. Results of fire behavior of PUR foams modified with POSS-Cl.

Sample Code	TTI (s)	pHRR (kW/m^2)	THR (MJ/m^2)	MAHRE (kW/m^2)	TSR (m^2/m^2)	Char Residue [%]
PUR-0	2 (0)	268 (8)	21.4 (1)	168 (8)	1490	18.9 (3)
PUR-POSS-Ar-Cl	5 (1)	172 (11)	20.2 (2)	128 (13)	1411	21.4 (5)
PUR-POSS-Pr-Cl	4 (0)	211 (13)	20.6 (2)	153 (16)	1472	19.6 (4)
PUR-POSS-Ar-Et-Cl	4 (0)	189 (13)	20.5 (3)	139 (11)	1367	18.2 (4)

3. Materials and Methods

3.1. Materials

The water-blown RPUFs used in this study were obtained from a two-component system supplied by Purinova Sp. z o. o., after mixing the polyol (Izopianol 30/10/C) and the diphenylmethane diisocyanate (Purocyn B). The polyol is a mixture of components containing polyester polyol (hydroxyl number ca. 230–250 mgKOH/g, functionality of 2), catalyst (*N,N*-Dimethylcyclohexylamine), flame retardant (Tris(2-chloro-1-methylethyl) phosphate), a chain extender (1,2-propanediol) and water as a blowing agent. PUR foams were modified with silsesquioxanes functionalized with halogen: chloropropylisobutyl-POSS (POSS-Pr-Cl), chlorobenzylisobutyl-POSS (POSS-Ar-Cl), chlorobenzylethylisobutyl-POSS (POSS-Ar-Et-Cl) from Hybrid Plastics, Inc (Hattiesburg, MS, USA). Their structural formulas are shown in Table 5. The POSS-Cl modifiers were added to the reaction system in an amount of 2 wt.% relative to the weight of the polyol.

Table 5. POSS-Cl compounds used for modification of PUR foams.

Coumpound	Abbreviation in the Text	Structure	Summary Formula
Chlorobenzyllsobutyl POSS	POSS-Ar-Cl	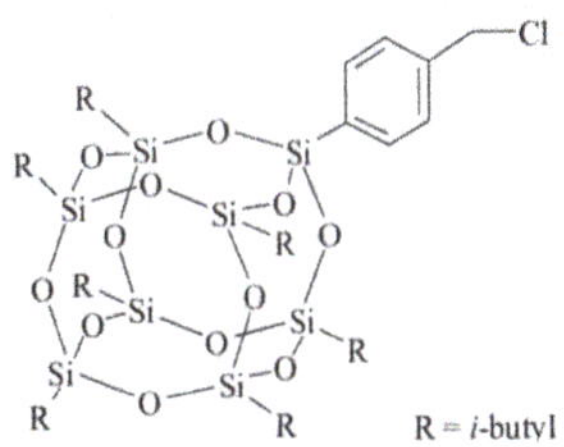	[$C_{35}H_{69}ClO_{12}Si8$]
Chloropropyllsobutyl POSS	POSS-Pr-Cl	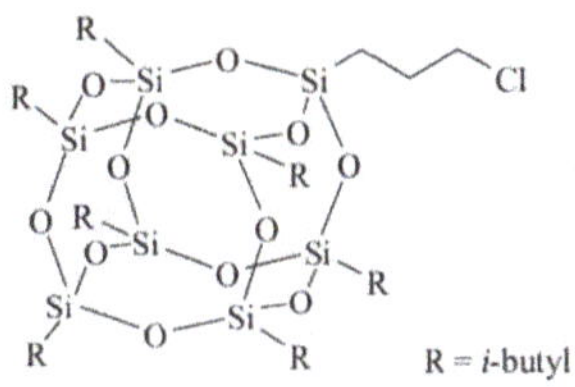	[$C_{31}H_{69}ClO_{12}Si_8$]
Chlorobenzylethyllsobutyl POSS	POSS-Ar-Et-Cl	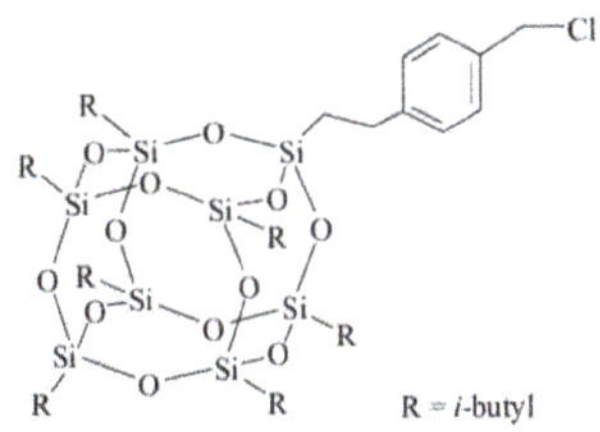	[$C_{37}H_{73}ClO_{12}Si_8$]

3.2. *Synthesis of PUR Composites Foams*

PUR composite foams were produced using the method described in the previous works. Briefly, the synthesis of PUR composite foams modified with POSS-Cl was as follows: polyol premix (Izopianol) was placed in a plastic cup and intensively mixed at 1000 rpm with a mechanical mixer for 60 s. The appropriate amount of POSS-Cl modifier was then added to the cup and mixed for another 60 s to obtain a homogeneous dispersion. The calculated amount of isocyanate (Purocyn) was added to the reaction mixture and thoroughly mixed for 30 s. The reaction mixture was allowed to grow freely by measuring the individual cream, growth and tack-free times, and the maximum reaction temperature (Tmax). The grown PUR foams were left at room temperature for 24 h to ensure complete curing of the composites.

3.3. *Methods and Instruments*

The viscosity of the polyol systems was evaluated using a Viscometer DVII+ (Brookfield, Hadamar-Steinbach, Germany) in the function of a shear rate according to ISO 2555. The measurement was performed at ambient temperature.

The apparent density of the analyzed foams was measured according to the standard ASTM D1622 (equivalent to ISO 845). The density was tested on five samples of each foam and expressed as an average.

Cell size distribution and foam morphology were examined based on the cellular structure pictures of foams taken using JEOL JSM-5500 LV scanning electron microscopy (JEOL LTD, Akishima, Japan). The microscopic research was carried out in a high-vacuum mode and at the accelerating voltage of 10 kV.

A three-point bending test was carried out accordingly to the standard ASTM D7264 (equivalent to ISO 178) using a Zwick Z100 Testing Machine (Zwick/Roell Group, Ulm, Germany). The analyzed samples were bent with a speed of 2 mm min−1. For each series of foams, at least five measurements were made. Obtained flexural stress at the break results for each sample was expressed as a mean value and averaged.

The compressive strength (σ10%) of the foams was determined according to the standard ASTM D1621 (equivalent to ISO 844). The measurement was conducted using a Zwick Z100 Testing Machine (Zwick/Roell Group, Ulm, Germany) with a load cell of 2 kN and a speed of 2 mm min^{-1}. The compression strength was examined as a ratio of the load causing 10% deformation of samples cross-section in both parallel and perpendicular direction to the square surface. The compressive strength was measured in five samples of foam (8 cm × 8 cm × 5 cm) and expressed as an average.

Surface hydrophobicity of PUR foams was measured using contact angle goniometer OEC-15EC (DataPhysics Instruments GmbH, Filderstadt, Germany) with software module SCA 20. Water absorption of PU foams was performed according to ISO 2896:2001.

The thermal stability of the foams was analyzed using a Mettler Toledo thermogravimetric analyzer TGA/DSC1. A thermal decomposition examination was conducted in air (flow 50 mL min^{-1}) and in the temperature range between 25 and 600 °C (heating rate 10 °C min^{-1}). The measurement included an analysis of the mass change as a function of temperature during thermal decomposition of the polyurethane foams. The initial temperatures of the following decomposition stages were noticed and designated as T5%, T10%, T50%. These temperature values corresponded to the percentage of weight loss.

The burning behavior and flame-retardant properties of the foams were analyzed using a cone calorimeter, according to the standard ISO 5660 in S.Z.T.K. TAPS (Maciej Kowalski Company, Saugus, Poland). The measurement for each foam was repeated on three samples and averaged. Each specimen with dimensions of 10 cm × 10 cm × 5 cm was wrapped with aluminum foil and burned at an external heat flux of 35 kW m^{-2}. The parameters were recorded during the time.

4. Conclusions

The paper presents the effect of POSS-Cl as modifiers of the functional properties of rigid PUR foams, with particular emphasis on the flame-retardant properties of silsesquioxanes. The conducted research shows that the introduction of the silicate cage modifier significantly influences the properties of the tested systems, starting from the rheological properties of the polyol premixes, which resulted in the subsequent characteristics of the obtained PUR-POSS-Cl foams. Starting from a change in structure to a more heterogeneous one with smaller and more irregular cells, they simultaneously resulted in a higher density of the molded foams. This, in turn, translated into an increase in the mechanical properties and hardness of the modified PUR foams. It was found that the addition of POSS-Cl had a significant effect on increasing the hydrophobicity of the system as compared to the PUR-0 reference sample. However, the most important aspect of the presented research is the improvement of thermal stability and reduction of flammability of modified PUR-POSS-Cl systems. Among all the silsesquioxanes used, POSS-Ar-Cl can be indicated as the best, as it had the greatest impact on increasing thermal and mechanical properties.

Author Contributions: Conceptualization, A.S.; methodology, A.S. and S.C.; validation, A.S. and S.C.; investigation, A.S., S.C. and K.M.; writing—original draft preparation, A.S.; writing—review and editing, A.S. and K.S.; visualization, A.S.; supervision, A.S. All authors have read and agreed to the published version of the manuscript.

Funding: This research received no external funding.

Institutional Review Board Statement: Not applicable.

Informed Consent Statement: Not applicable.

Data Availability Statement: The data presented in this study are available on request from the corresponding author.

Conflicts of Interest: The authors declare no conflict of interest.

References

1. Gama, N.V.; Ferreira, A.; Barros-Timmons, A. Polyurethane foams: Past, present, and future. *Materials* **2018**, *11*, 1841. [CrossRef]
2. Szycher, M. *Szycher's Handbook of Polyurethanes*, 2nd ed.; CRC Press: New York, NY, USA, 2006.
3. Engels, H.W.; Pirkl, H.G.; Albers, R.; Albach, R.W.; Krause, J.; Hoffmann, A.; Casselmann, H.; Dormish, J. Polyurethanes: Versatile materials and sustainable problem solvers for today's challenges. *Angew. Chem. Int. Ed.* **2013**, *52*, 9422–9441. [CrossRef]
4. Joshi, M.; Adak, B.; Butola, B.S. Polyurethane nanocomposite based gas barrier films, membranes and coatings: A review on synthesis, characterization and potential applications. *Prog. Mater. Sci.* **2018**, *97*, 230–282. [CrossRef]
5. Zhang, G.; Lin, X.; Zhang, Q.; Jiang, K.; Chen, W.; Han, D. Anti-flammability, mechanical and thermal properties of bio-based rigid polyurethane foams with the addition of flame retardants. *RSC Adv.* **2020**, *10*, 32156–32161. [CrossRef]
6. Kurańska, M.; Polaczek, K.; Auguścik-Królikowska, M.; Prociak, A.; Ryszkowska, J. Open-cell rigid polyurethane bio-foams based on modified used cooking oil. *Polymer* **2020**, *190*, 1–7. [CrossRef]
7. Członka, S.; Strakowska, A.; Strzelec, K.; Adamus-Włodarczyk, A.; Kairyte, A.; Vaitkus, S. Composites of rigid polyurethane foams reinforced with POSS. *Polymers* **2019**, *11*, 336. [CrossRef] [PubMed]
8. Wang, S.X.; Zhao, H.B.; Rao, W.H.; Huang, S.C.; Wang, T.; Liao, W.; Wang, Y.Z. Inherently flame-retardant rigid polyurethane foams with excellent thermal insulation and mechanical properties. *Polymer* **2018**, *153*, 616–625. [CrossRef]
9. Lee, J.; Jung, I. Tuning sound absorbing properties of open cell polyurethane foam by impregnating graphene oxide. *Appl. Acoust.* **2019**, *151*, 10–21. [CrossRef]
10. Kaur, R.; Kumar, M. Addition of anti-flaming agents in castor oil based rigid polyurethane foams: Studies on mechanical and flammable behaviour. *Mater. Res. Express* **2020**, *7*. [CrossRef]
11. Jiao, L.; Xiao, H.; Wang, Q.; Sun, J. Thermal degradation characteristics of rigid polyurethane foam and the volatile products analysis with TG-FTIR-MS. *Polym. Degrad. Stab.* **2013**, *98*. [CrossRef]
12. Strąkowska, A.; Członka, S.; Konca, P.; Strzelec, K. New flame retardant systems based on expanded graphite for rigid polyurethane foams. *Appl. Sci.* **2020**, *10*, 5817. [CrossRef]
13. Xu, W.; Wang, G.; Zheng, X. Research on highly flame-retardant rigid PU foams by combination of nanostructured additives and phosphorus flame retardants. *Polym. Degrad. Stab.* **2015**, *111*, 142–150. [CrossRef]
14. Ye, L.; Meng, X.-Y.; Liu, X.-M.; Tang, J.-H.; Li, Z.-M. Flame-Retardant and Mechanical Properties of High-Density Rigid Polyurethane Foams Filled with Decabrominated Dipheny Ethane and Expandable Graphite. *J. Appl. Polym. Sci.* **2009**, *111*, 2373–2380. [CrossRef]

15. Thirumal, M.; Khastgir, D.; Nando, G.B.; Naik, Y.P.; Singha, N.K. Halogen-free flame retardant PUF: Effect of melamine compounds on mechanical, thermal and flame retardant properties. *Polym. Degrad. Stab.* **2010**, *95*, 1138–1145. [CrossRef]
16. Shi, L.; Li, Z.-M.; Xie, B.-H.; Wang, J.-H.; Tian, C.-R.; Yang, M.-B. Flame retardancy of different-sized expandable graphite particles for high-density rigid polyurethane foams. *Polym. Int.* **2006**, *55*, 862–871. [CrossRef]
17. Lindholm, J.; Brink, A.; Wilen, C.-E.; Hupa, M. Cone Calorimeter Study of Inorganic Salts as Flame Retardants in Polyurethane Adhesive with Limestone Filler. *J. Appl. Polym. Sci.* **2012**, *123*, 1793–1800. [CrossRef]
18. Chen, L.; Wang, Y.Z. A review on flame retardant technology in China. Part I: Development of flame retardants. *Polym. Adv. Technol.* **2010**, *21*, 1–26. [CrossRef]
19. Zhao, P.; Guo, C.; Li, L. Exploring the effect of melamine pyrophosphate and aluminum hypophosphite on flame retardant wood flour/polypropylene composites. Constr. *Build. Mater.* **2018**, *170*, 193–199. [CrossRef]
20. Fanglong, Z.; Qun, X.; Qianqian, F.; Rangtong, L.; Kejing, L. Influence of nano-silica on flame resistance behavior of intumescent flame retardant cellulosic textiles: Remarkable synergistic effect? *Surf. Coat. Technol.* **2016**, *294*, 90–94. [CrossRef]
21. Ai, L.; Chen, S.; Zeng, J.; Yang, L.; Liu, P. Synergistic Flame Retardant Effect of an Intumescent Flame Retardant Containing Boron and Magnesium Hydroxide. *ACS Omega* **2019**, *4*, 3314–3321. [CrossRef]
22. Zhou, F.; Zhang, T.; Zou, B.; Hu, W.; Wang, B.; Zhan, J.; Ma, C.; Hu, Y. Synthesis of a novel liquid phosphorus-containing flame retardant for flexible polyurethane foam: Combustion behaviors and thermal properties. *Polym. Degrad. Stab.* **2020**, *171*, 109029. [CrossRef]
23. Hebda, E.; Ozimek, J.; Raftopoulos, K.N.; Michałowski, S.; Pielichowski, J.; Jancia, M.; Pielichowski, K. Synthesis and morphology of rigid polyurethane foams with POSS as pendant groups or chemical crosslinks. *Polym. Adv. Technol.* **2015**, *26*, 932–940. [CrossRef]
24. Fina, A.; Tabuani, D.; Frache, A.; Camino, G. Polypropylene-polyhedral oligomeric silsesquioxanes (POSS) nanocomposites. *Polymer* **2005**, *46*, 7855–7866. [CrossRef]
25. Shi, H.; Yang, J.; You, M.; Li, Z.; He, C. Polyhedral Oligomeric Silsesquioxanes (POSS)-Based Hybrid Soft Gels: Molecular Design, Material Advantages, and Emerging Applications. *ACS Mater. Lett.* **2020**, *2*, 296–316. [CrossRef]
26. Zatorski, W.; Sałasinska, K. Combustibility studies of unsaturated polyester resins modified by nanoparticles. *Polimery* **2016**, *61*, 815–823. [CrossRef]
27. Fina, A.; Abbenhuis, H.C.L.; Tabuani, D.; Camino, G. Metal functionalized POSS as fire retardants in polypropylene. *Polym. Degrad. Stab.* **2006**, *91*, 2275–2281. [CrossRef]
28. Gentiluomo, S.; Veca, A.D.; Monti, M.; Zaccone, M.; Zanetti, M. Fire behavior of polyamide 12 nanocomposites containing POSS and CNT. *Polym. Degrad. Stab.* **2016**, *134*, 151–156. [CrossRef]
29. Janowski, B.; Pielichowski, K. Thermo(oxidative) stability of novel polyurethane/POSS nanohybrid elastomers. *Thermochim. Acta* **2008**, *478*, 51–53. [CrossRef]
30. Zhang, W.; Camino, G.; Yang, R. Polymer/polyhedral oligomeric silsesquioxane (POSS)nanocomposites: An overview of fire retardance. *Prog. Polym. Sci.* **2017**, *67*, 77–125. [CrossRef]
31. Fan, X.; Cao, M.; Zhang, X.; Li, Z. Synthesis of star-like hybrid POSS-(PDMAEMA-b-PDLA)8 copolymer and its stereocomplex properties with PLLA. *Mater. Sci. Eng. C* **2017**, *76*, 211–216. [CrossRef] [PubMed]
32. Laoutid, F.; Bonnaud, L.; Alexandre, M.; Lopez-Cuesta, J.M.; Dubois, P. New prospects in flame retardant polymer materials: From fundamentals to nanocomposites. *Mater. Sci. Eng. R Rep.* **2009**, *63*, 100–125. [CrossRef]
33. Ye, X.; Zhang, W.; Yang, R.; He, J.; Li, J.; Zhao, F. Facile synthesis of lithium containing polyhedral oligomeric phenyl silsesquioxane and its superior performance in transparency, smoke suppression and flame retardancy of epoxy resin. *Compos. Sci. Technol.* **2020**, *189*, 108004. [CrossRef]
34. Marcinkowska, A.; Przadka, D.; Dudziec, B.; Szczesniak, K.; Andrzejewska, E. Anchor effect in polymerization kinetics: Case of monofunctionalized POSS. *Polymers* **2019**, *11*, 515. [CrossRef] [PubMed]
35. Sałasińska, K.; Leszczyńska, M.; Celiński, M.; Kozikowski, P.; Kowiorski, K.; Lipińska, L. Burning Behaviour of Rigid Polyurethane Foams with Histidine and Modified Graphene Oxide. *Materials* **2021**, *14*, 1184. [CrossRef] [PubMed]
36. Członka, S.; Strąkowska, A.; Kairytė, A.; Kremensas, A. Nutmeg filler as a natural compound for the production of polyurethane composite foams with antibacterial and anti-aging properties. *Polym. Test.* **2020**, *86*, 106479. [CrossRef]
37. Estravísa, S.; Tirado-Mediavillaa, J.; Santiago-Calvoa, M.; Ruiz-Herreroa, J.L.; Villafañeb, F.; Rodríguez-Pérez, M.A. Rigid polyurethane foams with infused nanoclays: Relationship between cellular structure and thermal conductivity. *Eur. Polym. J.* **2016**, *80*, 1–15. [CrossRef]
38. Strąkowska, A.; Kosmalska, A.; Zaborski, M. Surface modification of methylvinylsilicone rubber vulcanizates with polyhedral oligomeric silsesquioxanes functionalized using chloride groups (POSS-Cl). *Polimery* **2016**, *61*, 272–278. [CrossRef]
39. Septevani, A.A.; Evans, D.A.C.; Chaleat, C.; Martin, D.J.; Annamalai, P.K. A systematic study substituting polyether polyol with palm kernel oil based polyester polyol in rigid polyurethane foam. *Ind. Crops Prod.* **2015**, *66*. [CrossRef]
40. Acuña, P.; Li, Z.; Santiago-Calvo, M.; Villafañe, F.; Rodríguez-Perez, M.Á.; Wang, D.-Y. Influence of the characteristics of expandable graphite on the morphology, thermal properties, fire behaviour and compression performance of a rigid polyurethane foam. *Polymers* **2019**, *11*, 168. [CrossRef]
41. Chattopadhyay, D.; Webster, D.C. Thermal stability and flame retardancy of polyurethanes. *Prog. Polym. Sci.* **2009**, *34*, 1068–1133. [CrossRef]

42. Chua, M.H.; Zhou, H.; Xu, J. POSS as Fire Retardant: Preparation, Properties, Applications. In *Polymer/POSS Nanocomposites and Hybrid Materials*; Springer: Berlin, Germany, 2018; pp. 337–372. [CrossRef]
43. Szatylowicz, H.; Jezuita, A.; Krygowski, T.M. On the relations between aromaticity and substituent effect. *Struct. Chem.* **2019**, *30*, 1529–1548. [CrossRef]

molecules

MDPI

Article

Novel Polyhedral Silsesquioxanes [POSS(OH)$_{32}$] as Anthracycline Nanocarriers—Potential Anticancer Prodrugs

Kinga Piorecka *, Jan Kurjata and Wlodzimierz A. Stanczyk

Centre of Molecular and Macromolecular Studies, Polish Academy of Sciences, Sienkiewicza 112, 90-363 Lodz, Poland; jkurjata@cbmm.lodz.pl (J.K.); was@cbmm.lodz.pl (W.A.S.)
* Correspondence: kgradzin@cbmm.lodz.pl

Abstract: Anthracyclines belong to the anticancer drugs that are widely used in chemotherapy. However, due to their systemic toxicity they also exert dangerous side effects associated mainly with cardiovascular risks. The pathway that is currently often developed is their chemical and physical modification via formation of conjugated or complexed prodrug systems with a variety of nanocarriers that can selectively release the active species in cancer cells. In this study, six new nanoconjugates were synthesized with the use of polyhedral oligosilsesquioxanes [POSS(OH)$_{32}$] as nanocarriers of the anticancer drugs anthracyclines—doxorubicin (DOX) and daunorubicin (DAU). These prodrug conjugates are also equipped with poly(ethylene glycol) (PEG) moieties of different structure and molecular weight. Water-soluble POSS, succinic anhydride modified (SAMDOX and SAMDAU) with carboxylic function, and PEGs (PEG1, PEG2 and PEGB3) were used for the synthesis. New nanoconjugates were formed via ester bonds and their structure was confirmed by NMR spectroscopy (^{1}H-NMR, ^{13}C-NMR, ^{1}H-^{13}C HSQC, DOSY and ^{1}H-^{1}H COSY), FTIR and DLS. Drug release rate was evaluated using UV-Vis spectroscopy at pH of 5.5. Release profiles of anthracyclines from conjugates **4–9** point to a range of 10 to 75% (after 42 h). Additionally, model NMR tests as well as diffusion ordered spectroscopy (DOSY) confirmed formation of the relevant prodrugs. The POSS-anthracycline conjugates exhibited prolonged active drug release time that can lead to the possibility of lowering administered doses and thus giving them high potential in chemotherapy. Drug release from conjugate **7** after 42 h was approx. 10%, 33% for conjugate **4**, 47% for conjugate **5**, **6**, **8** and 75% for conjugate **9**.

Keywords: anticancer conjugates; polyhedral oligomeric silsesquioxanes; anthracyclines

Citation: Piorecka, K.; Kurjata, J.; Stanczyk, W.A. Novel Polyhedral Silsesquioxanes [POSS(OH)$_{32}$] as Anthracycline Nanocarriers—Potential Anticancer Prodrugs. *Molecules* **2021**, *26*, 47. https://dx.doi.org/10.3390/molecules26010047

Academic Editor: Sonia Trombino
Received: 9 December 2020
Accepted: 22 December 2020
Published: 24 December 2020

1. Introduction

We are currently witnessing a rapid development of anticancer drug delivery systems [1,2], exploiting such nanocarriers as, e.g., gold and mesoporous silica nanoparticles, polymers, and dendrimers [3]. Nanotechnology and nanomedicine [4,5] now offer a variety of systems that are intensively studied as delivery vehicles, modulating cytotoxicity and mediating sustained drug release in tumour tissues.

Anthracycline chemotherapy with antibiotics such as doxorubicin (DOX) and daunorubicin (DAU) plays an important role in treating many types of cancer. They exhibit wide action spectrum, high efficacy and are used in the form of soluble hydrochlorides.

They intercalate with DNA, inhibiting proliferation of cancer cells and leading to their apoptosis. Unfortunately, they are not selective and administration of high doses can often lead to heart failure [6]. Nanotechnology is one of the ways to increase the selectivity of chemotherapy via limiting systemic toxicity. The use of nanoparticles has been shown to improve the selective transport of drugs and their accumulation in the tumour, which in turn can lead to dose reduction [7]. Encapsulation, complexation or conjugation of these drugs with nanocarriers leads to slow release of the active drug, extending its time of action, leading to decreased doses. Our previous studies have been devoted to the use of polyhedral oligosilsesquioxanes as nanocarriers of anti-cancer drugs [8–11]. Silsesquioxanes are well known to facilitate cell membrane penetration [12], an important feature in drug delivery. Additionally, they are biocompatible [13] and nontoxic [14]. The current research presents results on application of hydrophilic silsesquioxanes, known in the literature as [POSS(OH)$_{32}$], as novel anthracycline nanocarriers, synthesized by hydrolytic condensation of a functionalized precursor—*N*,*N*-di(2,3-dihydroxypropyl)(aminopropyl)triethoxysilane [15]. Contrary to the previously used POSSs, the silsesquioxanes obtained from *N*,*N*-di(2,3-dihydroxypropyl)(aminopropyl)triethoxysilane have relatively narrow size distribution and a large number of functional OH moieties. The average particle diameter of these water soluble structures is on average ~3.0 nm and the synthesis of such a system is simple and effective. Additionally, due to partial coupling of hydroxyl groups of the carrier with carboxy terminated, methoxylated poly(ethylene glycol) (PEG), apart from esterification with succinic anhydride modified anthracyclines [16], the solubility in aqueous media can be largely improved. It also allows additional binding of the targeting moiety biotin (Figures 1 and 2).

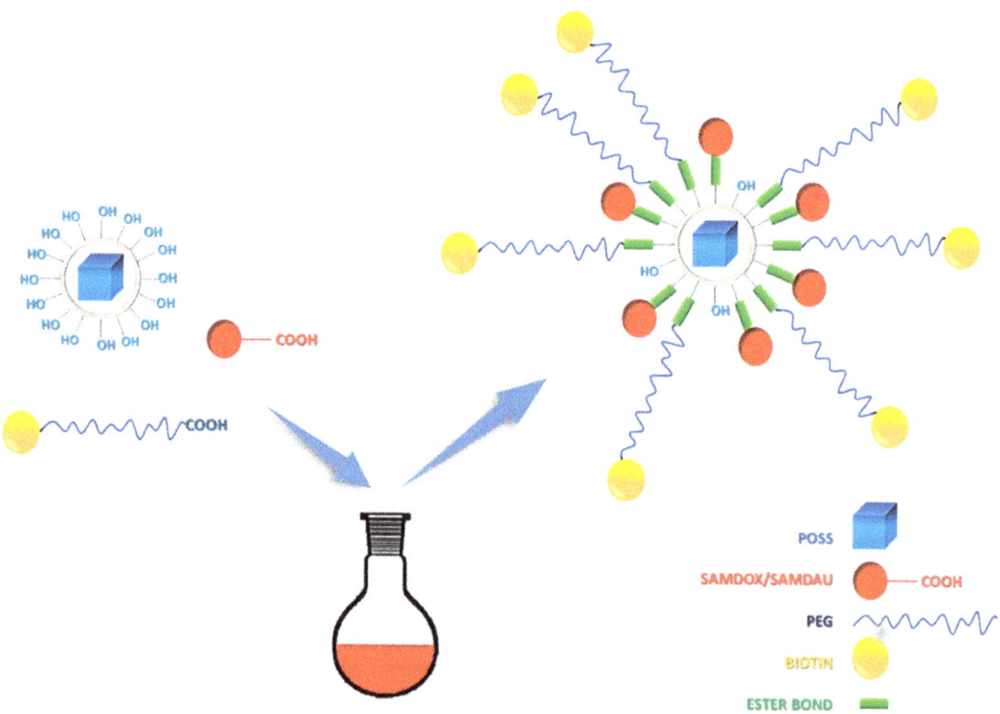

Figure 1. Synthetic scheme of [POSS(OH)$_{32}$] conjugates with anthracyclines, PEG and biotin.

In this work, 6 POSS-anthracycline conjugates were synthesized: POSS conjugate containing doxorubicin and PEG1 [*O*-(2-carboxyethyl)-*O*′-methyl-undecaethylene glycol; M_w 588.68]—PossDoxPEG1 (**4**), doxorubicin and PEG2 [*O*-Methyl-*O*′-succinylpolyethylene glycol 2′000; M_w ~2100]—PossDoxPEG2 (**5**), doxorubicin and PEGB3 [*O*-[2-(Biotinyl-amino)ethyl]-*O*′-(2-carboxyethyl)polyethylene glycol; M_w 3000]—PossDoxPEGB3 (**6**), daunorubicin and PEG1—PossDauPEG1 (**7**), daunorubicin and PEG2—PossDauPEG2 (**8**) and daunorubicin and PEGB3—PossDauPEGB3 (**9**) (Figure 2). The research was aimed to develop effective synthetic methods of silsesquioxane nanoconjugates with anticancer drugs. Such conjugates would be capable of releasing active drugs in cancer cells. The structure of all synthesized nanoconjugates was confirmed by NMR spectroscopy, FTIR, and DLS.

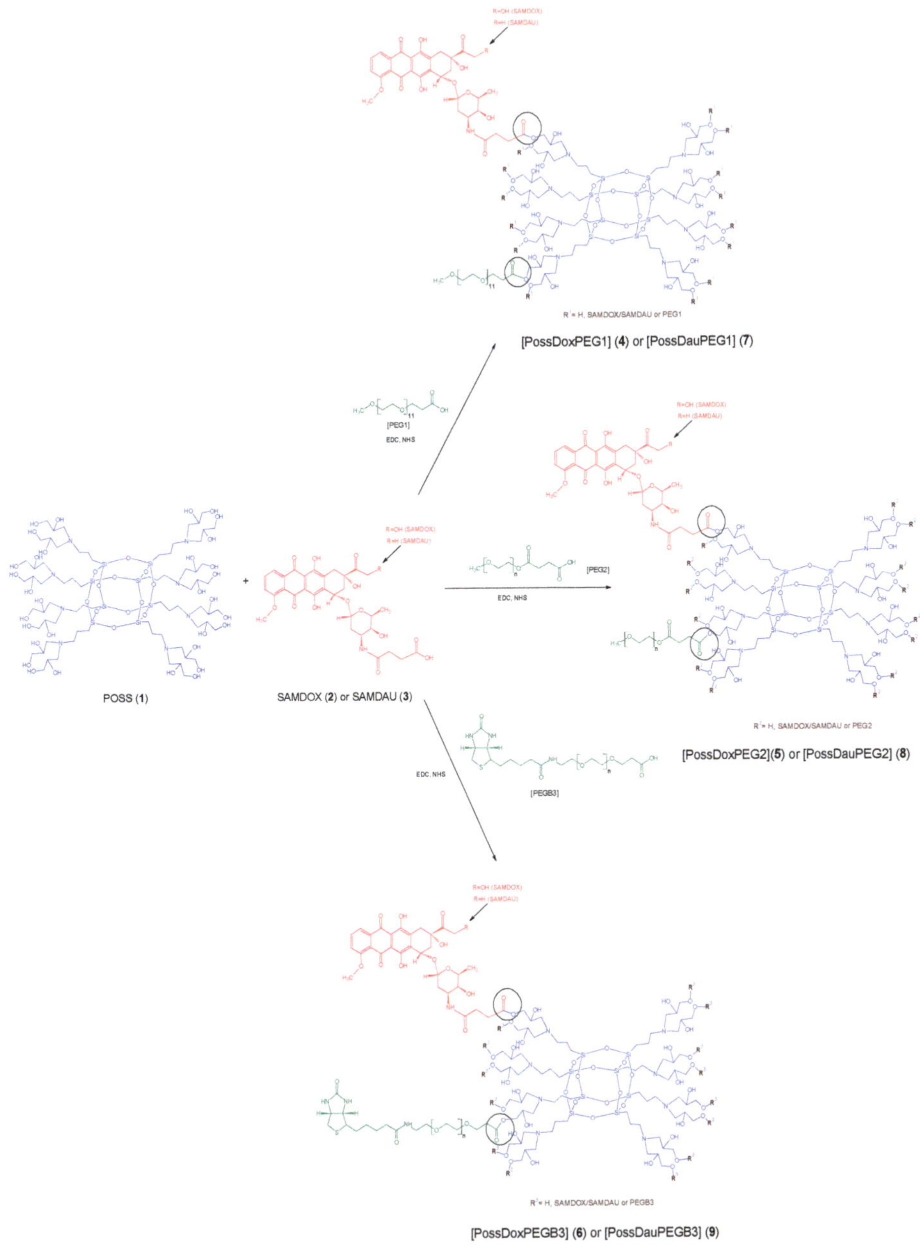

Figure 2. Synthesis of POSS conjugates **4–9**. PossDoxPEG1 (**4**): POSS conjugate containing doxorubicin and PEG1 [*O*-(2-Carboxyethyl)-*O*′-methyl-undecaethylene glycol; M_w 588.68), PossDoxPEG2 (**5**): POSS conjugate containing doxorubicin and PEG2 [*O*-Methyl-*O*′-succinylpolyethylene glycol 2′000; M_w ~2100], PossDoxPEGB3 (**6**): POSS conjugate containing doxorubicin and PEGB3 [*O*-[2-(Biotinyl-amino)ethyl]-*O*′-(2-carboxyethyl)polyethylene glycol; M_w 3000], PossDauPEG1 (**7**): POSS conjugate containing daunorubicin and PEG1, PossDauPEG2 (**8**): POSS conjugate containing daunorubicin and PEG2, PossDauPEGB3 (**9**): POSS conjugate containing daunorubicin and PEGB3.

2. Results and Discussion

2.1. Model Reaction

The model reaction (Figures S2 and S3 and Table S1) was carried out to allow interpretation of complex NMR spectra of nanoconjugates (**4–9**). The POSS structure—$T^8[(CH_2)_2S(CH_2)_2OH]_8$ applied in the model reaction has a well-defined structure, in contrast to $[POSS(OH)_{32}]$ being a mixture of cage silsesquioxanes. It was obtained by modifying commercial POSS-Vi with 2-mercapotetanol. In addition, PEG1 with a lower molecular weight was used, which simplifies interpretation of the NMR spectra. The post reaction mixture was pre-purified (see Section 3), and its composition was determined using ^{1}H-^{1}H COSY and ^{1}H-^{13}C HSQC NMR spectra of the substrates and the product. Thus, in the subsequent syntheses, it was possible to determine whether the purified products also contained free drugs and/or free PEGs.

2.2. Determination of Total Drug Content in Nanoconjugates 4–9

Total drug content was measured using UV-Vis spectroscopy (Figure 3). Nanoconjugates (**4–9**) (1 mg) were dissolved in H_2O/DMF (1 mL, 5:1), then 200 µL of this solution was taken and diluted with 3 mL of solvent (H_2O/DMF (5:1)), followed by mixing with 0.2 mL HCl (36%). It was kept at 50 °C for 2 h and then at room temperature for 24 h. A calibration curve for DOX/DAU was constructed and total drug content was calculated from UV-Vis spectra of DOX, DAU and all the nanoconjugates (**4–9**) at 480 nm (Figures S4 and S5).

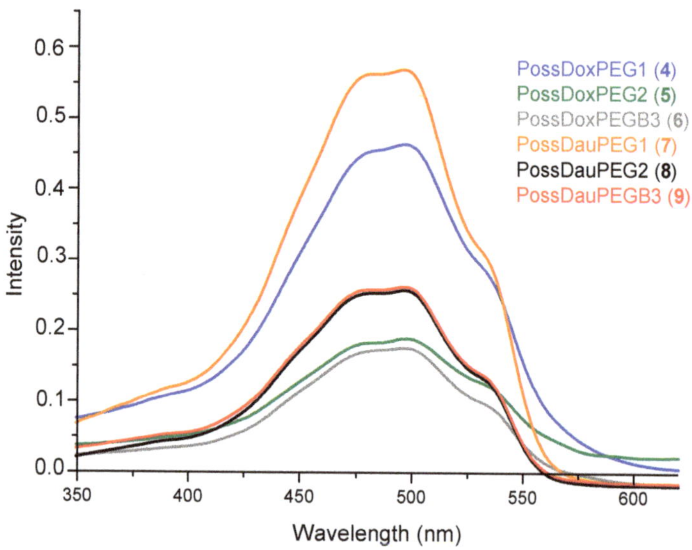

Figure 3. Measurement of maximum contents of DOX and DAU released from the relevant conjugates (**4–9**) (by mixing with HCl (36%) at 50 °C).

Table 1 shows that conjugate **4** (42.88%) has the highest anthracycline content and conjugate **6** (16.38%) the lowest. Conjugate **4** (88.0%) exhibits the highest efficiency of drug attachment, while conjugate **5** is characterized by the lowest (47.4%). The PEG1-containing conjugates (**4** and **7**) are most effective in conjugation with anthracyclines. In contrast, the conjugates containing PEGB3 in the molecule show the lowest efficiency of binding drugs. This may be related to the molecular mass and structure of PEGs—PEGB3 has the highest molecular mass, thus slowing down the ester bond formation.

Table 1. Characteristics of conjugates **4–9**. Total drug content refers to the weight percentage of the drug in the conjugate. The calculated maximum drug content (wt%) relates to the theoretical maximum drug content—calculated from the weight of drug taken for the reaction. The drug attachment efficiency was calculated on the basis of total drug content and calculated maximum drug content.

Type of Nanoconjugate	Total Drug Content (wt%)	Calculated Maximum Drug Contents (wt%)	Drug Attachment Efficiency (wt%)
PossDoxPEG1 (**4**)	42.88%	48.70%	88.0%
PossDoxPEG2 (**5**)	17.54%	37.02%	47.4%
PossDoxPEGB3 (**6**)	16.38%	27.00%	60.7%
PossDauPEG1 (**7**)	42.33%	54.50%	77.7%
PossDauPEG2 (**8**)	19.20%	33.93%	56.6%
PossDauPEGB3 (**9**)	19.52%	27.18%	71.8%

2.3. Drugs Release Study

DOX/DAU release from the nanoconjugates was determined in citrate buffer solution—(pH 5.5 0.1 M) at 310 K. Nanoconjugates, dissolved in DMF, were placed in a dialysis bag (MWCO: 2 kD, Spectrum Laboratories) and inserted into buffer solution (conjugate concentrations are described in the Table S3). The amount of released drug was calculated from the UV-Vis spectra. The drug release profile of PossDauPEGB3 (**9**) conjugate showed the fastest release of DAU in an acid environment. In this case, over 60% of POSS conjugated DAU was released within 21 h and over 70% within 42 h. Release profiles of anthracyclines from conjugates **4–6** and **8** point to a maximum amount after 42 h in the range of 33 to 47%. Conjugate **7** showed the lowest drug release profile (only ~10% after 21 h and 42 h). As can be seen in Figure 4, the drug release profiles from the conjugates with PEG1 indicate the lowest release values, and are not dependent only on the type of applied PEG. It can be also related to the complexation of drugs and PEGs (not covalently bound to POSS). Conjugate **9** has a biotynyl fragment that can form drug complexes with NH moieties via hydrogen bonds. Non-covalent systems involving anthracyclines are known to be formed and are cleaved easier than relevant conjugates [1]. The hydrogen bonded anthracyclines and $POSS(OH)_{32}$ complexes have been recently described [17]. Additionally, Figure S6 shows the graphs of absorbance intensity measured with UV-Vis for drug release from the relevant conjugates after 21 h and 42 h. It is evident that the drug release rate from conjugates is slow enough to make them potentially good candidates for anti-cancer therapy [17].

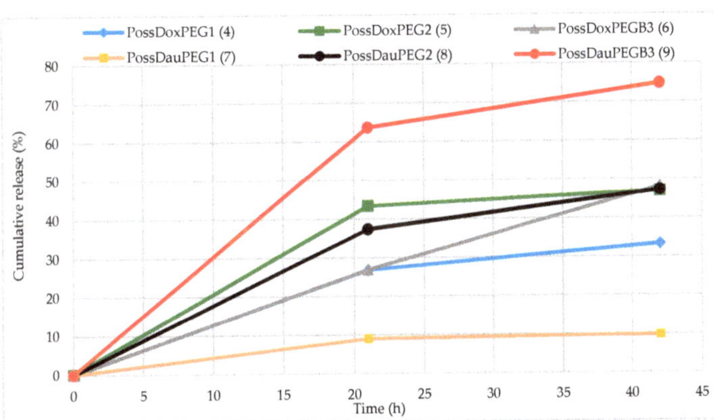

Figure 4. In vitro DOX/DAU release profiles from nanoconjugates in citrate buffer solution at 310 K quantified by UV-Vis method.

As the plots in Figure 4 prove, the conjugates (**4** and **7**) containing PEG1 in their structure show the slowest rate release of the drugs, while the PEG2-containing conjugates

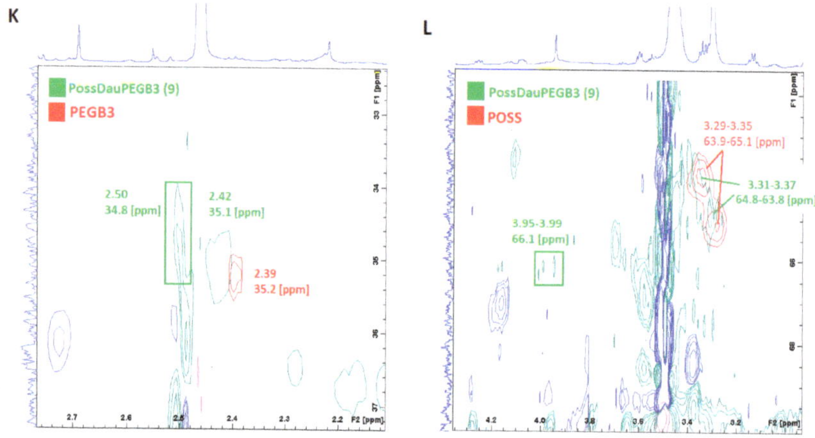

Figure 5. Superimposition of the ^{1}H-^{13}C HSQC spectra of the substrates (PEG and POSS) on the ^{1}H-^{13}C HSQC spectra of the products (conjugates **4–9**). (**A**,**B**) PossDoxPEG1 (**4**); (**C**,**D**) PossDoxPEG2 (**5**); (**E**,**F**) PossDoxPEGB3 (**6**); (**G**,**H**) PossDauPEG1 (**7**); (**I**,**J**) PossDauPEG2 (**8**); (**K**,**L**) PossDauPEGB3 (**9**).

Figure 6 shows the superimposed ^{1}H-^{13}C HSQC spectra of the conjugates on the SAMDOX and SAMDAU spectra to investigate differences in chemical shifts of the anthracycline protons adjacent to the ester/carboxylic functions after (red) and before (green—SAMDOX and SAMDAU) the conjugation. The formation of an ester bond is associated with a change in the chemical shifts of adjacent protons (POSS-antracycline-$\underline{CH_2CH_2}$-ester bond) towards the lower field. For conjugates 5 and 8 containing PEG2 in their molecule, this is a marked change from ~2.3 ppm to ~2.5 ppm. In contrast to the ^{1}H-^{13}C HSQC spectra shown after superimposing PEG/POSS on the spectra of the conjugates (Figure 5), the ^{1}H-^{13}C HSQC spectra (Figure 6) did not give an unequivocal answer as to whether SAMDOX and SAMDAU are incorporated into the product.

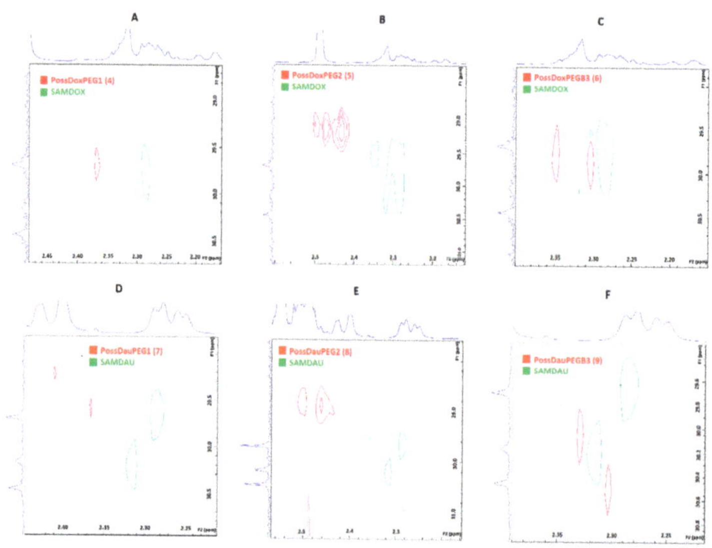

Figure 6. Superimposition of the ^{1}H-^{13}C HSQC spectra of the substrates (SAMDOX/SAMDAU) on the ^{1}H-^{13}C HSQC spectra of the products (conjugates **4–9**). (**A**) PossDoxPEG1 (**4**); (**B**) PossDoxPEG2 (**5**); (**C**) PossDoxPEGB3 (**6**); (**D**) PossDauPEG1 (**7**); (**E**) PossDauPEG2 (**8**); (**F**) PossDauPEGB3 (**9**).

2.5. Diffusion NMR Spectroscopy

Diffusion NMR spectroscopy (DOSY) was performed to further confirm the formation of the conjugates **4–9**. The results were determined by superimposing the DOSY spectra of the substrates (SAMDOX/SAMDAU, PEG1/PEG2/PEGB3 and POSS(OH)$_{32}$) on the DOSY spectrum of the products (conjugates **4–9**) (Figure 7). Self-diffusion coefficients (D) of the SAMDOX, SAMDAU, POSS(OH)$_{32}$, PEG1, PEG2, PEGB3 and conjugates **4–9** were determined from the resonance signals: POSS (1.45 ppm), SAMDOX (7.84 ppm), SAMDAU (7.79 ppm), PEG1 (2.43 ppm), PEG2 (3.49 ppm), PEGB3 (3.48 ppm), PossDox-PEG1 (1.42 ppm; 7.85 ppm; 2.43 ppm), PossDoxPEG2 (1.42 ppm; 7.83 ppm; 3.49 ppm), PossDoxPEGB3 (1.47 ppm; 7.84 ppm; 3.43 ppm), PossDauPEG1 (1.45 ppm; 7.79 ppm; 2.42 ppm), PossDauPEG2 (1.49 ppm; 7.78 ppm; 3.52 ppm) and PossDauPEGB3 (1.49 ppm; 7.82 ppm; 3.51 ppm). The results are presented in Figure 8. Differences in diffusion rates of substrates and conjugates (**4–9**) point to the formation of larger conjugated molecules, as they migrate at a smaller rate than the relevant substrates, due to a lower value of self-diffusion coefficients.

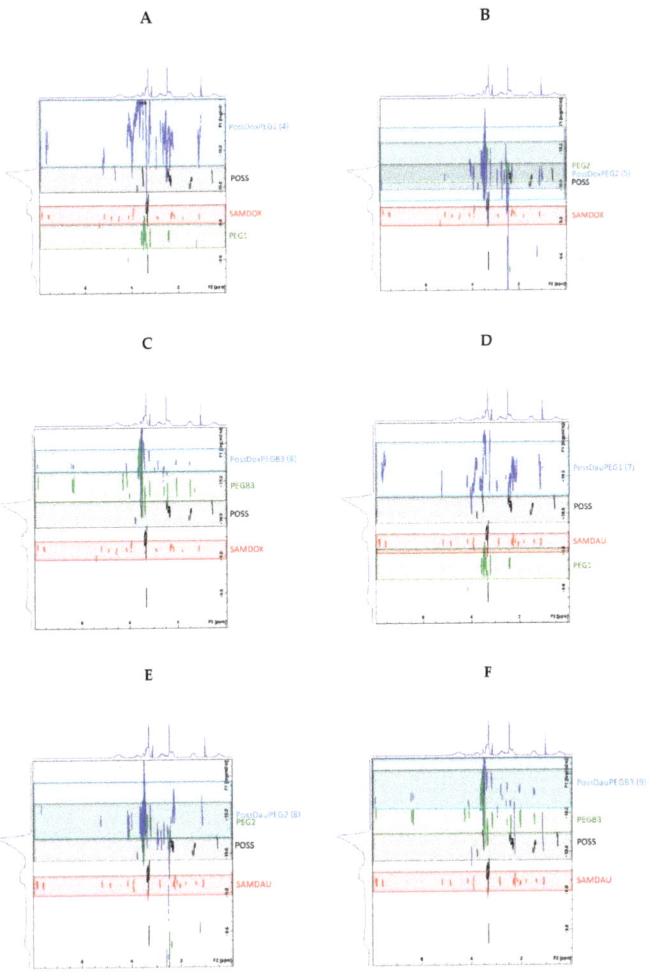

Figure 7. Superimposition of the DOSY spectra of the substrates (SAMDOX/SAMDAU, PEG1/PEG2/PEGB3 and POSS(OH)$_{32}$) on the DOSY spectra of the products (conjugates **4–9**). (**A**) PossDoxPEG1 (**4**); (**B**) PossDoxPEG2 (**5**); (**C**) Poss-DoxPEGB3 (**6**); (**D**) PossDauPEG1 (**7**); (**E**) PossDauPEG2 (**8**); (**F**) PossDauPEGB3 (**9**).

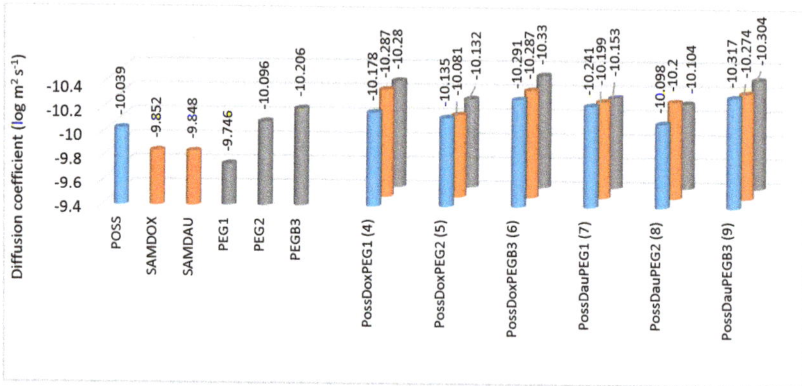

Figure 8. Self-diffusion coefficients (log m^2s^{-1}) of the SAMDOX, SAMDAU, POSS$(OH)_{32}$, PEG1, PEG2, PEGB3 and conjugates **4–9** from the resonance signals: POSS (1.45 ppm), SAMDOX (7.84 ppm), SAMDAU (7.79 ppm), PEG1 (2.43 ppm), PEG2 (3.49 ppm), PEGB3 (3.48 ppm), PossDoxPEG1 (1.42 ppm; 7.85 ppm; 2.43 ppm), PossDoxPEG2 (1.42 ppm; 7.83 ppm; 3.49 ppm), PossDoxPEGB3 (1.47 ppm; 7.84 ppm; 3.43 ppm), PossDauPEG1 (1.45 ppm; 7.79 ppm; 2.42 ppm), PossDauPEG2 (1.49 ppm; 7.78 ppm; 3.52 ppm) and PossDauPEGB3 (1.49 ppm; 7.82 ppm; 3.51 ppm).

2.6. Fourier Transform Infrared Spectroscopy

FTIR spectroscopy was additionally used for structural characterization of the conjugates formed via ester bond between drugs and the POSS carrier. FTIR spectra of conjugates differed from that of SAMDOX/SAMDAU [11]. The spectra of conjugates **4–9** (Figure 9) show a new absorption frequency at ~1734 cm^{-1} ($\nu_{C=O}$, ester bond) indicating formation of an ester bond between SAMDOX/SAMDAU and the POSS hydroxyl groups.

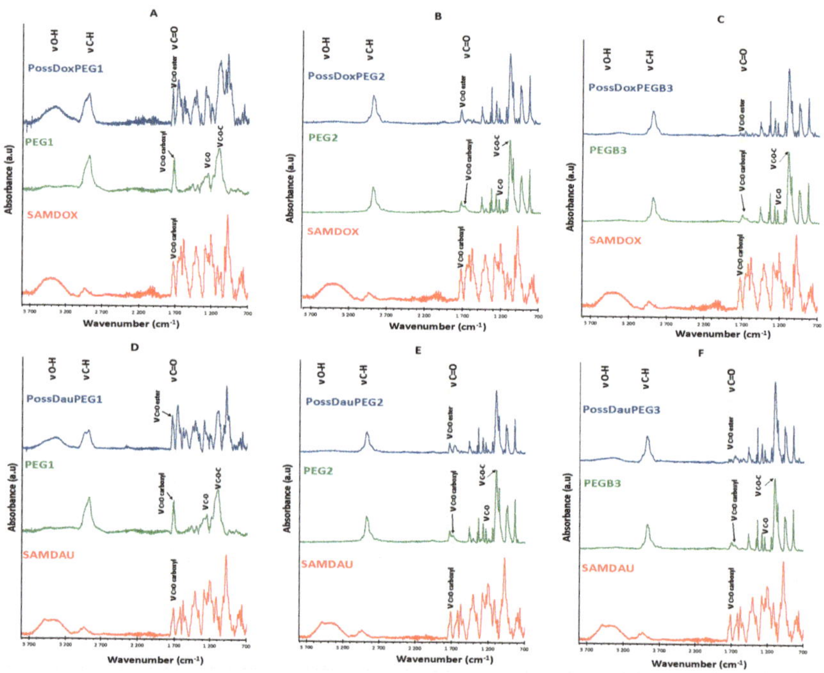

Figure 9. Fourier transform infrared spectra of conjugates **4–9** compared with those of antracyclines and PEGs. Blue color—range 2700–3700 cm^{-1} ($\nu_{O\text{-}H}$, $\nu_{C\text{-}H}$); green color—range 1700–1750 cm^{-1} ($\nu_{C=O}$, ester). (**A**) PossDoxPEG1 (**4**); (**B**) PossDoxPEG2 (**5**); (**C**) PossDoxPEGB3 (**6**); (**D**) PossDauPEG1 (**7**); (**E**) PossDauPEG2 (**8**); (**F**) PossDauPEGB3 (**9**).

Conjugate **4** (Figure 10A) exhibits the new absorption band at 1732 cm^{-1} ($\nu_{C=O}$, ester bond), clearly indicating the formation of ester bond between SAMDOX/PEG1 and $POSS(OH)_{32}$. There is no absorption in the spectrum indicating the presence of unreacted carboxyl groups (SAMDOX 1721 cm^{-1} ($\nu_{C=O}$ carboxylic bond) and PEG1 1722 cm^{-1} ($\nu_{C=O}$ carboxylic bond)). The same characteristic absorption is also evident for conjugate **6** (Figure 10C).

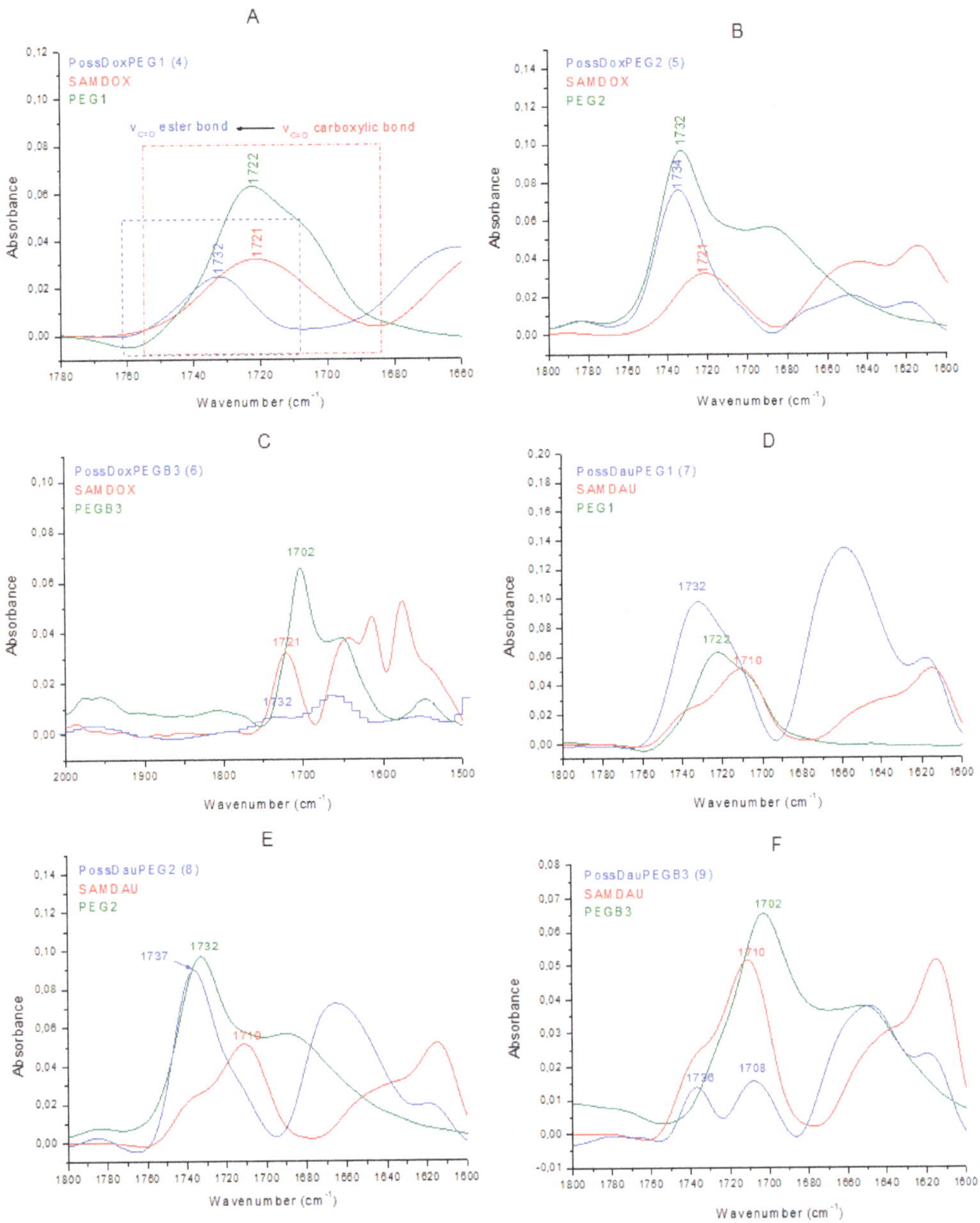

Figure 10. FTIR spectra of conjugates **4–9** compared with the ones of anthracyclines and PEGs (vibration range—$\nu_{C=O}$, ester bond). (**A**) PossDoxPEG1 (**4**); (**B**) PossDoxPEG2 (**5**); (**C**) PossDoxPEGB3 (**6**); (**D**) PossDauPEG1 (**7**); (**E**) PossDauPEG2 (**8**); (**F**) PossDauPEGB3 (**9**).

The spectrum of conjugate **5** (Figure 10B) is more difficult to interpret as PEG2 also has an ester linkage. We can assume that the ester bond in conjugate **5** has been formed, as shown by the shift of the $\nu_{C=O}$ PEG2 (ester bond) absorption from 1732 cm^{-1} to 1734 cm^{-1}. There is no absorption in the spectrum that would indicate the presence of unreacted carboxyl groups. The same conclusion can be drawn from the spectrum of conjugate **8** (Figure 10E).

In contrast, in the spectrum of conjugate **7** (Figure 10D), the absorption at 1732 cm^{-1} ($\nu_{C=O}$ ester bond), indicating the formation of ester bond, also appears, but there are still unreacted carboxyl groups from SAMDAU or PEG1 as shown by widening of the band at 1732 cm^{-1}. The situation is similar for conjugate **9** (Figure 10F), however, here there is a clear absorption at 1708 cm^{-1} indicating unreacted –COOH moieties from SAMDAU or PEGB3.

2.7. *Hydrodynamic Diameters of Conjugates* **4–9**

The hydrodynamic diameter of the POSS(OH)$_{32}$ and conjugates **4–9** was investigated by DLS. Figure 11 shows that POSS has the highest hydrodynamic diameter, which may suggest that its hydroxyl groups form a network of hydrogen bonds. The involvement of the POSS hydroxyl groups in the formation of ester bonds in conjugates **4–9** resulted in a reduction in their hydrodynamic diameter.

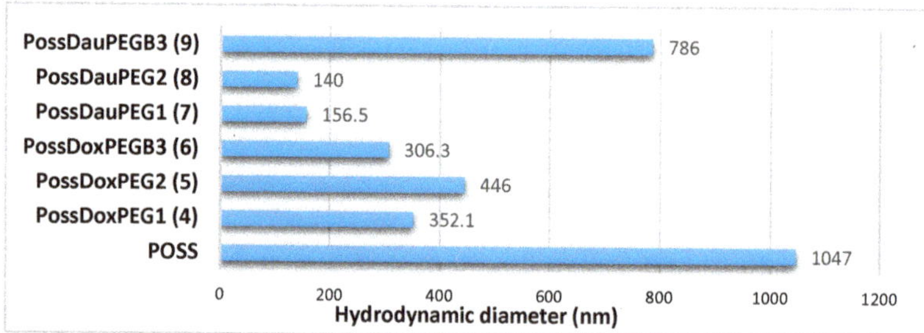

Figure 11. Comparison of hydrodynamic diameters of POSS(OH)$_{32}$ and conjugates **4–9**.

3. Materials and Methods

3.1. *Materials*

Doxorubicin (DOX) and daunorubicin (DAU) (Beijing Packbuy M&C, Beijing, China), succinic anhydride (Sigma-Aldrich, Saint Louis, MO, USA), *N*-hydroxysuccinimide (NHS) (Sigma-Aldrich), 1-ethyl-3-(3-dimethylaminopropyl) carbodiimide hydrochloride (EDC) (Sigma-Aldrich), PEG1: *O*-(2-Carboxyethyl)-*O*′-methyl-undecaethylene glycol [Molecular Weight 588.68; n = 11], PEG2: *O*-Methyl-*O*′-succinylpolyethylene glycol 2′000 [M_r ~2100], PEGB3: *O*-[2-(Biotinyl-amino)ethyl]-*O*′-(2-carboxyethyl)polyethylene glycol [M_p 3000] (Sigma-Aldrich) were used as supplied. Triethylamine (Et_3N, CHEMPUR, Piekary Slaskie, Poland), methylene chloride (POCh) and *N,N*-dimethylformamide (DMF, POCh, Gliwice, Poland) were purified as described in the literature [18]. Hydroxyl functionalized silsesquioxane cage POSS(OH)$_{32}$ was synthesized according to the method described previously in the literature [19]. The reproducibility of synthesized structures was proven by three independent experiments [17]. Syntheses of succinic anhydride-modified daunorubicin (SAMDAU) and succinic anhydride-modified doxorubicin were performed according to the methods described in the literature [16] (Figure S1).

3.2. *General Remarks*

Ultraviolet–visible (UV–Vis) measurements were carried out using the Specord S600 spectrophotometer using 10 mm path length quartz cuvettes. Solutions of conjugates at concentration $C_{conjugates}$ = 0.05882 mg/mL were prepared in water/DMF (5:1). DOX/DAU release from the nanoconjugates was determined at a pH of 5.5 (citrate buffer solution—0.1 M) at 310 K (Table S3).

^{1}H, ^{13}C, ^{1}H-^{13}C HSQC and DOSY nuclear magnetic resonance (NMR) spectra were recorded using Bruker Avance III 500 MHz instrument (Bruker BioSpin GmbH, Rheinstetten, Germany). Chemical shifts are reported in ppm downfield from TMS using DMSO-d6 as a solvent and at 295 K (^{1}H, ^{13}C, ^{1}H-^{13}C HSQC, ^{1}H-^{1}H COSY) and 298 K (DOSY).

FTIR (Fourier transform infrared spectroscopy) spectra were obtained using a Nicolet 6700 spectrometer equipped with a deuterated triglycine sulfate detector and using attenuated total reflectance (ATR) for 64 scans at a 2cm^{-1} resolution.

Hydrodynamic diameters of conjugates **4–9** measured by dynamic light scattering (DLS) in deionized water (with 5% v/v DMF) at 298 K. DLS studies were performed using a ZetaSizer Nano ZS (Malvern Instruments, Malvern, UK) equipped with HeNe red laser (λ = 633 nm) at a measurement angle of 173°. DLS measurements were carried out at the concentration of pure $POSSOH_{32}$ (C_{POSS} = 1.934 × 10^{-3} mg/mL) and of the conjugates **4–9** ($C_{conjugates}$ = 1.187 × 10^{-3} mg/mL).

3.3. General Synthesis of Nanoconjugates **4–9**

All of the synthetic steps were carried out in the dark (Figure 2). Table S2 shows the concentrations of the reagents used in the reaction. SAMDOX/SAMDAU, NHS and EDC were placed in a flask, dissolved in DMF and stirred under nitrogen. At the same time, a PEG, NHS and EDC mixture was prepared in DMF and stirred under nitrogen. Both reaction mixtures were stirred for 3 h at room temperature. Then both mixtures were added dropwise to the POSS solution in DMF and the reaction mixture was stirred for 5 days at room temperature, filtered, concentrated under reduced pressure (rotary evaporator) and dialised in molecular porous membranę tubing (MWCO: 3.5 kD, Standard RC Tubing, Spectrum Laboratories) in DMF for 1 week (DMF was changed three times). At the end the final products were dried on a vacuum line.

3.4. Model Reaction

Synthesis of POSS $T_8[(CH_2)_2S(CH_2)_2OH]_8$. Octa(vinyl)silsesquioxane (T_8-Vi) (2.00 g, 3.1593 × 10^{-3} mol) and 2-mercaptoethanol (2.5 mL), 2,2-dimethoxy-2-phenylacetophenone (DMPA) (0.1295 g) were introduced into a quartz reactor and dissolved in THF (30 mL). The reaction mixture was irradiated with a UV lamp at 350 nm for 1.5 h. Volatiles were removed under vacuum. The residue was dissolved in THF (5 mL), precipitated in pentane (10 × 30 mL), separated, washed with pentane (100 mL) and dried under vacuum.

Synthesis of POSSDAU-MR. The reaction carried out in the dark. SAMDAU, PEG1, NHS and EDC were placed in a flask, dissolved in DMF and stirred under nitrogen for 17 h at room temperature. After this time the mixture was added dropwise to the POSS solution in DMF and the reaction mixture was stirred for 3 days at room temperature. Then the reaction mixture was concentrated under reduced pressure and the product was purified on a column Sephadex LH20 (length 55 cm, ø (diameter of the column) 1 cm, eluent: dry DMF). At the end the product was dried under vacuum.

4. Conclusions

Six new nanoconjugates were synthesized using polyhedral oligosilsesquioxanes (POSS) as nano-carriers for anthracyclines (DOX and DAU) and PEGs (water solubilizing agent). The new conjugates contain an ester bond capable of hydrolysis under the conditions of lowered pH (5.5), characteristic for cancer cells. The structure of the new prodrugs were confirmed by NMR (^{1}H-NMR, ^{13}C-NMR, ^{1}H-^{13}C HSQC and DOSY), FTIR and DLS. The analytical methods applied in this work can serve as the important tools and model approach in studies of other nanocarrier-anthracycline conjugates. This is evidenced by the disappearance of the signal at ~2.5 ppm ($\underline{CH_2}OH$) associated with the formation of an ester bond (in the NMR analysis) and the appearance of the 1730 cm^{-1} peak in the FTIR analysis. Using simple and efficient NMR techniques, it was possible to detect formation of the prodrug conjugates. The conjugates are larger in size, compared to anthracycline antibiotics themselves, which is favorable due to the presumably limited normal cell penetration during chemotherapy. In addition, they show longer release time, which makes them potential candidates for biomedical applications in anticancer therapy. POSS type nanocarriers were proven again [10,11] to be useful systems in formation of nanoconjugates and nanocomplexes with anthracycline drugs. The present work shall be expanded by in vitro studies as soon as the test laboratory reopens after COVID-19 closure.

Supplementary Materials: The following are available online, Figure S1: Scheme for the synthesis of SAMDOX and SAMDAU; Figure S2: Scheme for the synthesis of POSSDAU-MR.; Figure S3: Structure of POSSDAU-MR; Table S1: NMR results for POSSDAU-MR.; Table S2: Concentrations of the reagents used in conjugation reaction **4–9**; Table S3: Concentrations of the conjugates **4–9** in drugs release study; Figure S4: Calibration curves of: (A) DOX in H_2O/DMF (5:1) (B) DAU in H_2O/DMF; Figure S5: Dependence of anthracycline concentration (A. DOX, B. DAU) on the absorbance intensity in the UV-Vis spectrum; Figure S6: A. Study of DOX/DAU release from nanoconjugates at pH 5.5 at 310 K quantified by UV-Vis method after 21 h (A) and after 42 h (B); ^{1}H NMR spectra of **4–9** (500 MHz, 295 K, DMSO-d6).

Author Contributions: K.P. has written the research proposal, carried out part of the synthetic work and interpreted NMR, UV-Vis and FTIR results, J.K. interpreted the DLS and UV-Vis results, W.A.S. supervised the studies and finally corrected the paper. All authors have read and agreed to the published version of the manuscript.

Funding: This research was funded by National Science Center, grant number UMO-2016/21/N/ST5/03360.

Acknowledgments: The financial support for this study was provided by the Centre of Molecular and Macromolecular Studies. K. Piorecka wishes to acknowledge the Preludium Grant UMO-2016/21/N/ST5/03360.

Conflicts of Interest: The authors declare no conflict of interest.

Sample Availability: Samples of all compounds (POSS, SAMDAU, SAMDOX, compounds **4–9**) are available from the authors.

References

1. Piorecka, K.; Kurjata, J.; Stanczyk, M.; Stanczyk, W.A. Synthetic routes to nanomaterials containing anthracyclines: Noncovalent systems. *Biomater. Sci.* **2018**, *6*, 2552–2565. [CrossRef] [PubMed]
2. Piorecka, K.; Smith, D.; Kurjata, J.; Stanczyk, M.; Stanczyk, W.A. Synthetic routes to nanoconjugates of anthracyclines. *Bioorg. Chem.* **2020**, *96*, 103617. [CrossRef] [PubMed]
3. Ma, P.; Mumper, R.J. Anthracycline Nano-Delivery Systems to Overcome Multiple Drug Resistance: A Comprehensive Review. *Nano Today* **2013**, *8*, 313–331. [CrossRef] [PubMed]
4. Carvalho, M.R.; Reis, R.L.; Oliveira, J.M. Dendrimer nanoparticles for colorectal cancer applications. *J. Mater. Chem. B* **2020**, *8*, 1128–1138. [CrossRef] [PubMed]
5. Avramović, N.; Mandić, B.; Savić-Radojević, A.; Simić, T. Polymeric Nanocarriers of Drug Delivery Systems in Cancer Therapy. *Pharmaceutics* **2020**, *12*, 298. [CrossRef] [PubMed]
6. McGowan, J.V.; Chung, R.; Maulik, A.; Piotrowska, I.; Walker, J.M.; Yellon, D.M. Anthracycline Chemotherapy and Cardiotoxicity. *Cardiovasc. Drugs Ther.* **2017**, *31*, 63–75. [CrossRef] [PubMed]
7. Graham, U.M.; Jacobs, G.; Yokel, R.A.; Davis, B.H.; Dozier, A.K.; Birch, M.E.; Tseng, M.T.; Oberdörster, G.; Elder, A.; DeLouise, L. From Dose to Response: In Vivo Nanoparticle Processing and Potential Toxicity. *Adv. Exp. Med. Biol.* **2017**, *947*, 71–100. [PubMed]
8. Janaszewska, A.; Gradzinska, K.; Marcinkowska, M.; Klajnert-Maculewicz, B.; Stanczyk, W.A. In Vitro Studies of Polyhedral Oligo Silsesquioxanes: Evidence for Their Low Cytotoxicity. *Materials* **2015**, *8*, 6062–6070. [CrossRef] [PubMed]
9. Sobierajska, E.; Konopka, M.; Janaszewska, A.; Piorecka, K.; Blauz, A.; Klajnert-Maculewicz, B.; Stanczyk, M.; Stanczyk, W.A. Unusual Enhancement of Doxorubicin Activity on Co-Delivery with Polyhedral Oligomeric Silsesquioxane (POSS). *Materials* **2017**, *10*, 559. [CrossRef] [PubMed]
10. Piorecka, K.; Radzikowska, E.; Kurjata, J.; Rozga-Wijas, K.; Stanczyk, W.A.; Wielgus, E. Synthesis of the first POSS cage—Anthracycline conjugates via amide bonds. *New J. Chem.* **2016**, *40*, 5997–6000. [CrossRef]
11. Piorecka, K.; Kurjata, J.; Bak-Sypien, I.; Cypryk, M.; Steinke, U.; Stanczyk, W.A. Reasons for enhanced activity of doxorubicin on co-delivery with octa(3-aminopropyl)silsesquioxane. *RSC Adv.* **2020**, *10*, 15579–15585. [CrossRef]
12. Shanmugan, S.; Cani, D.; Pescarmona, P.P. The design and synthesis of an innovative octacarboxy-silsesquioxane building block. *Chem. Commun.* **2014**, *50*, 11008–11011. [CrossRef] [PubMed]

13. Ghanbari, H.; de Mel, A.; Seifalian, A. Cardiovascular application of polyhedral oligomeric silsesquioxane nanomaterials: A glimpse into prospective horizons. *Int. J. Nanomed.* **2011**, *6*, 775.
14. Rosenholm, J.M.; Linden, M. Towards establishing structure-activity relationships for mesoporous silica in drug delivery applications. *J. Control. Release* **2008**, *28*, 996. [CrossRef] [PubMed]
15. Mori, H.; Lanzendorfer, M.G.; Muller, A.H.E. Silsesquioxane-Based Nanoparticles Formed via Hydrolytic Condensation of Organotriethoxysilane Containing Hydroxy Groups. *Macromolecules* **2004**, *37*, 5228–5238. [CrossRef]
16. Piorecka, K.; Stanczyk, W.; Florczak, M. NMR analysis of antitumor drugs: Doxorubicin, daunorubicin and their functionalized derivatives. *Tetrahedron Lett.* **2017**, *58*, 152–155. [CrossRef]
17. Piorecka, K.; Janaszewska, A.; Majkowska, M.; Marcinkowska, M.; Kurjata, J.; Kazmierski, S.; Radzikowska-Cieciura, E.; Kost, B.; Klajnert-Maculewicz, B.; Stanczyk, W.A. Hydrophilic polyhedral oligomeric silsesquioxane—POSS$(OH)_{32}$ as a complexing nanocarrier for doxorubicin and daunorubicin. *Materials* **2020**, *13*, 5512. [CrossRef] [PubMed]
18. Armarego, W.L.F.; Chai, C.L.L. *Purification of Laboratory Chemicals*; Elsevier: Amsterdam, The Netherlands, 2009.
19. Ni, C.; Wu, G.; Zhu, C.; Yao, B. The Preparation and Characterization of Amphiphilic Star Block Copolymer Nano Micelles Using Silsesquioxane as the Core. *J. Phys. Chem. C* **2010**, *114*, 13471–13476. [CrossRef]

molecules

MDPI

Article

Synthesis of Silsesquioxanes with Substituted Triazole Ring Functionalities and Their Coordination Ability †

Monika Rzonsowska [1,2,*], Katarzyna Kozakiewicz [1], Katarzyna Mituła [1,2], Julia Duszczak [1,2], Maciej Kubicki [3] and Beata Dudziec [1,2,*]

1 Department of Organometallic Chemistry, Faculty of Chemistry, Adam Mickiewicz University in Poznań, Uniwersytetu Poznańskiego 8, 61-614 Poznań, Poland; k.kozakiewicz96@gmail.com (K.K.); katarzyna.mitula@gmail.com (K.M.); julia.duszczak@amu.edu.pl (J.D.)
2 Centre for Advanced Technologies, Adam Mickiewicz University in Poznań, Uniwersytetu Poznańskiego 10, 61-614 Poznań, Poland
3 Faculty of Chemistry, Adam Mickiewicz University in Poznań, Uniwersytetu Poznańskiego 8, 61-614 Poznań, Poland; mkubicki@amu.edu.pl
* Correspondence: mrzons@amu.edu.pl (M.R.); beata.dudziec@gmail.com (B.D.); Tel.: +48-618291878 (B.D.)
† Dedicated to Professor Julian Chojnowski on the occasion of his 85th birthday.

Abstract: A synthesis of a series of mono-T_8 and difunctionalized double-decker silsesquioxanes bearing substituted triazole ring(s) has been reported within this work. The catalytic protocol for their formation is based on the copper(I)-catalyzed azide-alkyne cycloaddition (CuAAC) process. Diverse alkynes were in the scope of our interest—i.e., aryl, hetaryl, alkyl, silyl, or germyl—and the latter was shown to be the first example of terminal germane alkyne which is reactive in the applied process' conditions. From the pallet of 15 compounds, three of them with pyridine-triazole and thiophenyl-triazole moiety attached to T_8 or DDSQ core were verified in terms of their coordinating properties towards selected transition metals, i.e., Pd(II), Pt(II), and Rh(I). The studies resulted in the formation of four SQs based coordination compounds that were obtained in high yields up to 93% and their thorough spectroscopic characterization is presented. To our knowledge, this is the first example of the DDSQ-based molecular complex possessing bidentate pyridine-triazole ligand binding two Pd(II) ions.

Keywords: polyhedral oligomeric silsesquioxane (SQs); click chemistry; CuAAC; coordination compounds; bidentate ligand

Citation: Rzonsowska, M.; Kozakiewicz, K.; Mituła, K.; Duszczak, J.; Kubicki, M.; Dudziec, B. Synthesis of Silsesquioxanes with Substituted Triazole Ring Functionalities and Their Coordination Ability . *Molecules* **2021**, 26, 439. https://doi.org/10.3390/molecules26020439

Academic Editors: Sławomir Rubinsztajn, Marek Cypryk and Wlodzimierz Stanczyk

Received: 21 December 2020
Accepted: 12 January 2021
Published: 15 January 2021

1. Introduction

Polyhedral oligomeric silsesquioxanes (SQs) are a large family of compounds that feature diverse structures with Si-O-Si linkages and tetrahedral Si vertices—i.e., random, amorphous, ladder, and cage-like—and the architecture of the latter has attracted considerable scientific interest. It is due to the presence of the inorganic, rigid core (thermal stability, chemical resistance) and organic moieties attached to it (tunable processability) which is the essence of hybrid materials. Functionalized SQs derivatives may be regarded as their nanosized, smallest fragments and precursors that affect and drive the directions of their potential applications [1–6]. Significant development of catalytic protocols for effective and selective anchoring of respective organic functionality to the SQs core has been observed during the last years. The crucial aspect of this is the presence of a proper prefunctional moiety at the Si-O-Si framework, enabling its modification, e.g., Si-H, Si-OH, Si–CH=CH_2 units, etc. This, in turn, influences the selection of a respective catalytic procedure for this purpose, e.g., hydrosilylation, cross-metathesis, *O*-silylation, Friedel-Crafts, silylative, Heck, Suzuki, or Sonogashira coupling reactions [2,7–20]. Among these methods, the copper(I)-catalyzed azide-alkyne cycloaddition (CuAAC) may be an alternative but the only route to yield substituted 1,4-triazole ring functionalities regioselectively [21–23]. This is a powerful synthetic tool to build covalent connections between dissimilar units, and since its discovery in 2002 (independently by Sharpless, Fokin, and Meldal groups), it is widely utilized in (bio-)organic, medicinal and for some time now, also in surface/materials chemistry [24–27]. Despite the simplicity of the reaction, the immense development in the catalyst design, incl. stabilizing ligands, has be observed in the past few years. This may be visible in the aspect of functional group diversity in both reagents and their resilience to the CuAAC reaction conditions, i.e., solvent type, temperature, time, etc. The variety of Cu-based catalysts for this process requires the presence of the Cu(I) species at the highest concentration whether it is introduced in this state or generated in situ [26]. On the other hand, the Cu(I) favors the Glasser coupling of terminal alkynes [28], so the CuAAC conditions should be optimized to avoid the formation of by-products. For this, the in situ creation of Cu(I) ions seems reasonable [24]. As a result of this evolvement, the process gained popularity in the chemistry of silsesquioxanes as well. There are some examples of CuAAC methodology applied for silsesquioxanes to introduce the triazole moiety(-ies) substituted at 1,4 positions with the SQs core and organic group. The process may be applied in the case of mono- and octa-substituted T_8 SQs, linked with the Si-O-Si core via mainly alkyl [29–36] but there are also reports on a phenyl group [37,38]. There are also few examples of using di- substituted double-decker (DDSQ) [39,40] or ladder [41] silsesquioxanes [42].

The application of the resulting products with 1,4-triazole ring(s) depends on the kind of alkyne moiety used as a reagent. Due to the interesting photoelectronic properties of SQs-based systems with aryl-triazole groups, they could be possibly used as luminescent materials with enhanced thermal resistance [29,41,43–45]. One of the promising branches of their application is materials chemistry, e.g., as polymers or dendrimers modifiers/synthons [30,39,42,46–54]. They may be also found in macromolecular SQs-based surfactants of amphiphilic and self-assembly or encapsulation properties [31,55,56]. Additionally, the CuAAC methodology serves as a tool for the formation of SQs functionalized peptide dendrimers [35] or glycoconjugates and in targeted bioimaging [57,58].

Catalysis has become a very prospective direction of employing the SQs with 1,4-substituted triazole rings, as catalysts by themselves, e.g., in asymmetric Michael or Aldol reactions [32,33]. Even though there have been still very few reports in this area of interest. One of them reported on the potential coordinating character of a pyridine-triazole moiety attached to the SQs core and revealed its use as a ligand for Pd-complexation. Interestingly, this compound was found to be active in Suzuki–Miyaura cross-coupling [59].

Herein, we present our studies on the copper(I)-catalyzed azide-alkyne cycloaddition process of two different types of silsesquioxanes with the variation of alkynes bearing aryl, hetaryl, alkyl, silyl, siloxyl, or even germyl groups. As a result, we report on the preparation of mono-T_8 and difunctionalized double-decker silsesquioxanes with substituted triazole ring(s). The second part of the paper is focused on the application of selected SQs with hetaryl substituted triazole moieties as potential ligands in complexing reaction with transition metals (Pd, Rh, Pt), resulting in the formation of respective SQs-based coordination systems (Figure 1).

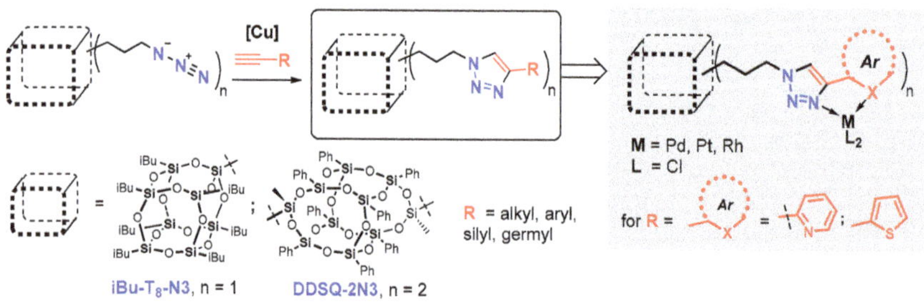

Figure 1. Mono-T_8 and difunctionalized DDSQ silsesquioxanes with Substituted Triazole Ring and the coordinating ability of the pyridine- and thiophenyl-derivatives towards selected TM ions, presented in this work.

2. Results and Discussion

2.1. *The Copper(I)-Catalyzed Azide-Alkyne Cycloaddition (CuAAC) Using* $\mathbf{iBuT_8}$**-N3** *and* **DDSQ-2N3**

In the first step, the starting precursors, i.e., the azidopropyl-derivative(s) of mono-$\mathbf{iBuT_8}$**-N3** and di-**DDSQ-2N3** were prepared in a sequence of hydrolytic condensation of respective silanol precursor of SQs and chlorosilane followed by nucleophilic substitution with NaN_3 [60,61] (Figure 2). The idea of the synthetic path is presented below.

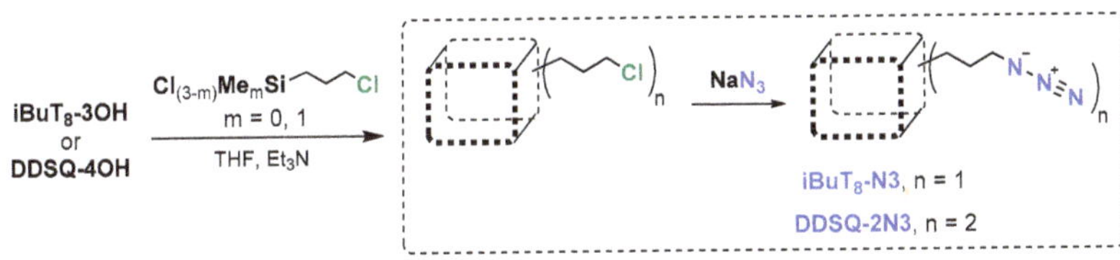

Figure 2. General route for the synthesis of azide-derivatives $\mathbf{iBuT_8}$**-N3** and **DDSQ-2N3**.

The two SQs-based azides $\mathbf{iBuT_8}$**-N3** and **DDSQ-2N3** were used as reagents in CuAAC coupling process with a variety of alkynes bearing aryl, alkyl, silyl, and germyl functionalities. The reaction progress was monitored by FT-IR, due to the large mass of the product eliminating the possibility of using GC or GC-MS and confirmed by ^{1}H-NMR. The representative FT-IRs are presented in Figure 3. For all alkynes tested nearly complete conversion of SQs azides was observed within up to 3 days which depended on the type of reaction conditions and Cu catalyst.

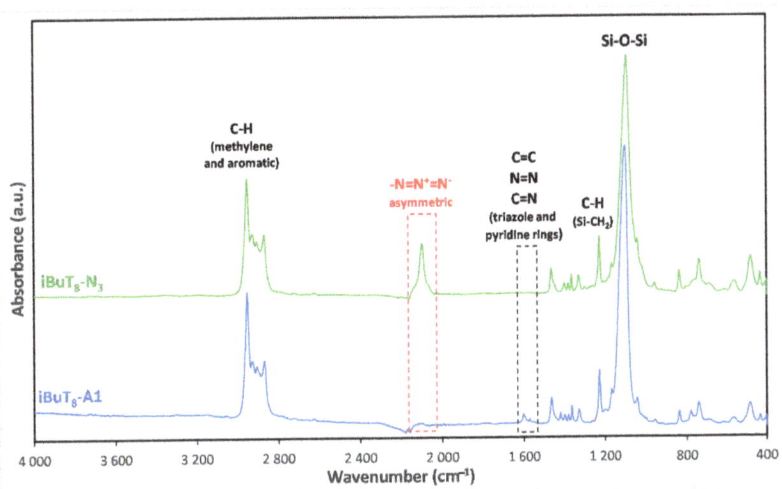

Figure 3. FT-IR spectra of $\mathbf{iBuT_8}$**-N3** and $\mathbf{iBuT_8}$**-A1** after completion of CuAAC coupling reaction.

The stacked FT-IR spectra of starting material $\mathbf{iBuT_8}$**-N3** and the selected product with 4-pyridine-triazole group $\mathbf{iBuT_8}$**-A1** are depicted in Figure 3. The established reaction conditions resulted in the complete conversion of azide (-N=N$^+$=N-) group in $\mathbf{iBuT_8}$**-N3**, confirmed by the disappearance of respective bands attributed to stretching asymmetric vibrations of -N=N- at ca. $\bar{\nu}$ = 2098 cm^{-1} (marked in Figure 3). For the CuAAC reaction product, i.e., $\mathbf{iBuT_8}$**-A1**, there are new bands in the spectrum, characteristics of C=C, C=N, N=N stretching vibrations from triazole as well as pyridine ring at ca. $\bar{\nu}$ = 1603 cm^{-1} and $\bar{\nu}$ = 1571 cm^{-1} that confirm the formation of the desired product.

We based on two types of Cu sources, i.e., $CuSO_4$ with sodium ascorbate [40,59] and CuBr with PMDTA [33]. At first, special conditions were created for the reduction of Cu(II) in situ to Cu(I) and then to maintain the introduction of Cu(I) in this oxidation state into the reaction. The main target was to perform the reaction until full conversion of SQ-based azides to avoid the problematic isolation issues of resulting SQ-products with substituted triazole rings(s) from unreacted SQ-based azides. The results of the reactions conducted to obtain products with substituted triazole ring(s) are collected in Table 1 for T_8-derivatives and in Table 2 for DDSQ-derivatives.

Table 1. Copper-catalyzed azide-alkyne cycloaddition using **iBuT$_8$-N3** and alkynes.

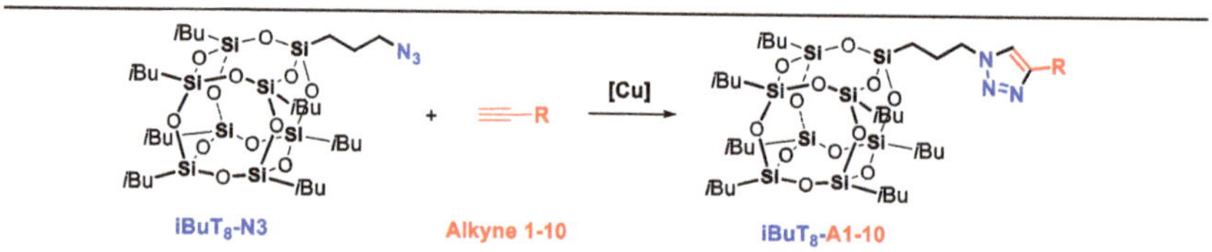

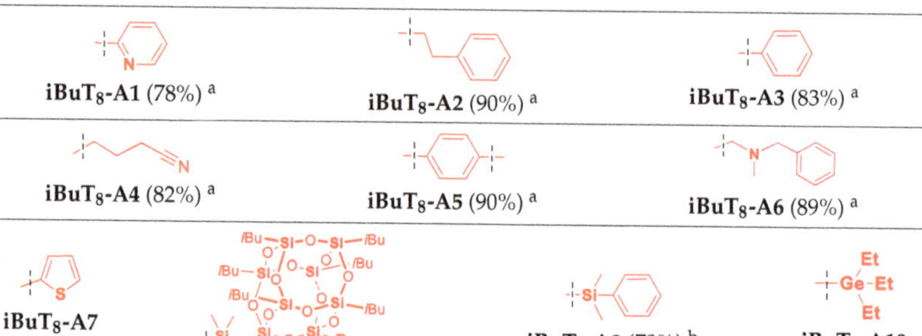

iBuT$_8$-A1 (78%) [a] | **iBuT$_8$-A2** (90%) [a] | **iBuT$_8$-A3** (83%) [a]

iBuT$_8$-A4 (82%) [a] | **iBuT$_8$-A5** (90%) [a] | **iBuT$_8$-A6** (89%) [a]

iBuT$_8$-A7 (81%) [a] | **iBuT$_8$-A8** (84%) [b] | **iBuT$_8$-A9** (70%) [b] | **iBuT$_8$-A10** (85%) [c]

[a] Reaction conditions for Cu(II) source: [azide]:[alkyne]:[$CuSO_4$][sodium ascorbate] = 1:1.4–8:0.025–0.25:0.3–5; 25–60 °C; 72–96 h. [b] Reaction conditions for Cu(I) source: [azide]:[alkyne]:[CuBr][PMDTA] = 1:1.45:0.1:0.1; 25 °C; 24 h. [c] additional 12 h at 45 °C. > 99% conversion of **iBuT$_8$-N3** was confirmed by FT-IR in situ and ^{1}H-NMR analyses. Value in parenthesis is given for isolation yield (%).

Table 2. Copper-catalyzed azide-alkyne cycloaddition using **DDSQ-2N3** [a].

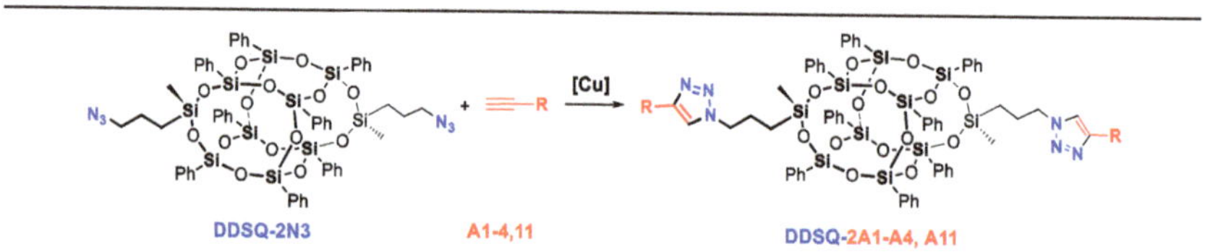

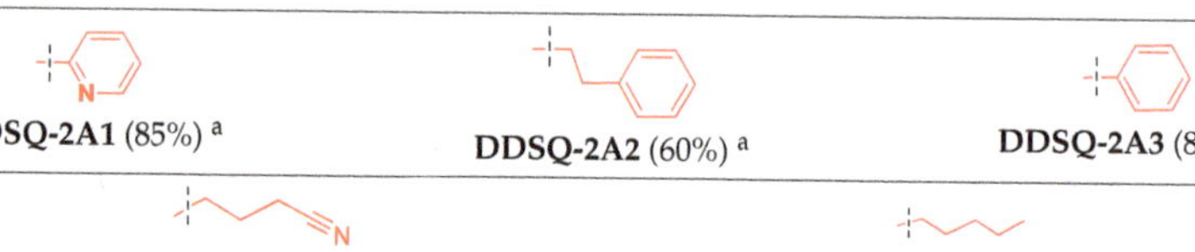

DDSQ-2A1 (85%) [a] | **DDSQ-2A2** (60%) [a] | **DDSQ-2A3** (82%) [a]

DDSQ-2A4 (78%) [a] | **DDSQ-2A11** (84%) [a]

[a] Reaction conditions for Cu(II) source: [azide]:[alkyne]:[$CuSO_4$][sodium ascorbate] = 1:1.4–8:0.025–0.25:0.3–5; 25–60 °C; 72–96 h. > 99% conversion of **DDSQ-2N3** was confirmed by FT-IR in situ and ^{1}H-NMR analyses. Value in parenthesis is given for isolation yield (%).

The results of DDSQ-based systems with substituted triazole rings are collected in Table 2 and involves the selective formation of DDSQ-compounds with the two above-mentioned triazole moieties. The spectrum of used alkynes varies as they contain aryl, hetaryl, alkyl, and silyl derivatives of commercial availability. Additionally, we tested ethynyl(triethyl)germane (A9) and ethynylsiloxysubstituted-iBuT$_8$ (A8) to verify their potential in the CuAAC process.

The tested reaction conditions based on Cu(II) and Cu(I) catalysts seem to be analogous in the case of less demanding alkynes, i.e., simple aryl or alkyl derivatives. Interestingly, for the 5-hexynenitrile, the applied catalytic conditions did not affect the present -CN moiety that in general may also be reactive and susceptible to alkyne-azide coupling reaction conditions to form respective 5-substituted tetrazoles [62]. For this, the presence of a reactive -CN moiety could be used in further modifications of the obtained products: **iBuT$_8$-A4** and **DDSQ-2A4**. The reactivity of ethynylsilane (A9), ethynylgermane (A10) and also ethynylsiloxysubstituted iBuT$_8$ (A8) compounds was tested with positive results. However, the use of silyl (A9) or germyl (A10) alkyne proceeded with >99% conversion of SQs-based azides (**iBuT$_8$-N3** and **DDSQ-2N3**) only when modified reaction conditions with Cu(I) [33] were applied (heating at 45 °C). Even though, for ethynylsiloxysubstituted-iBuT$_8$ (A8) up to 10% of unreacted **iBuT$_8$-N3** was observed. It could be separated from the resulting product **iBuT$_8$-A8** during the purification with the use of chromatography column and proper eluent selection (hexane:DCM 3:1 for separation of **iBuT$_8$-N3** from **iBuT$_8$-A8**). Lower reactivity of A8 may derive from the presence of oxygen as the silicon atom in the vicinity of ethynyl-moiety and its electron-withdrawing impact. It should be noted that ethynylsilanes exhibit reactivity in this process, but conditions created by us seem to be milder for lower reaction temperature [63]. On the other hand, it would be the first example for ethynylgermane (A10) to exhibit high reactivity in the CuAAC reaction. One report on the formation of 4-germyl-substituted triazole ring derivative concerns using internal alkyne, i.e., 3-(trimethylgermyl)-2-propynal [64]. Additionally, the reports on the reactivity of the ethynylsiloxy-moiety (meaning A8) in the CuAAC process are very scarce [65].

An interesting relationship was found for ^{1}H-NMR analyses of DDSQs bearing triazole ring substituted at 4-positition with an aryl (**DDSQ-2A1**) and alkyl (**DDSQ-2A4**) group. The resonance line of a very significant triazole proton N=C-H^t at 5*H*-position of triazole ring depends on the type of the moiety at 4-position of the latter. The crucial aspect may be its electronic property and the respective shielding effect of alkyl and deshielding effect characteristic for the aryl moiety presence. It affects the N=C-H^t signal shift and it is upfield for **DDSQ-2A1** to be present at 6.75 ppm and downfield for **DDSQ-2A4**, to appear at 7.84 ppm, which gives a total change in resonance lines of 1.09 ppm (Figure 4). Due to the presence of a triazole, aromatic ring, this effect is also insensibly perceptible for -CH$_2$- group at 1*N*-position of this ring (for **DDSQ-2A1** δ = 4.15 ppm and **DDSQ-2A4** δ = 4.21 ppm) (Figure 4). This is a notable difference in chemical shifts of N=C-H^t at triazole ring for its alkyl and aryl derivatives when compared with analogous compounds of iBu-SQs, i.e., **iBuT$_8$-A4** (alkyl δ = 7.31 ppm) and **iBuT$_8$-A1** (aryl δ = 8.12 ppm) that equals 0.82 ppm (Figure 5). It is even more significant when comparing analogous products with alkyl groups at triazole ring but with diverse Si-O-Si cores, i.e., **DDSQ-2A4**, N=C-H^t proton present at 6.75 ppm with **iBuT$_8$-A4**, =C-H^t at 7.31 ppm. These differences in result may be explained by the presence and electronic effect of the DDSQ core with phenyl substituents.

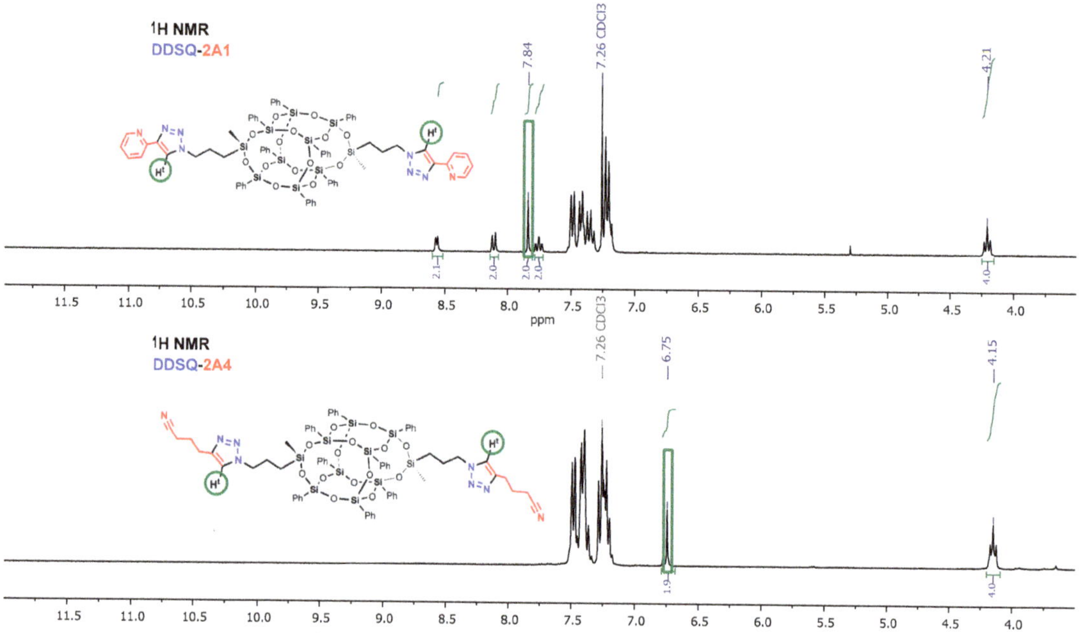

Figure 4. Selected range of stacked ^{1}H-NMR spectra of **DDSQ-2A1** and **DDSQ-2A4**.

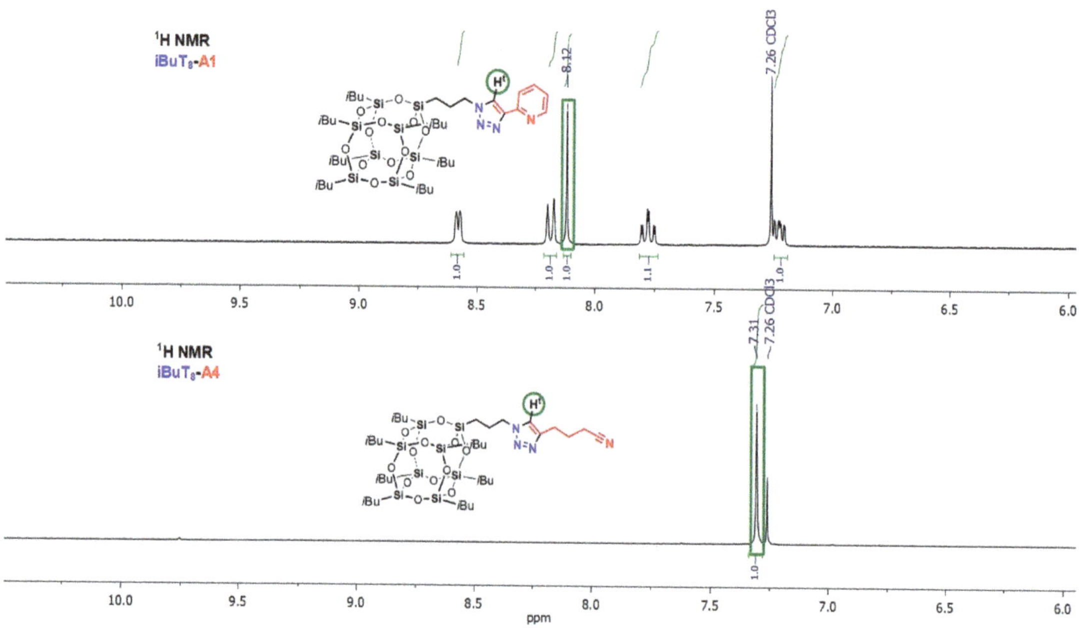

Figure 5. Selected range of stacked ^{1}H-NMR spectra of **iBuT$_8$-A1** and **iBuT$_8$-A4**.

2.2. *X-ray Analysis of* **DDSQ-2A1**

A DDSQ-based pyridine-triazole derivative, i.e., **DDSQ-2A1** proved to be a solid and acquired the form of crystals amenable to X-ray crystal structure determination (Figure 6). The molecule is C_i-symmetrical, as it lies across the center of inversion in the space group $P2_1/c$. The structure of the core may be described as built of four rings, two 8-membered (four Si, four O), and two 10-membered (five Si, five O), which can be noted as $8^2 10^2$. The geometry of the core of the molecule is determined by two factors: one rigid—Si-O distance, which has a very narrow spread (mean value 1.615(8) Å), and one flexible Si-O-Si angles (140.77(16)°–162.43(16)°). Similar tendencies were noted in similar molecules [16,66]. The architecture of the crystal is determined by weak but numerous interactions (C-H···O, C-H···π, π···π etc.). These multiple interactions give rise to quite significant interaction energies. Calculations with PIXEL method give results as high as −160.5, −95.7, and −85.4 kJ/mol for the three highest interaction energies between molecules, and −555.5 kJ/mol as total packing energy [67,68].

All of the T_8 and DDSQ-based compounds with substituted triazole ring(s) were isolated in high, up to 90% yields. They are air-stable white or light-yellow solids with good solubility in DCM, $CHCl_3$, THF, toluene. The solubility in MeOH, MeCN and for hexane depends on the type of SQ's core, i.e., **iBuT$_8$** derivatives are more soluble than **DDSQs**.

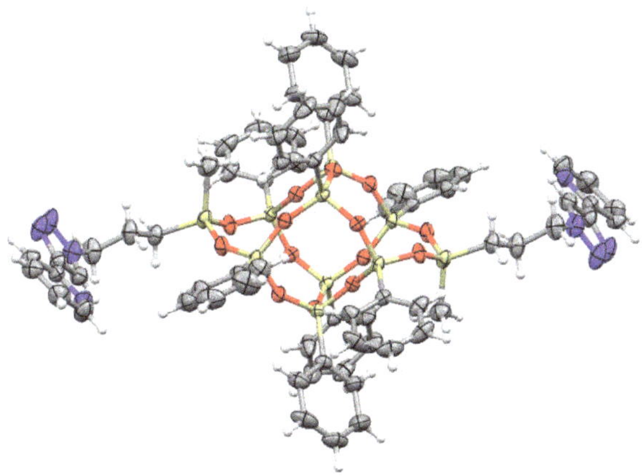

Figure 6. A perspective view of the molecule. Ellipsoids are drawn at the 50% probability level, hydrogen atoms are shown as spheres of arbitrary radii (grey-C, white-H, blue-N, red-O, yellow-Si).

2.3. *SQs-Based Pyridyl- and Thiophenyl-Triazole Derivatives* (**iBuT$_8$-A1, DDSQ-2A1, iBuT$_8$-A7**) *as Bidentate Ligands in the Formation of Coordination Complexes with Selected Transition Metals (TM = Pd, Pt, Rh)*

The next step was to verify the coordination properties of selected T_8 and DDSQ products type possessing heteroatom at 4C triazole ring, i.e., N (**iBuT$_8$-A1, DDSQ-2A1**) and S (**iBuT$_8$-A7**). Using 2-ethynylpyridine and 2-ethynylthiophene derivatives created the possibility to form bidentate ligands of N^N and N^S kind donation. For this purpose, we chose TM metals that are known to form coordination compounds with SQs-based ligands, i.e., Pd(II) [59,69], Rh(I) [70], and Pt(II) [71].The general scheme for using T_8-type ligands, i.e., **iBuT$_8$-A1** and **iBuT$_8$-A7** is disclosed in Figure 7 for Pd(II), Pt(II), and Rh(I) and DDSQ-based ligand with Pd(II) in Figure 8.

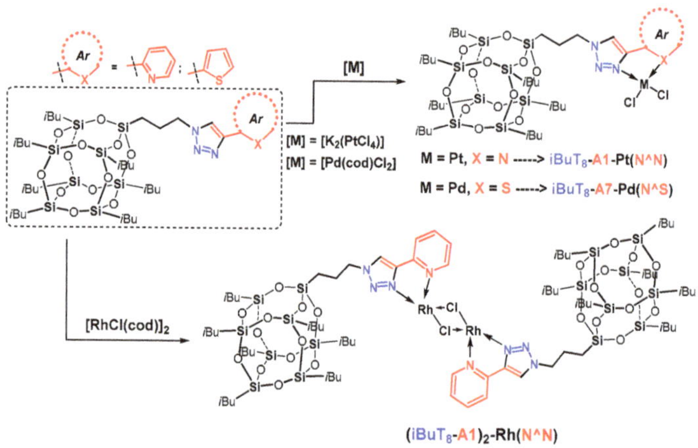

Figure 7. General procedure for the synthesis of T_8-based N^N and N^S type mononuclear coordination compounds with Pd(II) (**iBuT$_8$-A7-Pt(N^S)**), Pt(II) (**iBuT$_8$-A1-Pt(N^N)**), and binuclear with Rh(I) ((**iBuT$_8$-A1)$_2$-Rh(N^N)**).

Figure 8. General procedure for the synthesis of DDSQ-based N^N coordination compound with Pd(II) (**DDSQ-A1-[Pd(N^N)]$_2$**).

The analogous verification was performed in terms of **DDSQ-2A1** possessing bidentate N^N ligand Pd(II). The 1:2 (ligand: metal) stoichiometry of the reaction enabled the formation of a molecular system with two Pd(II) ions captured to the opposite parts of the DDSQ core (Figure 8).

To our knowledge, this is the first example of the DDSQ-based molecular complex possessing a bidentate pyridine-triazole ligand with coordination TM Pd(II) ion. Furthermore, it is an interesting example of using difunctionalized DDSQ compounds to anchor metal ions and the reports on these systems have been still profoundly limited [71,72].

For the reaction aiming at palladium and rhodium complexes, their cyclooctadiene precursors were used and for platinum, the tetrachloroplatinate(II) was applied. The mononuclear compounds **iBuT$_8$-A7-Pt(N^S)** and **iBuT$_8$-A1-Pt(N^N)** are air-stable, pale yellow solids. The dinuclear Rh(I) based complex ((**iBuT$_8$-A1)$_2$-Rh(N^N)**) is rather an air- and moisture sensitive orange solid and its synthesis was performed with the use of the Schlenk technique. The iBuT$_8$-derivatives are soluble in DCM, $CHCl_3$, THF, toluene, and of very low solubility in methanol. The DDSQ-based Pd(II) complex **DDSQ-A1-[Pd(N^N)]$_2$** is an air-stable pale yellow solid with very limited solubility in DCM, $CHCl_3$, and THF and soluble in DMF and DMSO. The four coordination SQ-based compounds were isolated in yields 55%–93% and characterized using spectroscopic analysis proving their formation (for details see ESI). The respective comparison of the ^{1}H-NMR stacked spectra of ligand **DDSQ-A1** and respective complex **DDSQ-A1-[Pd(N^N)]$_2$** are presented below (Figure 9).

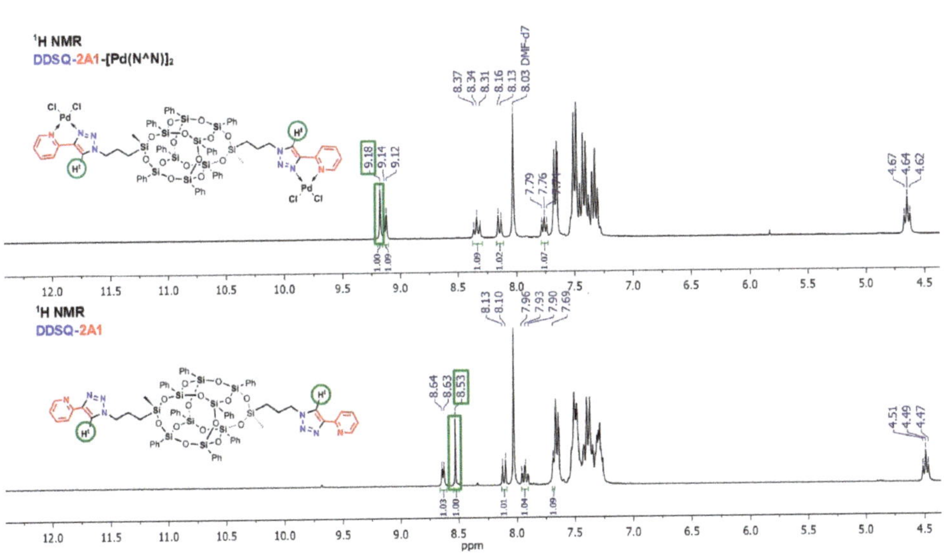

Figure 9. Selected range of stacked ^{1}H-NMR spectra of **DDSQ-A1** and **DDSQ-A1-[Pd(N^N)]$_2$**.

The presence of Pd with its chloro-ligands in **DDSQ-A1-[Pd(N^N)]$_2$** affects the polarity of the complex and restricts its solubility in a common, less polar solvent and for this reason DMF-d_7 was selected in order to compare ^{1}H-NMR spectra. As expected from the results obtained for the iBuT$_8$-based Pd, Pt, and Rh complexes and from the literature reports [59], the placement of resonance line of the triazole proton N=C-H^t is susceptible to the chemical surrounding and presence of a different type of TM ion. However, in general, in each complex its shift is downfield significantly. For **DDSQ-A1-[Pd(N^N)]$_2$** N=C-H^t there is a notable difference in its chemical shift to appear at δ = 9.18 ppm when compared with a bare ligand, i.e., **DDSQ-A1**: N=C-H^t at δ = 8.53 ppm (Figure 9). Additionally, the resonance lines derived from the pyridine ring are also shifted downfield due to the changes in the electron density on the hetaryl moiety while coordinating to Pd ion, especially for the =C-H^5.

3. Materials and Methods

3.1. Materials

The chemicals were purchased from the following sources: Hybrid Plastics (Hybrid Plastics, Hattiesburg, MS, USA) for DDSQ tetrasilanol form ($C_{48}H_{44}O_{14}Si_8$) (DDSQ-4OH), trisilanol ($C_{28}H_{66}O_{12}Si_7$)(iBuT$_8$-3OH); Sigma-Aldrich (Saint Louis, MO, USA) for: dichloromethane (DCM), tetrahydrofuran (THF), dimethylformamide (DMF), toluene, methanol, acetonitrile (MeCN), chloroform ($CHCl_3$), hexane, chloroform-*d*, dimethyl sulfoxide-d_6 (DMSO-d_6), dichloromethane-d_2 (DCM-d_2), (dimethylphenylsilyl)acetylene, 2-ethynylpyridine, *N*-methyl-*N*-propargylbenzylamine, 3-butynylbenzene, phenylacetylene, n-heptyne, 1,4-diethynylbenzene, 5-hexynenitrile, 2-ethynylthiophene; ABCR (ABCR, Karlsruhe, Germany) for dichloro(3-chloropropyl)methylsilane, molecular sieves, triethylamine, and silica gel 60. Chemat (Gdansk, Poland) for: sodium L-ascorbate crystalline, ammonium chloride, copper(II) sulfate pentahydrate, copper(I) bromide, *N,N,N′,N′,N″*-pentamethyldiethylenetriamine (PMDTA), sodium azide, sodium sulfate anhydrous, dichloro(1,5-cyclooctadiene)palladium(II), chloro(1,5-cyclooctadiene)rhodium(I) dimer, potassium tetrachloroplatinate(II). Tetrahydrofuran (THF) was refluxed over sodium/benzophenone and distilled. Triethylamine (Et_3N) was distilled over calcium hydride before use. DMF was stored under argon. Ethynyl(triethyl)germane (A10) and ethynyl (dimethylsiloxy)hepta(i-butyl)octasilsesquioxane (A8) was synthesized according to the literature procedures [73,74]. (3-chloropropyl)hepta(i-butyl)octasilsesquioxane (iBuT_8-Cl) and DDSQ-2Cl were obtained via corner- and side-capping hydrolytic condensation procedure, and mono- and diazido-functionalized silsesquioxanes (**iBuT_8-N3**, **DDSQ-2N3**) were synthesized via nucleophilic substitution according to the literature procedures [60,61]. All syntheses were conducted under argon atmosphere using standard Schlenk-line and vacuum techniques.

3.2. Methods

Nuclear magnetic resonance spectroscopy (NMR) measurements (^{1}H, ^{13}C, and ^{29}Si NMR) were conducted using spectrometers: Bruker Ultrashield 300 MHz and 400 MHz respectively (Bruker, Faellanden, Switzerland) with $CDCl_3$, CD_2Cl_2, DMF-d_7, and DMSO-d_6 as a solvent. Chemical shifts are reported in ppm with reference to the residual solvent signal peaks for ^{1}H and ^{13}C and to TMS for ^{29}Si.

Fourier transform-infrared (FT-IR) spectra were recorded on a Nicolet iS5 (Thermo Scientific, Waltham, MA, USA) spectrophotometer equipped with a SPECAC Golden Gate, diamond ATR unit with a resolution of 2 cm^{-1}. In all cases, 16 scans were collected to record the spectra in a range of 4000–430 cm^{-1}.

Elemental analyses (EA) were performed using a Vario EL III instrument (Elementar Analysensysteme GmbH, Langenselbold, Germany).

High-resolution mass spectra (HRMS) were obtained using Impact HD mass spectrometerQ-TOF type instrument equipped with electrospray ion source (Bruker Daltonics, GmbH, Bremen, Germany). The sample solutions (DCM:MeOH) were infused into the ESI source by a syringe pump (direct inlet) at the flow rate of 3 μL/min. The instrument was operated under the following optimized settings: endplate voltage 500 V; capillary voltage 4.2 kV; nebulizer pressure 0.3 bar; dry gas (nitrogen) temperature 200 °C; dry gas flow rate 4 L/min. The spectrometer was previously calibrated with the standard tune mixture.

X-ray crystallography. Diffraction data were collected by the ω-scan technique, using graphite-monochromated MoKα radiation (λ = 0.71073 Å), at 100(1) on Rigaku XCalibur (Rigaku OD, Neu-Isenburg, Germany) four-circle diffractometer with EOS CCD detector. The data were corrected for Lorentz-polarization as well as for absorption effects [75]. Precise unit-cell parameters were determined by a least-squares fit of the 6861 reflections of the highest intensity, chosen from the whole experiment. The structures were solved with SHELXT [76] and refined with the full-matrix least-squares procedure on F2 by SHELXL [77]. All non-hydrogen atoms were refined anisotropically. Hydrogen atoms were placed in idealized positions and refined as 'riding model' with isotropic displacement parameters set at 1.2 (1.5 for CH_3) times Ueq of appropriate carrier atoms.

Crystallographic data for the structural analysis has been deposited with the Cambridge Crystallographic Data Centre, no. CCDC–2045899. Copies of this information may be obtained free 589 of charge from: The Director, CCDC, 12 Union Road, Cambridge, CB2 1EZ, UK; e-mail: deposit@ccdc.cam.ac.uk, or www.ccdc.cam.ac.uk.

Crystal data. $C_{70}H_{68}N_8O_{14}Si_{10}$, M_r = 1526.22, monoclinic, $P2_{1/c}$, a = 14.8188(5) Å, b = 22.6566(9) Å, c = 11.1567(3) Å, β = 101.194(3)°, V = 3674.5(2) Å^3, Z = 2, dx = 1.379 g·cm^{-3}, F(000) = 1592, μ = 0.248 mm^{-1}, 16,862 reflections collected, 6455 symmetry-independent (Rint = 2.40%), 5402 with I > 2σ(I). Final R(F) [I > 2σ(I)] = 0.0580, wR2 [I > 2σ(I) = = 0.1257, R(F) [all data] = 0.0709, wR2 [all data] = 0.1298, Δρmax/min = 1.398/−0.627 e/Å^{-3}.

3.3. General Procedure for Copper(I)-Catalyzed Azide-Alkyne Cycloaddition (CuAAC)

3.3.1. Synthetic Procedure with the Use of $CuSO_4$ as Cu(II) Ion Source

The exemplary synthetic procedure is presented for **iBuT_8-A1**. To a solution of **iBuT_8-N3** (300 mg, 0.33 mmol) in THF (15 mL), sodium L-ascorbate crystalline (in general 0.3–5 eq., herein 5 eq.), **A1** (in general 1.4–8 eq., herein 7.85 eq.) and copper(II) sulfate pentahydrate (in general 0.025–0.25eq., herein 0.25 eq.) diluted in water, were added respectively. The reaction was conducted in a closed system until full conversion of **iBuT_8-N3**, confirmed by FT-IR analysis (typically 72–96 h depending on alkyne used). The crude product was filtered off by column chromatography (silica gel 60, THF) to remove solid impurities and the solvent was evaporated. It was extracted with DCM and water. Organic layer was dried with anhydrous sodium sulfate. Then the solvent was removed under reduced pressure, and the product was precipitated in methanol as white solid. The product (83 mg, i.e., 78% isolated yield) was analyzed by ^{1}H, ^{13}C, and ^{29}Si NMR and EA to confirm its structure. For the spectroscopic analysis please see Supplementary Materials.

3.3.2. Synthetic Procedure with Use of CuBr as Cu(I) Ion Source

The exemplary synthetic procedure is presented for **iBuT$_8$-A7** [33]. To a solution of **iBuT$_8$-N3** (200 mg, 0,22 mmol) and **A7** (35 µL, 0.33 mmol) in THF (3 mL) stirring under argon, CuBr (3.2 mg, 0.02 mmol) and PMDTA (4.6 µL, 0.02 mmol) were added. The reaction mixture was stirred at room temperature for 30 h. After the reaction was completed (FT-IR analysis), the crude product was filtered off by column chromatography (silica gel 60, THF) to remove solid impurities and solvent was evaporated. The resulted solid was washed with methanol and dried in vacuo. Second option for isolation is the extraction in DCM and water. Organic layer was dried with sodium anhydrous sulfate, solvent was removed under reduced pressure, and product was precipitated in methanol as a pale yellow solid in 81% yield (181 mg). Product was analyzed by ^{1}H, ^{13}C, and ^{29}Si NMR and EA to confirm its structure.

3.4. *General Procedure for Using SQs-Based Pyridyl-Triazole- and Thiophenyl-Triazole Derivatives as Ligands in the Formation of Coordination Complexes with Selected Transition Metals (Pd, Pt, Rh)*

3.4.1. Procedure for the Synthesis of **iBuT$_8$-A7-Pd(N^S)**, **DDSQ-A1-[Pd(N^N)]$_2$**, and **(iBuT$_8$-A1)$_2$-Rh(N^N)**

The procedure for the synthesis of **iBuT$_8$-A7-Pd(N^S)** is described as an example.

A mixture of 1 equiv. of **iBuT8-A7** ligand and a stoichiometric amount of Pd(cod)Cl_2 was dissolved in dichloromethane and stirred at room temperature for 24 h. After this time, a solvent was evaporated. The crude product was dissolved in hexane and filtrated off via cannula. The solvent was evaporated and afforded in pure **iBuT$_8$-A7-Pd(N^S)** as yellow solid in 60% yield (91 mg). It was dried in vacuo. Complex **DDSQ-A1-[Pd(N^N)]$_2$** was obtained analogously, however, it precipitated from the DCM solution. After 24 h, the solvent was evaporated and washed with hexane and dried in vacuo. Obtained products were yellow **DDSQ-A1-[Pd(N^N)]$_2$** (93%, 228 mg) and orange **(iBuT$_8$-A1)$_2$-Rh(N^N)** (for Rh, the complexation was performed within 96 h) (55%, 70 mg) solids. **DDSQ-A1-[Pd(N^N)]$_2$** exhibits very restricted solubility in DCM, chloroform, THF or hexane and is soluble in DMF and DMSO.

3.4.2. Procedure for Synthesis of **iBuT$_8$-A1-Pt(N^N)**

The complex was synthesized as described by Galanski and Keppler et al. with slight modifications [78]. To a solution of ligand **iBuT$_8$-A1** (0.051 g, 0.05 mmol, 1.005 eq.) in THF, a solution of K_2PtCl_4 in water-MeOH (1:1) was added. The mixture was stirred overnight in a light-protected flask at 40 °C. After this time, the mixture was in a form of suspension and the addition of MeOH resulted in crude product precipitation. It was washed with methanol and afforded in pure **iBuT$_8$-A1-Pt(N^N)** as pale pale-yellow solid in 87% yield (55 mg) and then dried in vacuo.

4. Conclusions

In conclusion, we reported on the synthesis and characterization of a series of T_8- and DDSQ-based double-decker silsesquioxanes bearing 4-substituted triazole ring with aryl, hetaryl, alkyl, silyl, and germyl groups via copper(I)-catalyzed azide-alkyne cycloaddition (CuAAC). From this group of compounds hetaryl-triazole i.e., pyridine- and thiophenyl-derivatives were selected and verified in terms of their coordinating properties towards Pd(II), Pt(II), and Rh(I) ions. As a result of performed tests, four types of complexes, i.e., two mononuclear iBuT$_8$-based Pt- and Pd- with N^N and N^S ligands, dinuclear iBuT$_8$-based Rh with N^N ligand as well as DDSQ-based with two Pd ions coordinated with N^N bidentate ligand were obtained and fully characterized. For the **DDSQ-2A1** ligand, this is the first example of using the pyridine-triazole moiety to anchor TM ion in the chemistry of DDSQ-compounds. These may be potentially valuable systems of catalytic activity that will be tested in our future studies.

Supplementary Materials: The following are available online, ^{1}H, ^{13}C, ^{29}Si NMR spectra of all obtained compounds (Figures S1–S53).

Author Contributions: Designed and supervised, B.D.; Synthesis and characterization of the reported compounds, K.K., M.R., K.M., and J.D.; X-ray analysis and description M.K.; Writing—original draft preparation, M.R. and B.D. All authors have read and agreed to the published version of the manuscript.

Funding: This research was funded by National Science Centre (Poland) Project OPUS UMO-2016/23/B/ST5/00201.

Data Availability Statement: The data presented in this study are available in this article or in a Supplementary Materials.

Acknowledgments: The authors are grateful to Rafał Januszewski for donation of iBu-SQ-Cl and to Jan Jarożek for the project of graphical abstract.

Conflicts of Interest: The authors declare no conflict of interest.

Sample Availability: Not available.

References

1. Hartmann-Thompson, C. *Applications of Polyhedral Oligomeric Silsesquioxanes*; Springer: Berlin/Heidelberg, Germany, 2011; ISBN 9789048137862.
2. Du, Y.; Liu, H. Cage-like Silsesquioxanes-based Hybrid Materials. *Dalt. Trans.* **2020**, *49*, 5396–5405. [CrossRef]
3. Dong, F.; Lu, L.; Ha, C.S. Silsesquioxane-Containing Hybrid Nanomaterials: Fascinating Platforms for Advanced Applications. *Macromol. Chem. Phys.* **2019**, *220*, 1800324. [CrossRef]
4. John, Ł. Selected developments and medical applications of organic–inorganic hybrid biomaterials based on functionalized spherosilicates. *Mater. Sci. Eng. C* **2018**, *88*, 172–181. [CrossRef] [PubMed]
5. Cordes, D.B.; Lickiss, P.D.; Rataboul, F. Recent developments in the chemistry of cubic polyhedral oligosilsesquioxanes. *Chem. Rev.* **2010**, *110*, 2081–2173. [CrossRef]
6. Ahmed, N.; Fan, H.; Dubois, P.; Zhang, X.; Fahad, S.; Aziz, T.; Wan, J. Nano-engineering and micromolecular science of polysilsesquioxane materials and their emerging applications. *J. Mater. Chem. A* **2019**, *7*, 21577–21604. [CrossRef]
7. Kaźmierczak, J.; Kuciński, K.; Hreczycho, G. Highly Efficient Catalytic Route for the Synthesis of Functionalized Silsesquioxanes. *Inorg. Chem.* **2017**, *56*, 9337–9342. [CrossRef]
8. Dudziec, B.; Zak, P.; Marciniec, B. Synthetic routes to silsesquioxane-based systems as photoactive materials and their precursors. *Polymers* **2019**, *11*, 504. [CrossRef]
9. Brick, C.M.; Ouchi, Y.; Chujo, Y.; Laine, R.M. Robust Polyaromatic Octasilsesquioxanes from Polybromophenylsilsesquioxanes, Br x OPS, via Suzuki Coupling. *Macromolecules* **2005**, *38*, 4661–4665. [CrossRef]
10. Walczak, M.; Januszewski, R.; Dutkiewicz, M.; Franczyk, A.; Marciniec, B. A facile approach for the synthesis of novel silsesquioxanes with mixed functional groups. *New J. Chem.* **2019**, *43*, 18141–18145. [CrossRef]
11. Żak, P.; Bołt, M.; Grzelak, M.; Rachuta, K.; Dudziec, B.; Januszewski, R.; Marciniec, B.; Marciniak, B. Synthesis and properties of chromophore-functionalized monovinylsilsesquioxane derivatives. *New J. Chem.* **2020**, *44*, 7659–7664. [CrossRef]
12. Grzelak, M.; Frąckowiak, D.; Januszewski, R.; Marciniec, B. Introduction of organogermyl functionalities to cage silsesquioxanes. *Dalt. Trans.* **2020**, *49*, 5055–5063. [CrossRef] [PubMed]
13. Kaźmierczak, J.; Hreczycho, G. Copper(II) triflate-mediated synthesis of functionalized silsesquioxanes via dehydrogenative coupling of POSS silanols with hydrosilanes. *Dalt. Trans.* **2019**, *48*, 6341–6346. [CrossRef] [PubMed]
14. Jung, J.H.; Furgal, J.C.; Goodson, T.; Mizumo, T.; Schwartz, M.; Chou, K.; Laine, R.M. 3-D Molecular Mixtures of Catalytically Functionalized [vinylSiO1.5]10/[vinylSiO1.5]12. Photophysical Characterization of Second Generation Derivatives. *Chem. Mater.* **2012**, *24*, 1883–1895. [CrossRef]
15. Vautravers, N.R.; André, P.; Slawin, A.M.Z.; Cole-Hamilton, D.J. Synthesis and characterization of photoluminescent vinylbiphenyl decorated polyhedral oligomeric silsesquioxanes. *Org. Biomol. Chem.* **2009**, *7*, 717–724. [CrossRef] [PubMed]
16. Żak, P.; Dudziec, B.; Kubicki, M.; Marciniec, B. Silylative Coupling versus Metathesis-Efficient Methods for the Synthesis of Difunctionalized Double-Decker Silsesquioxane Derivatives. *Chem. A Eur. J.* **2014**, *20*, 9387–9393. [CrossRef] [PubMed]

17. Asuncion, M.Z.; Roll, M.F.; Laine, R.M. Octaalkynylsilsesquioxanes, Nano Sea Urchin Molecular Building Blocks for 3-D-Nanostructures. *Macromolecules* **2008**, *41*, 8047–8052. [CrossRef]
18. Araki, H.; Naka, K. Syntheses and properties of dumbbell-shaped POSS derivatives linked by luminescent Pi-conjugated units. *J. Polym. Sci. Part A Polym. Chem.* **2012**, *50*, 4170–4181. [CrossRef]
19. Cho, H.J.; Hwang, D.H.; Lee, J.I.J.; Jung, Y.K.; Park, J.H.; Lee, J.I.J.; Lee, S.K.; Shim, H.K. Electroluminescent polyhedral oligomeric silsesquioxane-based nanoparticle. *Chem. Mater.* **2006**, *18*, 3780–3787. [CrossRef]
20. Guan, J.; Arias, J.J.R.; Tomobe, K.; Ansari, R.; Marques, M. de F.V.; Rebane, A.; Mahbub, S.; Furgal, J.C.; Yodsin, N.; Jungsuttiwong, S.; et al. Unconventional Conjugation via vinylMeSi(O−)2 Siloxane Bridges May Imbue Semiconducting Properties in [vinyl(Me)SiO(PhSiO 1.5) 8 OSi(Me)vinyl-Ar] Double-Decker Copolymers. *ACS Appl. Polym. Mater.* **2020**, *2*, 3894–3907. [CrossRef]
21. Hein, J.E.; Fokin, V.V. Copper-catalyzed azide-alkyne cycloaddition (CuAAC) and beyond: new reactivity of copper(I) acetylides. *Chem. Soc. Rev.* **2010**, *39*, 1302–1315. [CrossRef]
22. Ma, J.; Ding, S. Transition Metal-Catalyzed Cycloaddition of Azides with Internal Alkynes. *Asian J. Org. Chem.* **2002**. [CrossRef]
23. Huo, J.; Lin, C.; Liang, J. A brief minireview of poly-triazole: Alkyne and azide substrate selective, metal-catalyst expansion. *React. Funct. Polym.* **2020**, *152*, 104531. [CrossRef]
24. Rostovtsev, V.V.; Green, L.G.; Fokin, V.V.; Sharpless, K.B. A stepwise huisgen cycloaddition process: Copper(I)-catalyzed regioselective "ligation" of azides and terminal alkynes. *Angew. Chem. Int. Ed.* **2002**, *41*, 2596–2599. [CrossRef]
25. Tornøe, C.W.; Christensen, C.; Meldal, M. Peptidotriazoles on solid phase: [1,2,3]-Triazoles by regiospecific copper(I)-catalyzed 1,3-dipolar cycloadditions of terminal alkynes to azides. *J. Org. Chem.* **2002**, *67*, 3057–3064. [CrossRef] [PubMed]
26. Singh, M.S.; Chowdhury, S.; Koley, S. Advances of azide-alkyne cycloaddition-click chemistry over the recent decade. *Tetrahedron* **2016**, *72*, 5257–5283. [CrossRef]
27. Liang, L.; Astruc, D. The copper(I)-catalyzed alkyne-azide cycloaddition (CuAAC) "click" reaction and its applications. An overview. *Coord. Chem. Rev.* **2011**, *255*, 2933–2945. [CrossRef]
28. Sindhu, K.S.; Anilkumar, G. Recent advances and applications of Glaser coupling employing greener protocols. *RSC Adv.* **2014**, *4*, 27867–27887. [CrossRef]
29. Ervithayasuporn, V.; Abe, J.; Wang, X.; Matsushima, T.; Murata, H.; Kawakami, Y. Synthesis, characterization, and OLED application of oligo(p-phenylene ethynylene)s with polyhedral oligomeric silsesquioxanes (POSS) as pendant groups. *Tetrahedron* **2010**, *66*, 9348–9355. [CrossRef]
30. Wang, X.; Ervithayasuporn, V.; Zhang, Y.; Kawakami, Y. Reversible self-assembly of dendrimer based on polyhedral oligomeric silsesquioxanes (POSS). *Chem. Commun.* **2011**, *47*, 1282–1284. [CrossRef]
31. Han, J.; Zheng, Y.; Zheng, S.; Li, S.; Hu, T.; Tang, A.; Gao, C. Water soluble octa-functionalized POSS: All-click chemistry synthesis and efficient host–guest encapsulation. *Chem. Commun.* **2014**, *50*, 8712–8714. [CrossRef]
32. Zhou, Y.; Yang, G.; Lu, C.; Nie, J.; Chen, Z.; Ren, J. POSS supported C2-symmetric bisprolinamide as a recyclable chiral catalyst for asymmetric Aldol reaction. *Catal. Commun.* **2016**, *75*, 23–27. [CrossRef]
33. Zheng, W.; Lu, C.; Yang, G.; Chen, Z.; Nie, J. POSS supported diarylprolinol silyl ether as an efficient and recyclable organocatalyst for asymmetric Michael addition reactions. *Catal. Commun.* **2015**, *62*, 34–38. [CrossRef]
34. Zhu, Y.K.; Guang, S.Y.; Xu, H.Y. A versatile nanobuilding precursor for the effective architecture of well-defined organic/inorganic hybrid via click chemistry. *Chin. Chem. Lett.* **2012**, *23*, 1095–1098. [CrossRef]
35. Pu, Y.J.; Yuan, H.; Yang, M.; He, B.; Gu, Z.W. Synthesis of peptide dendrimers with polyhedral oligomeric silsesquioxane cores via click chemistry. *Chin. Chem. Lett.* **2013**, *24*, 917–920. [CrossRef]
36. Schäfer, S.; Kickelbick, G. Simple and high yield access to octafunctional azido, amine and urea group bearing cubic spherosilicates. *Dalt. Trans.* **2017**, *46*, 221–226. [CrossRef]
37. Ak, M.; Gacal, B.; Kiskan, B.; Yagci, Y.; Toppare, L. Enhancing electrochromic properties of polypyrrole by silsesquioxane nanocages. *Polymer (Guildf).* **2008**, *49*, 2202–2210. [CrossRef]
38. Ervithayasuporn, V.; Wang, X.; Gacal, B.; Gacal, B.N.; Yagci, Y.; Kawakami, Y. Formation of trimethylsilylated open-cage oligomeric azidophenylsilsesquioxanes. *J. Organomet. Chem.* **2011**, *696*, 2193–2198. [CrossRef]
39. Wei, K.; Wang, L.; Zheng, S. Organic-inorganic copolymers with double-decker silsesquioxane in the main chains by polymerization via click chemistry. *J. Polym. Sci. Part A Polym. Chem.* **2013**, *51*, 4221–4232. [CrossRef]
40. Liu, Y.; Kigure, M.; Koizumi, K.; Takeda, N.; Unno, M.; Ouali, A. Synthesis of Tetrachloro, Tetraiodo, and Tetraazido Double-Decker Siloxanes. *Inorg. Chem.* **2020**, *59*, 15478–15486. [CrossRef]
41. Nowacka, M.; Makowski, T.; Kowalewska, A. Hybrid fluorescent poly(Silsesquioxanes) with amide-and triazole-containing side groups for light harvesting and cation sensing. *Materials* **2020**, *13*, 4491. [CrossRef]
42. Li, Y.; Dong, X.H.; Zou, Y.; Wang, Z.; Yue, K.; Huang, M.; Liu, H.; Feng, X.; Lin, Z.; Zhang, W.; et al. Polyhedral oligomeric silsesquioxane meets "click" chemistry: Rational design and facile preparation of functional hybrid materials. *Polymer (Guildf).* **2017**, *125*, 303–329. [CrossRef]
43. Pérez-Ojeda, M.E.; Trastoy, B.; Lõpez-Arbeloa, Í.; Bañuelos, J.; Costela, Ú.; García-Moreno, I.; Chiara, J.L. Click Assembly of Dye-Functionalized Octasilsesquioxanes for Highly Efficient and Photostable Photonic Systems. *Chem. A Eur. J.* **2011**, *17*, 13258–13268. [CrossRef] [PubMed]

44. Sekiya, R.; Uemura, Y.; Naito, H.; Naka, K.; Haino, T. Chemical Functionalisation and Photoluminescence of Graphene Quantum Dots. *Chem. A Eur. J.* **2016**, *22*, 8198–8206. [CrossRef] [PubMed]
45. Zhao, G.; Zhu, Y.; Guang, S.; Ke, F.; Xu, H. Facile preparation and investigation of the properties of single molecular POSS-based white-light-emitting hybrid materials using click chemistry. *New J. Chem.* **2018**, *42*, 555–563. [CrossRef]
46. Namvari, M.; Du, L.; Stadler, F.J. Graphene oxide-based silsesquioxane-crosslinked networks-synthesis and rheological behavior. *RSC Adv.* **2017**, *7*, 21531–21540. [CrossRef]
47. Gungor, E.; Bilir, C.; Hizal, G.; Tunca, U. Multiarm Star Polymers with POSS at the Periphery EDA. *J. Polym. Sci. Part A Polym. Chem.* **2010**, *48*, 4835–4841. [CrossRef]
48. Arslan, I.; Tasdelen, M.A. POSS-based hybrid thermosets via photoinduced copper-catalyzed azide-alkyne cycloaddition click chemistry. *Des. Monomers Polym.* **2016**, *19*, 155–160. [CrossRef]
49. Niu, M.; Li, T.; Xu, R.; Gu, X.; Yu, D.; Wu, Y. Synthesis of PS-g-POSS hybrid graft copolymer by click coupling via "graft onto" strategy. *J. Appl. Polym. Sci.* **2013**, *129*, 1833–1844. [CrossRef]
50. Uner, A.; Doganci, E.; Tasdelen, M.A. Non-covalent interactions of pyrene end-labeled star poly(ε-caprolactone)s with fullerene. *J. Appl. Polym. Sci.* **2018**, *135*, 1–8. [CrossRef]
51. Bach, L.G.; Islam, M.R.; Nga, T.T.; Binh, M.T.; Hong, S.S.; Gal, Y.S.; Lim, K.T. Chemical modification of polyhedral oligomeric silsesquioxanes by functional polymer via azide-alkyne click reaction. *J. Nanosci. Nanotechnol.* **2013**, *13*, 1970–1973. [CrossRef]
52. Li, L.; Zhang, C.; Zheng, S. Synthesis of POSS-terminated polycyclooctadiene telechelics via ring-opening metathesis polymerization. *J. Polym. Sci. Part A Polym. Chem.* **2017**, *55*, 223–233. [CrossRef]
53. Gauthier, M.; Aridi, T. Synthesis of arborescent polystyrene by "click" grafting. *J. Polym. Sci. Part A Polym. Chem.* **2019**, *57*, 1730–1740. [CrossRef]
54. Chang, P.; Xu, S.; Zhao, B.; Zheng, S. A design of shape memory networks of poly(ε-caprolactone)s via POSS-POSS interactions. *Polym. Adv. Technol.* **2019**, *30*, 713–725. [CrossRef]
55. Wang, Z.; Li, Y.; Dong, X.H.; Yu, X.; Guo, K.; Su, H.; Yue, K.; Wesdemiotis, C.; Cheng, S.Z.D.; Zhang, W. Bin Giant gemini surfactants based on polystyrene-hydrophilic polyhedral oligomeric silsesquioxane shape amphiphiles: Sequential "click" chemistry and solution self-assembly. *Chem. Sci.* **2013**, *4*, 1345–1352. [CrossRef]
56. Yue, K.; Liu, C.; Guo, K.; Yu, X.; Huang, M.; Li, Y.; Wesdemiotis, C. Sequential " Click " Approach to Polyhedral Oligomeric Silsesquioxane-Based Shape Amphiphiles. *Macromolecules* **2012**, *45*, 8126–8134. [CrossRef]
57. Trastoy, B.; Eugenia Pérez-Ojeda, M.; Sastre, R.; Chiara, J.L. Octakis(3-azidopropyl)octasilsesquioxane: A versatile nanobuilding block for the efficient preparation of highly functionalized cube-octameric polyhedral oligosilsesquioxane frameworks through click assembly. *Chem. A Eur. J.* **2010**, *16*, 3833–3841. [CrossRef]
58. Pérez-Ojeda, M.E.; Trastoy, B.; Rol, Á.; Chiara, M.D.; García-Moreno, I.; Chiara, J.L. Controlled click-assembly of well-defined hetero-bifunctional cubic silsesquioxanes and their application in targeted bioimaging. *Chem. A Eur. J.* **2013**, *19*, 6630–6640. [CrossRef]
59. Ervithayasuporn, V.; Kwanplod, K.; Boonmak, J.; Youngme, S.; Sangtrirutnugul, P. Homogeneous and heterogeneous catalysts of organopalladium functionalized-polyhedral oligomeric silsesquioxanes for Suzuki-Miyaura reaction. *J. Catal.* **2015**, *332*, 62–69. [CrossRef]
60. Ervithayasuporn, V.; Wang, X.; Kawakami, Y. Synthesis and characterization of highly pure azido-functionalized polyhedral oligomeric silsesquioxanes (POSS). *Chem. Commun.* **2009**, *60*, 5130–5132. [CrossRef]
61. Vogelsang, D.F.; Dannatt, J.E.; Maleczka, R.E.; Lee, A. Separation of asymmetrically capped double-decker silsesquioxanes mixtures. *Polyhedron* **2018**, *155*, 189–193. [CrossRef]
62. Demko, Z.P.; Sharpless, K.B. Preparation of 5-substituted 1H-tetrazoles from nitriles in water. *J. Org. Chem.* **2001**, *66*, 7945–7950. [CrossRef] [PubMed]
63. Cuevas, F.; Oliva, A.I.; Pericàs, M.A. Direct copper(I)-catalyzed cycloaddition of organic azides with TMS-protected alkynes. *Synlett* **2010**, 1873–1877.
64. Demina, M.M.; Nguyen, T.L.H.; Shaglaeva, N.S.; Mareev, A.V.; Medvedeva, A.S. Highly efficient synthesis of 4-trialkylsilyl(germyl)-1H-1,2,3-triazole-5-carbaldehydes. *Russ. J. Org. Chem.* **2012**, *48*, 1582–1584. [CrossRef]
65. Ziarani, G.M.; Hassanzadeh, Z.; Gholamzadeh, P.; Asadi, S.; Badiei, A. Advances in Click Chemistry for the Silica based Material Construction. *RSC Adv.* **2016**, *6*, 21979–22006. [CrossRef]
66. Walczak, M.; Januszewski, R.; Majchrzak, M.; Kubicki, M.; Dudziec, B.; Marciniec, B. The unusual cis- and trans-architecture of dihydrofunctional double-decker shaped silsesquioxane—Design and construction of its ethyl bridged π-conjugated arene derivatives. *New J. Chem.* **2017**, *41*, 3290–3296. [CrossRef]
67. Gavezzotti, A.; Filippini, G. Geometry of the intermolecular X-H···Y (X, Y = N, O) hydrogen bond and the calibration of empirical hydrogen-bond potentials. *J. Phys. Chem.* **1994**, *98*, 4831–4837. [CrossRef]
68. Gavezzotti, A. Are crystal structures predictable? *Acc. Chem. Res.* **1994**, *27*, 309–314. [CrossRef]
69. Piec, K.; Kostera, S.; Jędrzkiewicz, D.; Ejfler, J.; John, Ł. Mono-substituted amine-oligosilsesquioxanes as functional tools in Pd(II) coordination chemistry: synthesis and properties. *New J. Chem.* **2020**, *44*, 10786–10795. [CrossRef]
70. Marciniec, B.; Kownacki, I.; Franczyk, A.; Kubicki, M. Silsesquioxyl rhodium(i) complexes—Synthesis, structure and catalytic activity. *Dalt. Trans.* **2011**, *40*, 5073–5077. [CrossRef]

71. Au-Yeung, H.-L.; Leung, S.Y.-L.; Yam, V.W.-W. Supramolecular assemblies of dinuclear alkynylplatinum(II) terpyridine complexes with double-decker silsesquioxane nano-cores: the role of isomerism in constructing nano-structures. *Chem. Commun.* **2018**, *54*, 4128–4131. [CrossRef]
72. Kucuk, A.C.; Matsui, J.; Miyashita, T. Synthesis and photochemical response of Ru(II)-coordinated double-decker silsesquioxane. *RSC Adv.* **2018**, *8*, 2148–2156. [CrossRef]
73. Mei, Y.; Loth, M.A.; Payne, M.; Zhang, W.; Smith, J.; Day, C.S.; Parkin, S.R.; Heeney, M.; McCulloch, I.; Anthopoulos, T.D.; et al. High mobility field-effect transistors with versatile processing from a small-molecule organic semiconductor. *Adv. Mater.* **2013**, *25*, 4352–4357. [CrossRef] [PubMed]
74. Dudziec, B.; Rzonsowska, M.; Marciniec, B.; Brząkalski, D.; Woźniak, B. New mono- and diethynylsiloxysilsesquioxanes—Efficient procedures for their synthesis. *Dalt. Trans.* **2014**, *43*, 13201–13207. [CrossRef] [PubMed]
75. Rigaku Oxford Diffraction. *CrysAlisPro v1.171.40.81a*; Rigaku Corporation: Oxford, UK, 2020.
76. Sheldrick, G.M. SHELXT—Integrated space-group and crystal-structure determination. *Acta Crystallogr. Sect. A Found. Crystallogr.* **2015**, *71*, 3–8. [CrossRef] [PubMed]
77. Sheldrick, G.M. Crystal structure refinement with SHELXL. *Acta Crystallogr. Sect. C Struct. Chem.* **2015**, *71*, 3–8. [CrossRef]
78. Sommerfeld, N.S.; Gülzow, J.; Roller, A.; Cseh, K.; Jakupec, M.A.; Grohmann, A.; Galanski, M.; Keppler, B.K. Antiproliferative Copper(II) and Platinum(II) Complexes with Bidentate N,N-Donor Ligands. *Eur. J. Inorg. Chem.* **2017**, *2017*, 3115–3124. [CrossRef]

molecules

MDPI

Article

Limonene Derivative of Spherosilicate as a Polylactide Modifier for Applications in 3D Printing Technology

Dariusz Brząkalski [1], Bogna Sztorch [2], Miłosz Frydrych [1], Daria Pakuła [1], Kamil Dydek [3], Rafał Kozera [4], Anna Boczkowska [3], Bogdan Marciniec [1,2,*] and Robert E. Przekop [2,*]

1 Faculty of Chemistry, Adam Mickiewicz University in Poznań, 8 Uniwersytetu Poznańskiego, 61-614 Poznań, Poland; dariusz.brzakalski@amu.edu.pl (D.B.); frydrych@amu.edu.pl (M.F.); Darpak@amu.edu.pl (D.P.)

2 Centre for Advanced Technologies, Adam Mickiewicz University in Poznań, 10 Uniwersytetu Poznańskiego, 61-614 Poznań, Poland; bogna.sztorch@amu.edu.pl

3 Faculty of Materials Science and Engineering, Warsaw University of Technology, 141 Wołoska, 02-507 Warsaw, Poland; kamil.dydek@pw.edu.pl (K.D.); anna.boczkowska@pw.edu.pl (A.B.)

4 Technology Partners Foundation, 5A Adolfa Pawińskiego, 02-106 Warsaw, Poland; rafal.kozera@technologypartners.pl

* Correspondence: bogdan.marciniec@amu.edu.pl (B.M.); rprzekop@amu.edu.pl (R.E.P.)

Academic Editors: Sławomir Rubinsztajn, Marek Cypryk and Wlodzimierz Stanczyk
Received: 26 November 2020; Accepted: 10 December 2020; Published: 12 December 2020

Abstract: The first report of using limonene derivative of a spherosilicate as a modifier of polylactide used for 3D printing and injection moulding is presented. The paper presents the use of limonene-functionalized spherosilicate derivative as a functional additive. The study compared the material characteristics of polylactide modified with SS-Limonene (0.25–5.0% *w/w*) processed with traditional injection moulding and 3D printing (FFF, FDM). A significant improvement in the processing properties concerning rheology, inter-layer adhesion, and mechanical properties was achieved, which translated into the quality of the print and reduction of waste production. Moreover, the paper describes the elementary stages of thermal transformations of the obtained hybrid systems.

Keywords: spherosilicate; limonene; hydrosilylation; polylactide; 3D printing; FDM; FFF; injection moulding; rheology; thermal analysis

1. Introduction

Three-dimensional (3D) printing is one of the most dynamically developing modern technologies. It was discovered and patented in the 1980s by Charles Hull (SLA technique) and was protected by a patent for 20 years [1]. The development and growth of interest in additive technologies has been going on continuously for about 15 years and is mainly caused by the expiration of patents, but also a decrease in printer prices and their increased availability. At that time, technological solutions became commercially available, initially only to the largest enterprises, now it is increasingly used by the sector of small and medium-sized enterprises, but also by individual consumers. Additive technologies are among those of much demand constituting the main pillar of Industry 4.0 [2]. Initially, 3D printing technology did not arouse much interest, and in the Gartner report from 2012 it was included in the area of the so-called 'Trough of Disillusionment' [3]. However, the aforementioned development of new techniques was so rapid that, in 2013, the position of the 3D printing sector changed dramatically and in 2015 Gartner published a separate report dedicated to it [4]. Additive technologies can be divided into: photochemical DLP [5], SLA [6,7], laser SLS [8], thermal FDM [9] or LOM [10].

One of the most popular additive techniques is FDM (fused deposition modelling). The technology originally developed by Stratasys involves extrusion from a die heated above the polymer melting point and then applying it layer by layer in the direction of the Z axis [11]. The extruded material is in the form of a filament with a diameter of usually 1.75 mm. The advantages of the FDM method are its versatility and accessibility [12], as well as the ease of designing and making a model of any shape and geometry [13]. The disadvantages include, first of all, insufficient mechanical strength in the direction of the Z axis due to the appearance of air gaps between successively superimposed layers [14]. This effect does not occur in the case of other traditional methods, such as injection moulding, where solid objects are obtained.

Such air gaps can contribute to the appearance and initiation of cracks and material design defects. Therefore, attempts are continuously made to reduce poor quality of printed materials and improve the interfacial strength of printed models [15]. In addition, when compared to the traditional method such as injection moulding, 3D printing generates a number of technological problems related to insufficient process speed [16] or product quality (often requiring additional post-processing) [17].

Significant disadvantages hampering the future of 3D printing also include high waste generation, higher than in traditional techniques such as extrusion or injection moulding. An attempt to eliminate unfavourable features by improving the processing device, which is the printer, have encountered significant limitations. In the FDM technique, several thermoplastics are usually used with the greatest emphasis on such polymers/copolymers as: PLA, PA, ABS, TPU. The same types of plastics are mostly used in 3D printing and mature processing techniques. FDM is micro-processing and has significant differences when compared to classic thermoplastics processing techniques, such as lower extruder pressure or smaller cross-sections of the canals in which the molten material flows. Therefore, plastics dedicated for 3D printing should be designed to show the properties addressing these differences (e.g., higher MFI/lower viscosity).

In the FDM technique, polylactide is the most commonly used polymer, mainly due to its ease of its processing, low thermal shrinkage [18] and biodegradability. Degradation rate is low enough to make it resistant to mild weather conditions [19] which is why, next to ABS, it is most often used in medicine [20]. The melting point of PLA is around 150–170 °C and is lower than those of many other popular polymers, it also requires much less energy due to low heat of fusion, so it can be widely used in various processing techniques [21]. One of the disadvantages of PLA limiting its application is low mechanical strength, especially the impact resistance [22]. In order to improve these parameters, structural fillers and plasticizers have been used, e.g., glass and carbon fibres, ceramic or metallic fillers, and glycols [23].

Chemical modifiers that improve the functional properties of composites, such as organic and organosilicon compounds, can also be applied. Limonene (4-isopropenyl-1-methylcyclohexene)—is the main component of oils obtained from waste citrus peels (biomass). It can be obtained by natural and synthetic methods, e.g., using pyrolytic processes [24]. Simple distillation or steam distillation of citrus peels makes it possible to recycle the waste citrus peels from food industry and, at the same time, to obtain pure limonene with a small amount of toxic waste. Annually, these methods produce over 70,000 tons of this compound. In 3D printing technology, it is used as a solvent for support materials made of high impact polystyrene [25]. From the point of view of synthetic applications, limonene should be classified as a green olefin that is subjectable to hydrosilylation reaction [26]. The use of limonene as a building block is one of the many necessary steps in creating a chain of products based on raw materials of natural origin with a lower environmental impact. In its pure form, however, it has a low boiling point for PLA processing (176 °C), which may cause its boiling during processing and, as a result, introducing gases into the polymer, as well as emission of vapours to the environment. Nevertheless, attempts have been made to use it in the processing of PLA, when it showed a plasticizing effect on the polymer matrix, which is a great advantage due to the brittleness of the neat polymer [27].

Polyhedral oligomeric silsesquioxanes are well-known organosilicon compounds, mostly recognized for their high symmetry, excellent solubility, or unparalleled simplicity of synthetic protocols [28]. Among those, a subgroup called spherosilicates may be distinguished, sometimes identified as a different group of compounds, and in such case, both spherosilicates and silsesquioxanes are collectively called cage siloxanes [29]. Due to their high thermal stability and good dispersion properties, they are considered interesting functional additives for polymer processing. In our previous works we have presented approaches towards processing of low-density polyethylene (LDPE) with polyhedral oligomeric silsesquioxanes [30], as well as spherosilicates [31]. These studies allowed determination of the critical level of practical loading in the LDPE matrix, which was much lower than what can be found in numerous literature reports, and was parallelly confirmed by Romo-Uribe et al. [32–34]. One derivative from among the tested ones was found to be particularly interesting, it was SS-Limonene, a product of limonene hydrosilylation with Octahydrospherosilicate. The results suggested its mildly plasticizing effect on the polymer matrix, besides its content improved thermal and mechanical properties of the obtained materials. Therefore, in this study, we decided to assess the applicability of this derivative as a functional and processing additive for PLA for 3D printing.

2. Results and Discussion

2.1. Characterization of SS-Limonene (1,3,5,7,9,11,13,15-octa(Dimethyl((2-(4-methylcyclohex-3-en-1-yl)propyl) silyl)-Pentacyclo [9.5.1.1^{3,9}.1^{5,15}.1^{7,13}]Octasiloxane

SS-Limonene (Figure 1) was prepared according to the synthesis procedure described in Section 3.3 and the reaction completion was determined by FT-IR spectra analysis, the disappearance of the characteristic signals assigned to stretching and bending vibrations of the Si-H group was observed at 2141 and 889 cm^{-1}, respectively). Upon completion of the reaction, the hydrosilylation reached ~99% conversion. The structure and purity of the modifier was confirmed by NMR and MALDI-TOF-MS analyses.

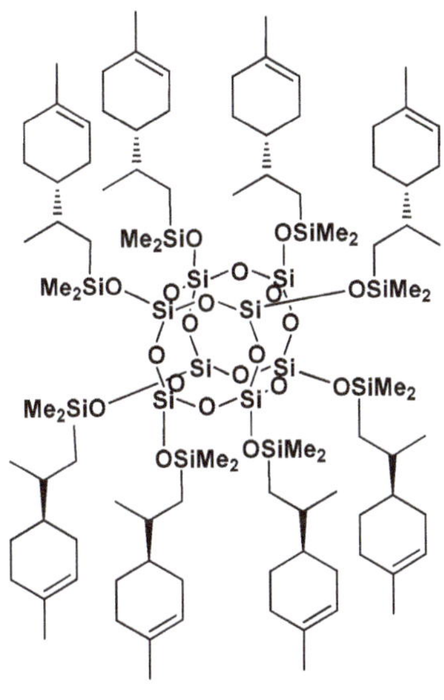

Figure 1. Structure of SS-Limonene.

The purity and chemical structure of the synthesized compound was confirmed by NMR spectroscopy, with the following signals assignment:

^{1}H-NMR (400 MHz, $CDCl_3$): δ (ppm) = 5.36 (s, 8H, ring position 3), 2.00–1.87 (m, 24 H, ring positions 1, 2, 5), 1.78–1.56 (m, 24 H, ring positions 2, 5, =CH- isopropenyl methylidene), 1.63 (s, 24 H,

-CH_3 methyl attached to ring position 4), 1.34–1.18 (m, 16 H, ring position 6), 0.90 (d, 24 H, isopropenyl methyl), 0.75–0.69 (m, 8H, isopropenyl -CH_aH_b-), 0.52–0.44 (m, 8H, isopropenyl -CH_aH_b-), 0.15 (s, 48H, $SiMe_2$); ^{13}C-NMR (101 MHz, $CDCl_3$): δ (ppm) = 133.96, 121.28, 121.26, 41.48, 41.33, 33.08, 32.93, 31.19, 31.12, 28.80, 28.24, 26.71, 25.71, 23.64, 23.63, 22.87, 22.48, 19.52, 19.22, 0.83, 0.72, 0.66; ^{29}Si-NMR (79.5 MHz, $CDCl_3$): δ (ppm) = 12.78 ($SiMe_2$), −109.10 (core). IR (ATR): 2980–2867, 1252, 1169–1069, 869–734, 549. MALDI-TOF-MS: $[M + Na]^+$: 2127.9581 (calc.), 2127.9606 (anal.)

2.2. Density of SS-Limonene/PLA Blends and Mass/Quantity Waste Factor of the Printed Samples

Densities of all samples were measured by the hydrostatic method. Measurements were performed for the samples of 1 cm in length. The average densities of all samples and the base sample of neat PLA were at the same level of around 1.24 g/cm^3. The waste factor of printed samples was also analysed for the bars printed with 100% infill. Data are collected in Table 1. On the basis of the obtained data, the mass and the volume waste factors were determined according to the formula:

$$Wf[m,q] = \frac{mass\ (quantity) of\ corect\ samples \times 100\%}{mass\ (quantity)\ of\ total\ samples} \quad (1)$$

Table 1. Masses and the numbers of models examined in order to establish waste factor values.

Sample	5%	2.5%	1%	0.5%	0.25%
			Total		
Mass [g]	111.11	84.43	91.15	105.52	59.95
Number	38	30	30	34	24
			Correct		
Mass [g]	53.55	51.69	57.44	66.85	44.16
Number	14	12	14	16	11
W_f [m%]	48	61	63	63	74
W_f [q%]	37	40	47	47	46

The calculated waste factors clearly showed that with decreasing content of the SS-Limonene modifier, the amount of 3D printing waste decreases. The issues with printing the samples with high loading of SS-Limonene can be explained on the basis of the additive polymerizing and curing during the polymer processing, which is further explained in Sections 2.4 and 2.5.

2.3. Rheology

The melt flow index (MFI) of pure PLA at 190 °C is 3.7075 g/10 min (Figure 2). For the PLA composites with SS-Limonene the MFI value increased with increasing concentrations of the filler. The composite samples containing 2.5% concentration of the additive are characterized by slightly higher MFR value. On the basis of the data analysis, it can be concluded that SS-Limonene as an additive to the polymer matrix will have a positive effect on its processing.

A capillary rheometer was used to determine the relation between shear rate and viscosity (Figure 3). At low shear rates, the viscosities of all compositions are lower than those of neat PLA, as a consequence of SS-Limonene acting as a plasticizer.

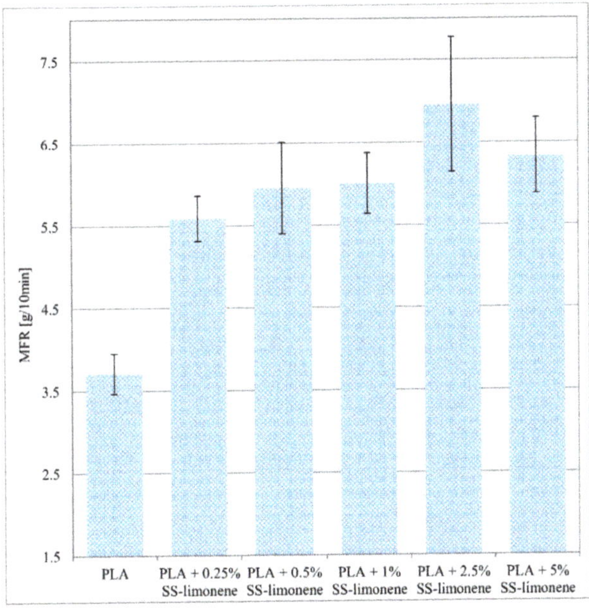

Figure 2. Result of MFR measurements.

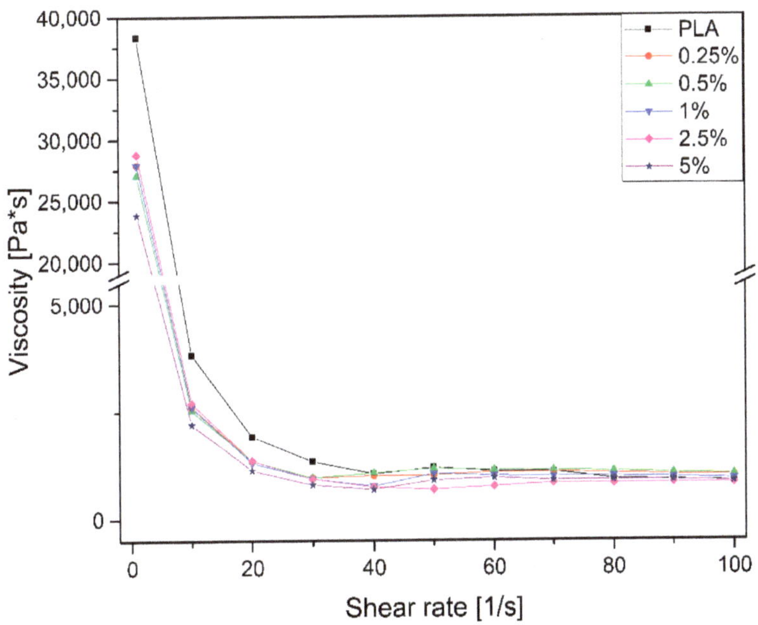

Figure 3. Viscosity measurement results.

2.4. Microscopy

Figures 4 and 5 present the outer and inner texture of 3D-printed (Figure 4) and injection-moulded samples (Figure 5) under a digital light microscope and SEM with an additional EDS analysis (Figure 4e). For the printed samples, different morphologies of the outer and inner regions of the fractured sample (Figure 4b,c) can be seen due to the printing pattern, as the infill has a grid pattern for improved mechanical strength, while the outer layers are made in a rectangular pattern and therefore can be seen as perpendicular to the fracture plane. This is a result of the standard printing process conditions, where the 3D printer outlines the outer shapes of the sample with straight lines first, and then infills the object with a chosen pattern. However, the outline of the outer layers is, in fact, of the highest interest, as it allows observation of the inter-layer interfaces between the individual extrudate strands (and, on this basis, an assessment of the layer-to-layer bonding), as well as the size of the air gaps. In Figure 4b, the inter-layer interfaces between extrudate strands of the neat PLA sample can be clearly seen, which contributes to the low mechanical resistance of the printed PLA objects. On the other hand, for 0.25% SS-Limonene/PLA sample (Figure 4c), this interface is hardly visible, which explains the improved inter-layer adhesion and mechanical rigidity of the samples, which made them mechanically more similar to the injection-moulded ones. However, the formation of air gaps was rather unaffected. Also, on closer inspection, all the samples containing SS-Limonene additive (either 3D printed or injection moulded) contained particles visible both under the light microscope (Figure 4d) and SEM (Figure 4f,g and Figure 5d,e). EDS analysis of silicon allowed identification of the particles to be agglomerates of the polymerized additive (Figure 4e). With increasing loading, more agglomerates of such particles were visible, which contributed to the issues with printing the samples containing high amounts of SS-Limonene. Polymerization of the additive is further discussed in Section 2.5. Additionally, for injection-moulded samples, together with the mentioned particles, small air gaps were observed (Figure 5d,e), which are not present in the sample made from neat PLA (Figure 5b,c). It may be due to difficulties with degassing of the polymer melt, or the generation of gas products either of evaporation or decomposition of SS-Limonene, as the thermogravimetric analysis thereof under the PLA processing temperatures revealed a small mass loss.

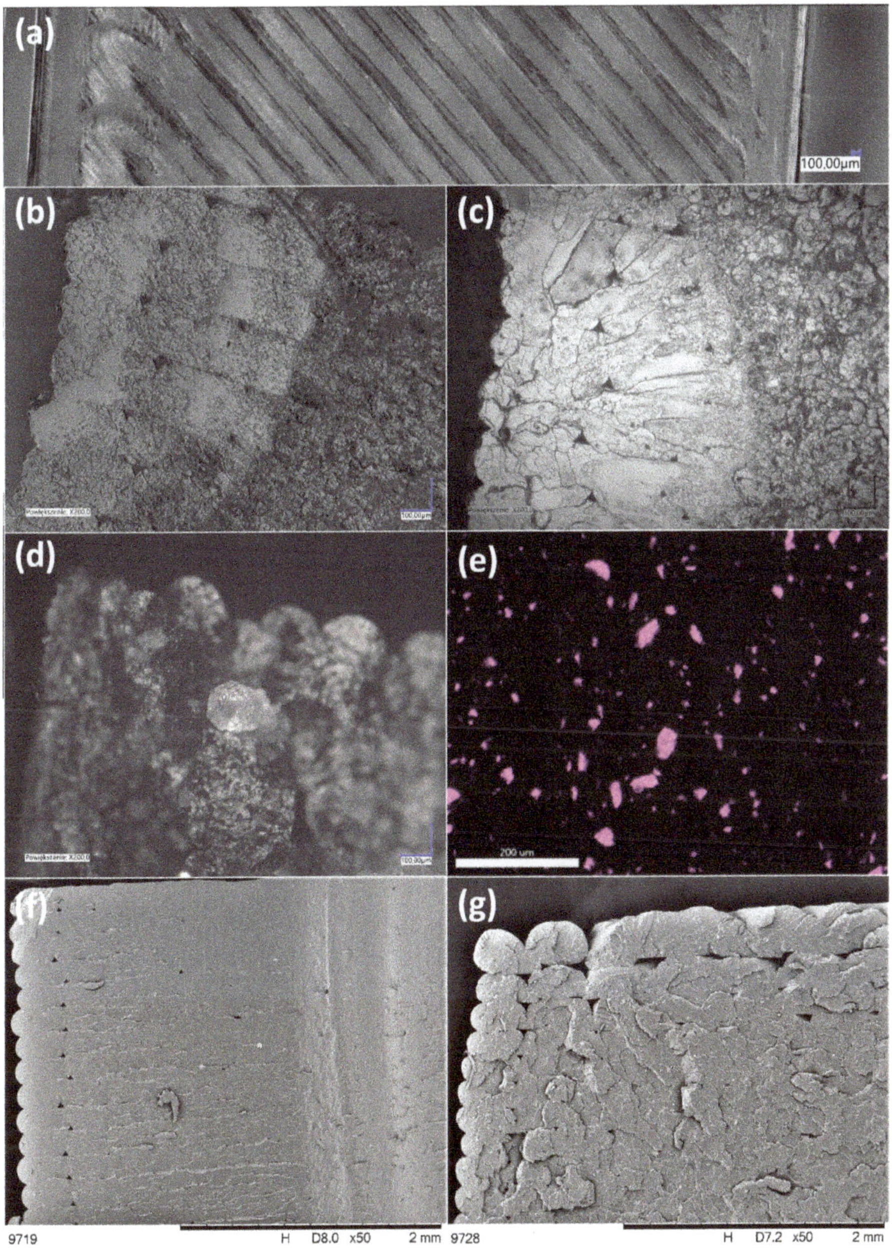

Figure 4. Optical microscopic images (**a**–**d**) and SEM images (**e**–**g**) of printed samples: outer surface (**a**), fractured sample from neat PLA (**b**), fractured sample from 0.25% SS-Limonene/PLA (**c**), a crystal of polymerized SS-Limonene (**d**), silicon EDS image of 0.25% SS-Limonene/PLA fractured sample (**e**), fractured sample from 0.25% SS-Limonene/PLA (**f**), fractured sample from 2.5% SS-Limonene/PLA (**g**).

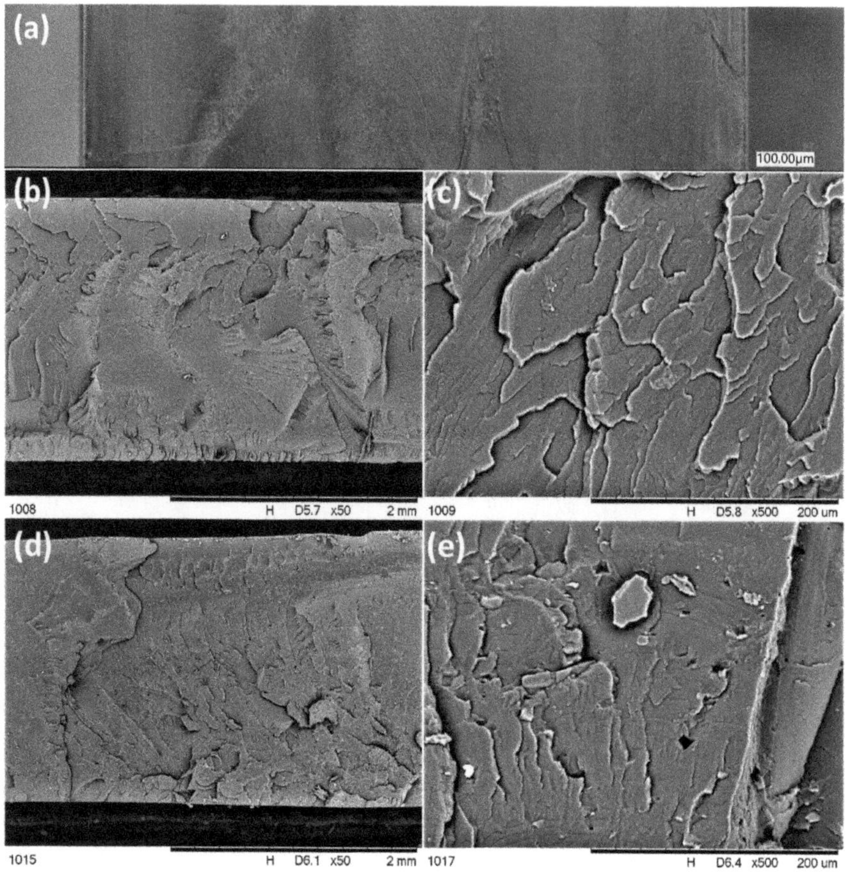

Figure 5. Optical microscopic image (**a**) and SEM images (**b–e**) of injection moulded samples: outer surface (**a**), fractured sample from neat PLA (**b**,**c**), fractured sample from 2.5% SS-Limonene/PLA (**d**,**e**).

2.5. Thermal Analysis Results

Thermal effects for SS-Limonene/PLA compositions were measured by differential scanning calorimetry (DSC) and thermogravimetric analysis (TGA).

DSC analysis was performed to determine the effect of SS-Limonene addition on the glass transition (T_g), crystallization (T_c) and melting (T_m) temperatures of the composites. The graphs are shown in Figure 6. In each case, a large peak of glass transition is noticeable at the first heating cycle, which is related to the low crystallinity of the extruded samples (high proportion of the amorphous phase), due to the rapid cooling of the polymer (in air and in a cooling bath during extrusion). The glass transition temperature (T_g) after the second heating cycle of the test samples containing the organosilicon additive is shifted towards lower values relative to that observed for the neat PLA. This is due to the plasticization of the polymer matrix. Similar conclusions can be drawn in the context of T_{cc}, as the presence of the plasticizing phase increases the freedom of the chains in the amorphous phase, accelerating the initiation of crystallization. The change in T_m and T_{cc} values also shows that SS-Limonene is at least partially mixed with PLA and the interaction of these two components occurs, despite the presence of the polymerized additive phase visible in the microscopic photos (SEM-EDS, light microscopy). On the basis of T_g, the strongest plasticizing effect was observed for the system containing 0.25% of the above-mentioned modifier, but the lowest melting point was observed for the system containing 5% of SS-Limonene. The DSC analysis of the modifier allowed observation of the polymerization of the compound on heating of the sample, which was also confirmed by microscopic analysis as an effect of agglomeration of the polymerized additive in the PLA matrix. A similar effect has been observed in our earlier work, however, most likely due to lower processing temperatures of LDPE, the agglomeration was far less severe [31].

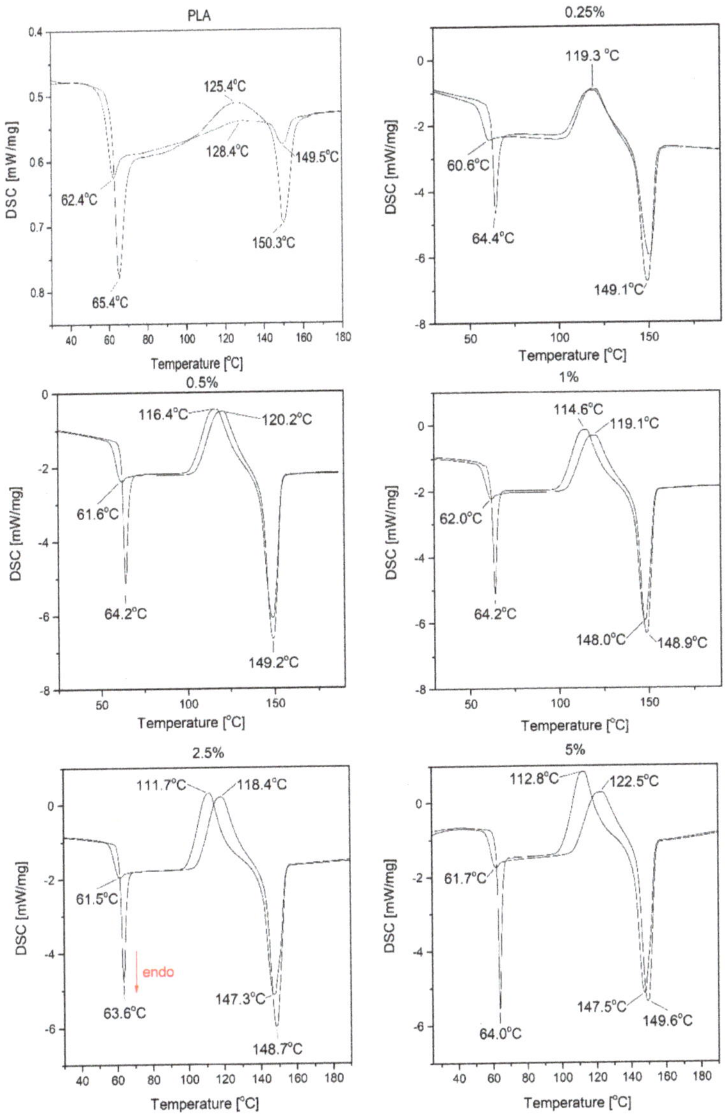

Figure 6. DSC curves recorded for samples of SS-Limonene/PLA composite.

All results of the TGA, DTG and c-DTA measurements are presented in Figure 7. The parameters determined, including the temperature of 1% mass loss, onset, and temperature at the maximum rate of mass loss are collected in Table 2. The process of thermal decomposition of samples was carried out in both nitrogen and air atmosphere. It should be remembered that thermal changes in thermoplastic systems at temperatures above the melting point of the matrix should not be defined as the 'thermal resistance' or 'thermal strength' of the composite, but refer to the influence of the applied additives on the thermal stability of the polymer system in molten state and the interaction between the system components in the melt. Based on the observation of the complete thermal analysis, i.e., the derivatographic curve, DTG and c-DTA results obtained both in air and nitrogen atmosphere, three stages of thermal transformation can be distinguished. The first step, observed in both air and nitrogen atmosphere, at 140–160 °C is related to the polymerization of SS-Limonene. The second stage, taking place in a nitrogen atmosphere at a temperature above 365.7 °C, is related to the cracking of PLA chains with the simultaneous endothermic distillation of mers, including SS-Limonene degradation products and lactides. According to the c-DTA curve determined in the air atmosphere, the endothermic distillation process overlaps with the exothermic oxidation process of the cracking products. In the last, third stage, in the air atmosphere, one more signal is observed at a temperature of 550–700 °C, which characterizes the process of burning coke, originating mainly from the organosilicon derivative. Stage 1 refers to the functional properties of the composite material, while stages 2 and 3 describe the irreversible processes of thermal decomposition of a mixture of molten PLA and an organosilicon derivative.

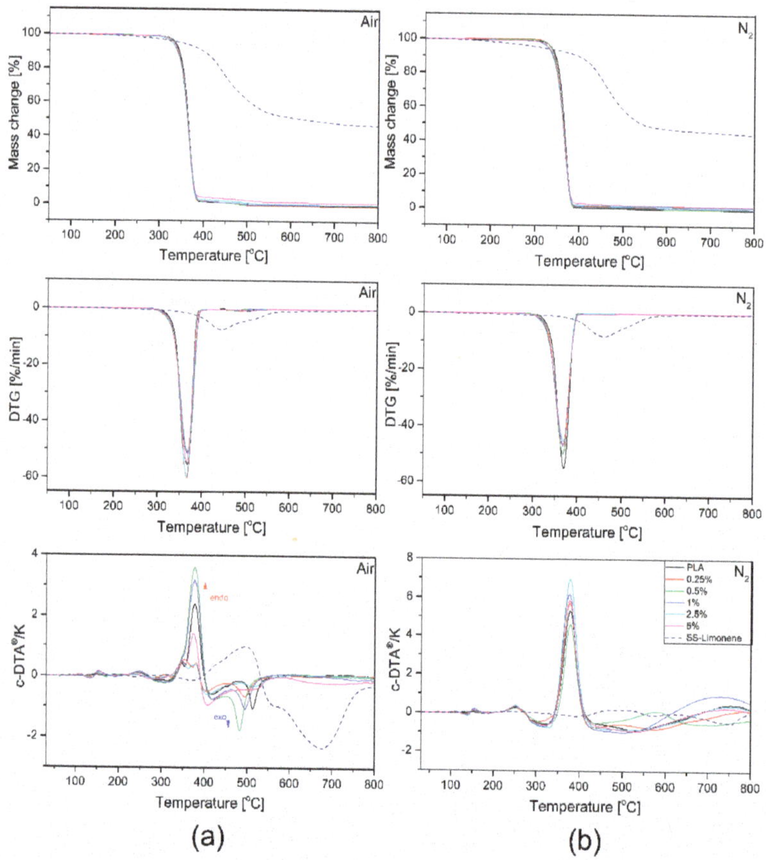

Figure 7. TGA, c-DTA and DTG curves of SS-Limonene/PLA composite in (**a**) air and (**b**) nitrogen atmosphere.

Table 2. Results of thermogravimetric analysis.

Conditions	1% Mass Loss [°C]		Onset Temperature [°C]		Temperature at the Maximum Rate of Mass Loss [°C]		Dry Mass Left [%]	
	N_2	Air	N_2	Air	N_2	Air	N_2	Air
SS-limonene	166.7	221.0	410.6	397.8	458.6	442.3	44.26	46.99
Neat PLA	304.8	307.9	351.1	348.3	370.1	367.0	0.00	0
PLA + 0.25% SS-Limonene	318.4	287.3	346.3	353.4	367.0	366.5	1.0	0
PLA + 0.5% SS-Limonene	316.3	308.2	348.3	346.1	368.9	366.8	0	0
PLA + 1% SS-Limonene	291.4	307.8	341.4	345.2	368.3	366.3	0.04	0
PLA + 2.5% SS-Limonene	294.9	287.9	342.4	348.5	368.4	365.5	1.03	0
PLA + 5% SS-Limonene	301.4	302.5	343.6	345.9	369.0	366.5	1.54	1.12

2.6. Contact Angle Measurements

The contact angle measurements were performed for SS-Limonene/PLA composites obtained by two different methods—FDM and injection moulding. For the measurements of 3D-printed samples, the samples were placed with the layer-by-layer deposition plane oriented parallelly to the plane of the goniometer stage. The contact angle of the neat PLA was 81.4° for the printed samples and 83.6° for the samples obtained in the injection process, in both cases the surface before modification showed hydrophilic properties. Modification of PLA with SS-Limonene increased the hydrophobicity of all the tested samples (see Table 3). For the printed samples, regardless of the modifier concentration, a hydrophobic surface effect was obtained (the value of the contact angle increased to above 90°). For the samples obtained by injection moulding, the increase in the value of the contact angle was insignificant and it remained at a similar level (max. by 4.3°). The difference in the degree of hydrophobicity of printed and injected samples is due to their surface structure and microstructure. 3D printing allows obtaining microstructures and surface roughness, which results in higher values of the contact angle. This thesis was confirmed by microscopic observations (see Section 2.4). On the other hand, injection moulding produces more smooth surfaces (if no modification of the mould surface is applied), which reduces the microstructure effect on the surface and therefore almost no effect of the organosilicon additive can be observed.

Table 3. Water contact angle [°].

	Injection Moulding [°]	3D Printing [°]
Neat PLA	83.6	81.4
PLA + 0.25% SS-Limonene	87.9	92.4
PLA + 0.5% SS-Limonene	87.8	95.4
PLA + 1% SS-Limonene	84.7	92.3
PLA + 2.5% SS-Limonene	85.7	95.1
PLA + 5% SS-Limonene	85.6	97.5

2.7. Mechanical Properties

2.7.1. Tensile Strength and Flexural Strength

Mechanical tests were carried out for the modified samples obtained by both 3D printing and traditional injection moulding. For the 3D-printed samples, the tensile load was applied parallelly to the plane of layer-by-layer deposition. The basic tensile strength values for neat PLA are 36.5 MPa for the samples obtained by the FDM method and 72.6 MPa for the samples obtained by the injection moulding (Figure 8). This difference is due to the technique of producing the dumbbells for tests. Lower values of mechanical parameters of printed samples are mainly related to low inter-layer adhesion between the extrudate strands and the presence of air gaps between the applied layers. In the caseof injection moulding, the materials are more solid with little to no structural defects. The addition of the SS-Limonene modifier increased the tensile strength of the printed samples, which brought them closer to the injection moulded ones. The highest value was observed for PLA with the content of 0.25% of the modifier, this value decreased with increasing concentration. This is mainly due to the improved flow of the polymer as shown in the capillary rheometry analysis of the samples (Section 3.1), as well as improved inter-layer adhesion (Section 2.4), which resulted in increased material consistency and improved fusing of the print paths (extrudate strands). For all the tested samples, high values of standard deviation were obtained, which is a characteristic feature of FDM printed objects due to the mentioned structural inconsistencies.

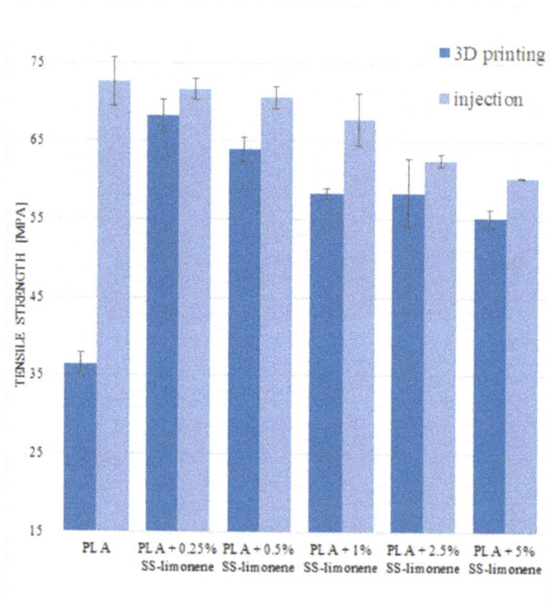

Figure 8. Tensile strength of SS-Limonene/PLA in 3D printing and injection moulding.

For the samples obtained by injection moulding, the tensile stress values are the highest for pure PLA and decrease with increasing loading of the modifier. This result can be explained by two reasons: one is the plasticizing effect of SS-Limonene, and the other is the presence of discontinuities in the polymer phase introduced together with the additive, that is the polymerized SS-Limonene agglomerates and additional air micropockets, as confirmed by the microscopy.

The elongation at maximum load for neat PLA samples obtained by 3D printing and injection moulding is characterized by similar values (2.43% and 2.29%, Figure 9). The addition of SS-Limonene to the samples obtained by the FDM method significantly improves the plasticity of the material. Higher elongation values in the case of modified samples indicate increased "mobility" of the polymer phase as a result of plasticization by SS-Limonene [35]. The plasticizer isolates the chains and spherulites of macromolecules, reducing the interaction between them. The highest value was obtained for PLA + 0.25% SS-Limonene, which was 4.21%. At higher loadings, the effect of the additive polymerization takes over and decreases the plasticizing effect. In the case of injection-moulded samples, the addition of the modifier gave negligible effects.

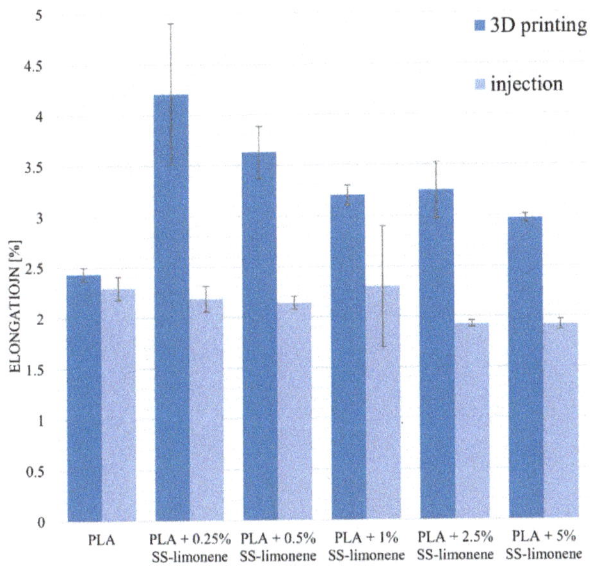

Figure 9. Elongation at maximum load of SS-Limonene/PLA in 3D printing and injection moulding.

2.7.2. Bending Tests

The basic flexural parameters of the samples were determined. For the 3D-printed samples, the bending load was applied perpendicularly to the plane of layer-by-layer deposition. The basic values of flexural strength for pure PLA are 77.98 MPa for the samples obtained by the FDM method and 99.98 MPa for the injection-moulded samples (Figure 10). The flexural strength is also significantly lower for the neat PLA obtained by 3D printing—this is due to the presence of air gaps and limited inter-layer adhesion, the structural discontinuities acting as stress concentration points.

In the case of the 3D printing technique, the addition of the SS-Limonene modifier significantly increased the value of the flexural strength. The samples obtained by printing with the modifier showed similar mechanical properties to those mould-injected. The highest values were obtained for the systems containing 0.25% and 0.5% of SS-Limonene (respectively 97.61 MPa; 98.46 MPa). Higher concentrations of the modifier caused a slight decrease in the strength values in relation to the 0.25% and 0.5% systems, but they were still higher than for the neat PLA samples.

The samples obtained by injection moulding technique were characterized by high values of flexural strength ranging from 88.46 MPa to 100.30 MPa (5% SS-Limonene/PLA and 1% SS-Limonene/PLA, respectively).

The values of flexural modulus for both types of samples were basically unchanged regardless of the additive loading (Figure 11).

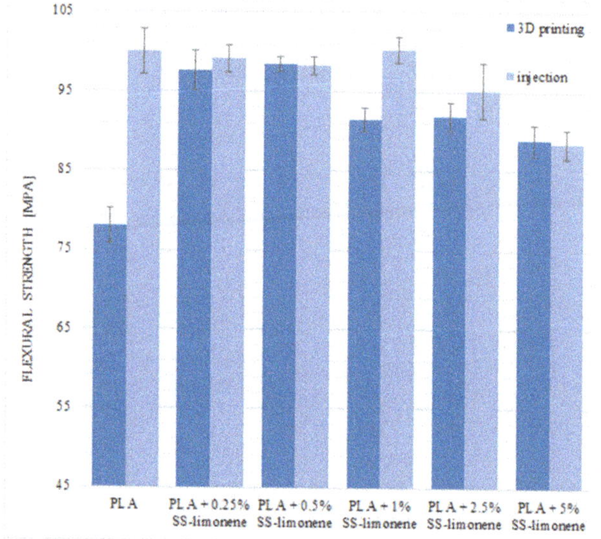

Figure 10. Flexural strength of SS-Limonene/PLA in 3D printing and injection moulding.

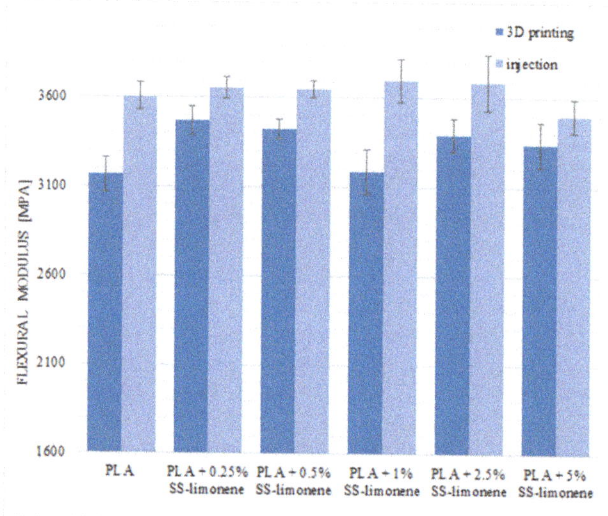

Figure 11. Flexural modulus of SS-Limonene/PLA in 3D printing and injection moulding.

2.7.3. Impact Strength and Hardness

Impact strength of the obtained composite samples was determined. For the 3D-printed samples, the impact direction was perpendicular to the plane of layer-by-layer deposition. Impact resistance tests confirmed the beneficial effect of SS-Limonene (especially at lower loadings) on the tested samples, regardless of the processing technique used (Figure 12). The modifier, as in the case of the previously discussed mechanical tests, acts as a plasticizer, the brittleness of the polymer is reduced therefore, the obtained composite is able to absorb more energy during an impact. High standard deviations are characteristic of the measurement method. The downward tendency along with the increase in the modifier content indicates a limited dispersion of the modifier in the polymer matrix and compatibility of the system components. Hardness, in Shore D scale, was determined to be virtually unaffected by the additive and on the level of 82–84 for all the samples regardless of the SS-Limonene loading or the processing method.

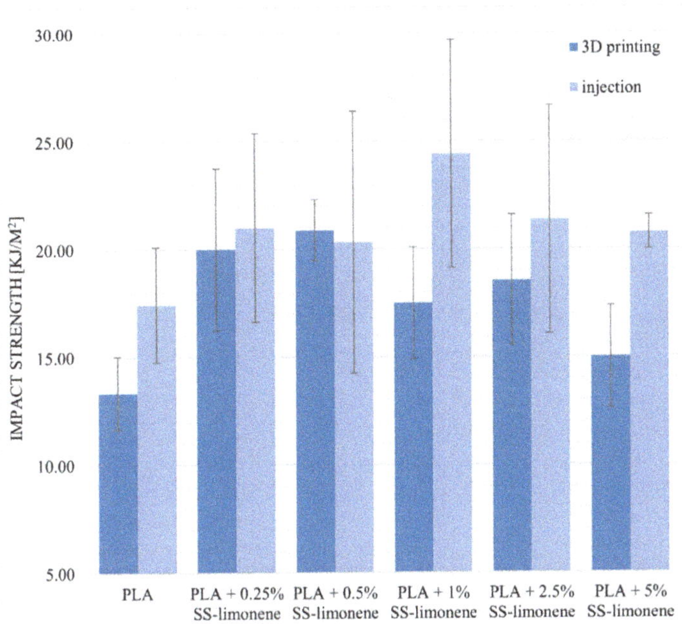

Figure 12. Impact strength of SS-Limonene/PLA in 3D printing and injection moulding.

3. Materials and Methods

3.1. Materials

Polylactide (PLA) type Ingeo 2003D was purchased from NatureWorks (Minnetonka, MN, USA). The chemicals were purchased from the following sources: Tetraethoxysilane (TEOS) from Unisil (Poland), chlorodimethylsilane, tetramethylammonium hydroxide (TMAH) 25% methanol solution from ABCR, (R)-(+)-limonene, toluene, chloroform-d, Karstedt's catalyst xylene solution from Aldrich, P_2O_5 from Avantor Performance Materials Poland S.A. Toluene was degassed and dried by distilling it from P_2O_5 under argon atmosphere.

3.2. Analyses

^1H, ^{13}C, and ^{29}Si Nuclear Magnetic Resonance (NMR) spectra were recorded at 25 °C on a Bruker Ascend 400 and Ultra Shield 300 spectrometers using $CDCl_3$ as a solvent. Chemical shifts are reported in ppm with reference to the residual solvent ($CHCl_3$) peaks for ^1H and ^{13}C.

MALDI-TOF mass spectra were recorded on a UltrafleXtreme mass spectrometer (Bruker Daltonics), equipped with a SmartBeam II laser (355 nm) in the 500–4000 *m*/*z* range. 2,5-Dihydroxybenzoic acid (DHB, Bruker Daltonics, Bremen, Germany) served as matrix. Mass spectra were measured in reflection mode. The data were analysed using the software provided with the Ultraflex instrument—FlexAnalysis (version 3.4).

Fourier Transform-Infrared (FT-IR) spectra were recorded on a Nicolet iS 50 Fourier transform spectrophotometer (Thermo Fisher Scientific) equipped with a diamond ATR unit with a resolution of 0.09 cm^{-1}.

Contact angle analyses were performed by the sessile drop technique at room temperature and atmospheric pressure, with a Krüss DSA100 goniometer (Hamburg, Germany). Three independent measurements were performed for each sample, each with a 5 μL water drop, and the obtained results were averaged to reduce the impact of surface nonuniformity.

Thermogravimetry (TG) was performed using a NETZSCH 209 F1 Libra gravimetric analyser (Selb, Germany). Samples of 5 ± 0.2 mg were cut from each granulate and placed in Al_2O_3 crucibles. Measurements were conducted under nitrogen (flow of 20 mL/min) in the range of 30–800 °C and a 20 °C/min heating rate. Differential scanning calorimetry (DSC) was performed using a NETZSCH 204 F1 Phoenix calorimeter Samples of 6 ± 0.2 mg were cut from each granulate and placed in an aluminium crucible with a punctured lid. The measurements were performed under nitrogen in the temperature range of −20–290 °C and at a 20 °C/min heating rate, and T_g was measured from the second heating cycle.

The effect of the modifier addition on the mass flow rate (MFR) was also determined. The measurements were made using a Instron plastometer (Norwood, MA, USA), model Ceast MF20 according to the applicable standard ISO 1133. The measurement temperature was 190 ± 0.5 °C, while the piston loading was 2.16 kg.

For flexural and tensile strength tests, the obtained materials were printed into type 1B dumbbell specimens in accordance with EN ISO 527:2012 and EN ISO 178:2006. Tests of the obtained specimens were performed on a universal testing machine INSTRON 5969 with a maximum load force of 50 kN. The traverse speed for tensile strength measurements was set at 2 mm/min, and for flexural strength was also set at 2 mm/min. Charpy impact test (with no notch) was performed on a Instron Ceast 9050 impact-machine according to ISO 179−1. For all the series, 6 measurements were performed.

Hardness of the composite samples was tested by the Shore method using a durometer Bareiss Prüfgerätebau GmbH.

A scanning electron microscope (SEM 3000, Hitachi, Japan) was used to analyse the microstructure and quality of the produced composite samples after 3D printing and injection moulding. Additionally, the effect of SS-Limonene addition on the microstructure of composite materials was investigated. Before the measurement, samples' cross-sections were coated with a thin layer of Au-Pd. The applied voltage for SEM observations was 15 kV.

Surface structure and breakthroughs were analysed under Digital Light Microscope Keyence VHX 7000 with 100× to 1000× VH-Z100T lens (Osaka, Japan). All of the pictures were recorded with a VHX 7020 camera.

3.3. *The Procedure for Synthesis of Octaspherosilicate Limonene Derivative (SS-Limonene)*

Octahydrospherosilicate was prepared according to a literature procedure [36]. The hydrosilylation reaction was performed accordingly to a previous report [31].

In a typical procedure, a 500 mL three-neck, round-bottom flask was charged with 25 g of Octahydrospherosilicate, 250 mL of toluene and 26.77 g of limonene, and a magnetic stirring bar was added. A thermometer and condenser equipped with an argon inlet and oil bubbler were attached, the flask placed in a heating mantle and the system was purged with argon. The reaction mixture was set on 110 °C and before reaching boiling, 25 μL of Karstedt's catalyst solution was added, which resulted in quick increase of temperature and the system starting to reflux. The reaction mixture was kept at reflux and samples were taken for FT-IR control until full Si-H group consumption was observed. Then, the solvent was evaporated under vacuum to dryness to obtain an analytically pure sample.

3.4. *Fabrication of Filaments*

3.4.1. Preparation of Granulates

The polymer and the filler were homogenized using a laboratory two-roll mill ZAMAK MERCATOR WG 150/280. A portion of 500 g PLA Ingeo™ 2003 D was mixed with SS-Limonene, until the final concentration of the additive of 5.0% *w/w*. The mixing was performed at the rolls temperature of 200 °C for 15 min., getting to full homogeneity of the concentrates. Masterbatch was granulated by a grinding mill WANNER C17.26 sv. The granulates were diluted with pure PLA up to the final filler concentrations of 0.25, 0.5, 1.0, 2.5 and 5.0% *w/w* upon extrusion moulding of a stream with consequent cold granulation on the twin-screw extrusion setup line HAAKE Rheomex OS, and then dried for 24 h at 40 °C.

3.4.2. Extrusion of Filaments

The granulates obtained as above were used for moulding of filaments of 1.75 mm in diameter by a single-screw extrusion setup HAAKE Rheomex OS.

3.4.3. 3D Printing (FDM)

Using a 3D printer FlashForge Finder two types of samples were printed by FDM: oars and bars, according to PN-EN-ISO 527-2. Parameters of printing are given in Table 4.

Table 4. Process parameters for sample printing.

Layer height	0.18 mm
Top layer height	0.27 mm
Shells	2
Top and bottom layers number	3
Bottom layers number	3
Infill density	100%
Infill pattern	Grid
Printing speed	60 mm/s
Idle speed	80 mm/s
Extruder temp.	220 °C

3.5. *Injection Moulding*

To compare the mechanical properties of composite materials made by 3D printing, the samples were produced by the injection moulding method. Specimens for static tensile, three-point bending, and impact tests were in accordance with the dimensions of the following standards: PN-EN ISO 527, PN-EN ISO 178, and PN-EN ISO 179, respectively. HAAKE Minijet Pro Piston Injection Moulding System (ThermoScientific, Bremen, Germany) equipped with a set of moulds was used to produce test samples. Parameters of the injection process are presented in Table 5

Table 5. Injection process parameters.

Cylinder temperature	225 °C
Mould temperature	45 °C
Injection pressure	750 bar
Injection time	10 s
Post-injection pressure	700 bar
Post-injection time	10 s

4. Conclusions

The obtained results confirm the effect of the addition of SS-Limonene on the improvement of rheological and mechanical properties of printed composites based on the PLA matrix. It is especially important for the FDM-based method for objects manufacturing, as usually objects prepared by such show much lower parameters than their injection-moulded counterparts, which can be explained by poor inter-layer adhesion and the presence of air gaps. On the basis of the obtained data, it can be concluded that the addition of a functionalized spherosilicate significantly improves such parameters as tensile strength, bending strength and impact strength of the samples obtained by 3D printing. SS-Limonene acted as a plasticizing additive for PLA. Additionally, the presence of SS-Limonene was found to increase hydrophobicity of the obtained composites. The addition of this modifier also facilitates the printing process itself, contributes to the improvement of rheological properties (reduces viscosity, increases MFR) and reduces production waste. It should be noted that in the conditions of processing the additive was found to undergo polymerization, leading to its secondary agglomeration observed under the microscope on increasing loading. It seems that the nature of the observed phenomena is complex, and the additive most likely exists in the system in two forms, namely a well-dispersed one and the agglomerated one, as suggested by previous research and the analysis of thermal data. The behaviour of the obtained composites is based on two types of interactions—between PLA and well-dispersed phase, as well as between PLA and the polymerized, agglomerated phase. Nevertheless, this complex interaction scheme should not be considered undesirable as it leads to a final improvement of the printing material system.

Author Contributions: Conceptualization, R.E.P.; Data curation, D.B. and B.S.; Formal analysis, M.F., D.P., and K.D.; Funding acquisition, D.B., B.S., and B.M.; Investigation, D.B. and B.S.; Methodology, B.S. and R.E.P.; Project administration, B.S.; Resources, D.B. and B.S.; Software, M.F., D.P., and K.D.; Supervision, B.M. and R.P.; Validation, R.K. and R.E.P.; Visualization, M.F., D.P., and K.D.; Writing—original draft, D.B., B.S., and R.E.P.; Writing—review & editing, R.K., A.B., and B.M. All authors have read and agreed to the published version of the manuscript.

Funding: This research was funded by the National Centre for Research and Development, Poland, grant number LIDER/01/0001/L-10/18/NCBR/2019, and the National Science Centre, Poland, grant number UMO-2018/29/N/ST5/00868.

Conflicts of Interest: The authors declare no conflict of interest.

References

1. Gibson, I.; Rosen, D.W.; Stucker, B. *Additive Manufacturing Technologies*; Springer: Berlin/Heidelberg, Germany, 2010; pp. 299–332.
2. Fawcett, S.E.; Waller, M.A. Supply chain game changers—Mega, nano, and virtual trends—And forces that impede supply chain design (i.e.; building a winning team). *J. Bus. Logist.* **2014**, *35*, 157–164. [CrossRef]
3. LeHong, H.; Fenn, J. Hype Cycle for Emerging Technologies. 2012. Available online: https://www.gartner.com/en/documents/2100915/hype-cycle-for-emerging-technologies-2012 (accessed on 26 July 2019).
4. Basiliere, P.; Shandler, M. Hype Cycle for 3D Printing. 2015. Available online: https://www.gartner.com/en/documents/3100228/hype-cycle-for-3d-printing-2015 (accessed on 26 July 2019).
5. Varghese, G.; Moral, M.; Castro-García, M.; López-López, J.J.; Marín-Rueda, J.R.; Yagüe-Alcaraz, V.; Hernandez-Afonso, L.; Ruiz-Morales, J.C.; Canales-Vázquez, J. Fabrication and characterisation of ceramics via low-cost DLP 3D printing. *Boletín de la Soc. Española de Cerámica y Vidr.* **2018**, *57*, 9–18. [CrossRef]
6. Schmidleithner, C.; Kalaskar, D.M. Stereolithography. In *3D Printing*; Cvetković, D., Ed.; IntechOpen: Hamilton, NJ, USA, 2018; pp. 1–22.
7. Ma, X.L. Research on Application of SLA Technology in the 3D Printing Technology. *Appl. Mech. Mater.* **2013**, *401*, 938–941. [CrossRef]
8. Fina, F.; Goyanes, A.; Gaisford, S.; Basit, A.W. Selective laser sintering (SLS) 3D printing of medicines. *Int. J. Pharm.* **2017**, *529*, 285–293. [CrossRef]
9. Liu, Z.; Wang, Y.; Wu, B.; Cui, C.; Guo, Y.; Yan, C. A critical review of fused deposition modeling 3D printing technology in manufacturing polylactic acid parts. *Int. J. Adv. Manuf. Technol.* **2019**, *102*, 2877–2889. [CrossRef]
10. Kalmanovich, G. "Curved-Layer" Laminated Object Manufacturing. Available online: https://repositories.lib.utexas.edu/handle/2152/70244 (accessed on 22 October 2020).

11. Grimm, T. *Fused Deposition Modeling: A Technology Evaluation;* T.A. Grimm & Associates Inc.: Atlanta, GA, USA, 2002.
12. Rayna, T.; Striukova, L. From rapid prototyping to home fabrication: How 3D printing is changing business model innovation. *Technol. Forecast. Soc. Chang.* **2016**, *102*, 214–224. [CrossRef]
13. A Third Industrial Revolution. Available online: https://www.economist.com/special-report/2012/04/21/a-third-industrial-revolution (accessed on 22 October 2020).
14. Sood, A.K.; Ohdar, R.K.; Mahapatra, S.S. Parametric appraisal of mechanical property of fused deposition modelling processed parts. *Mater. Des.* **2010**, *31*, 287–295. [CrossRef]
15. Wang, L.; Gardner, D.J. Contribution of printing parameters to the interfacial strength of polylactic acid (PLA) in material extrusion additive manufacturing. *Prog. Addit. Manuf.* **2018**, *3*, 165–171. [CrossRef]
16. Franchetti, M.; Kress, C. An economic analysis comparing the cost feasibility of replacing injection molding processes with emerging additive manufacturing techniques. *Int. J. Adv. Manuf. Technol.* **2017**, *88*, 2573–2579. [CrossRef]
17. Ngo, T.; Kashani, A.; Imbalzano, G.; Nguyen, K.; Hui, D. Additive manufacturing (3D printing): A review of materials, methods, applications and challenges. *Compos. Part B Eng.* **2018**, *143*, 172–196. [CrossRef]
18. Bai, J.; Goodridge, R.D.; Hague, R.J.M.; Okamoto, M. Processing and characterization of a polylactic acid/nanoclay composite for laser sintering. *Polym. Compos.* **2015**, *38*, 2570–2576. [CrossRef]
19. Mochizuki, M. Textile applications. In *Poly (Lactic Acid) Synthesis, Structures, Properties, Processing, and Applications*, 1st ed.; Auras, R.A., Lim, L.T., Selke, S.E., Tsuji, H., Eds.; John Wiley & Sons: Hoboken, NJ, USA, 2010; Volume 10, pp. 469–476. ISBN 978-0-470-29366-9.
20. Mardis, J.N. Emerging Technology and Applications of 3D Printing in the Medical Field. *Mo. Med.* **2018**, *115*, 368–373. [PubMed]
21. King, D.L.; Babasola, A.; Rozario, J.; Pearce, J.M. Mobile Open-Source Solar-Powered 3-D Printers for Distributed Manufacturing in Off-Grid Communities. *Chall. Sustain.* **2014**, *2*, 18–27. [CrossRef]
22. Rasal, R.M.; Hirt, D.E. Toughness decrease of PLA-PHBHHx blend films upon surface-confined photopolymerization. *J. Biomed. Mater. Res. A* **2019**, *88A*, 1079–1086. [CrossRef]
23. Liu, Z.; Lei, Q.; Xing, S. Mechanical characteristics of wood, ceramic, metal and carbon fiber-based PLA composites fabricated by FDM. *Environ. Sci. Technol.* **2019**, *53*, 1527–1535. [CrossRef]
24. Miranda, R.; Bustos-Martinez, D.; Blanco, C.S.; Villarreal, M.H.G.; Rodriguez, M.E.; Cantu, J. Pyrolysis of sweet orange (Citrus sinensis) dry peel. *Anal. Appl. Pyrolysis* **2009**, *86*, 245. [CrossRef]
25. Chae, M.P.; Rozen, W.M.; McMenamin, P.G.; Findlay, M.W.; Spychal, R.T.; Hunter-Smith, D.J. Emerging applications of bedside 3D printing in plastic surgery. *Front. Surg.* **2015**, *2*, 25. [CrossRef]
26. Drozdov, F.V.; Cherkaev, G.V.; Muzafarov, A.M. Synthesis of new functional siloxane derivatives of limonene. Part I: Combination of hydrosilylation and hydrothiolation reactions. *J. Organomet. Chem.* **2019**, *880*, 293–299. [CrossRef]
27. Arrieta, M.P.; López, J.; Ferrándiz, S.; Peltzer, M.A. Characterization of PLA-limonene blends for food packaging applications. *Polym. Test.* **2013**, *32*, 760–768. [CrossRef]
28. Cordes, D.B.; Lickiss, P.D.; Rataboul, F. Recent Developments in the Chemistry of Cubic Polyhedral Oligosilsesquioxanes. *Chem. Rev.* **2010**, *110*, 2081–2173. [CrossRef]
29. Liu, H.; Kondo, S.; Takeda, N.; Unno, M. Synthesis of Octacarboxy Spherosilicate. *J. Am. Chem. Soc.* **2008**, *130*, 10074–10075. [CrossRef] [PubMed]
30. Brząkalski, D.; Przekop, R.E.; Sztorch, B.; Jakubowska, P.; Jałbrzykowski, M.; Marciniec, B. Silsesquioxane Derivatives as Functional Additives for Preparation of Polyethylene-Based Composites: A Case of Trisilanol Melt-Condensation. *Polymers* **2020**, *12*, 2269. [CrossRef] [PubMed]
31. Brząkalski, D.; Przekop, R.; Dobrosielska, M.; Sztorch, B.; Marciniak, P.; Marciniec, B. Highly bulky spherosilicates as functional additives for polyethylene processing—Influence on mechanical and thermal properties. *Polym. Compos.* **2020**, *41*, 3389–3402. [CrossRef]
32. Romo-Uribe, A.; Lichtenhan, J.D. Melt extrusion and blow molding parts-per-million POSS interspersed the macromolecular network and simultaneously enhanced thermomechanical and barrier properties of polyolefin films. *Polym. Eng. Sci.* **2020**. [CrossRef]
33. Romo-Uribe, A.; Lichtenhan, J.D. Polyhedral Oligomeric Silsesquioxane Induced Thermomechanical Reinforcement in Polymer Films Using Only Parts-per-Million Content. *Macromol. Mater. Eng.* **2020**, *305*, 2000354. [CrossRef]

34. Romo-Uribe, A.; Lichtenhan, J.; Reyes-Mayer, A.; Calixto-Rodriguez, M.; Sarmiento-Bustos, E.; Yañez-Lino, M. Parts-per-million polyhedral oligomeric silsesquioxane loading induced mechanical reinforcement in polyethylene nanocomposites. When small and well-dispersed yields big. *Polym. Adv. Technol.* **2020**, *31*, 2453–2465. [CrossRef]
35. Fortunati, E.; Luzi, F.; Puglia, D.; Dominici, F.; Santulli, C.; Kenny, J.M.; Torre, L. Investigation of thermo-mechanical, chemical and degradative properties of PLA-limonene films reinforced with cellulose nanocrystals extracted from Phormium tenax leaves. *Eur. Polym. J.* **2014**, *56*, 77–91. [CrossRef]
36. Filho, N.L.D.; De Aquino, H.A.; Pires, G.; Caetano, L. Relationship between the Dielectric and Mechanical Properties and the Ratio of Epoxy Resin to Hardener of the Hybrid Thermosetting Polymers. *J. Braz. Chem. Soc.* **2006**, *17*, 533–541. [CrossRef]

Sample Availability: Samples of the compounds are available from the authors.

molecules

MDPI

Article

An Approach to the Use of Glycol Alkoxysilane–Polysaccharide Hybrids in the Conservation of Historical Building Stones

Miguel Meléndez-Zamudio [†], Ileana Bravo-Flores, Eulalia Ramírez-Oliva, Antonio Guerra-Contreras, Gilberto Álvarez-Guzmán, Ramón Zárraga-Nuñez, Antonio Villegas, Merced Martínez-Rosales and Jorge Cervantes *

División de Ciencias Naturales y Exactas, Departamento de Química, Universidad de Guanajuato, Guanajuato 36000, Mexico; melendem@mcmaster.ca (M.M.-Z.); ie.bravoflores@ugto.mx (I.B.-F.); eraoliva@ugto.mx (E.R.-O.); ja.guerra@ugto.mx (A.G.-C.); g.alvarez@ugto.mx (G.Á.-G.); rzarraga@ugto.mx (R.Z.-N.); vigaja@ugto.mx (A.V.); mercedj@ugto.mx (M.M.-R.)

* Correspondence: jauregi@ugto.mx; Tel.: +52-473-732-0006

† Current address: Department of Chemistry & Chemical Biology, McMaster University, Hamilton, ON 45015, Canada.

Abstract: Stone consolidants have been widely used to protect historical monuments. Consolidants and hydrophobic formulations based on the use of tetraethoxysilane (TEOS) and alkylalkoxysilanes as precursors have been widely applied, despite their lack of solubility in water and requirement to be applied in organic media. In the search for a "greener" alternative based on silicon that has potential use in this field, the use of tetrakis(2-hydroxyethyl)silane (THEOS) and tris(2-hydroxyethyl)methyl silane (MeTHEOS) as precursors, due their high water solubility and stability, is proposed in this paper. It is already known that THEOS and MeTHEOS possess remarkable compatibility with different natural polysaccharides. The investigated approach uses the water-soluble silanes THEOS–chitosan and MeTHEOS–chitosan as a basis for obtaining hybrid consolidants and hydrophobic formulations for the conservation of siliceous and calcareous stones. In the case of calcareous systems, their incompatibility with alkoxysilanes is known and is expected to be solved by the developed hybrid consolidant. Their application in the conservation of building stones from historical and archeological sites from Guanajuato, México was studied. The evaluation of the consolidant and hydrophobic formulation treatment was mainly conducted by determining the mechanical properties and contact angle measurements with satisfactory results in terms of the performance and compatibility with the studied stones.

Keywords: THEOS; MeTHEOS; chitosan; stone conservation; siliceous and calcareous stones

check for updates

Citation: Meléndez-Zamudio, M.; Bravo-Flores, I.; Ramírez-Oliva, E.; Guerra-Contreras, A.; Álvarez-Guzmán, G.; Zárraga-Nuñez, R.; Villegas, A.; Martínez-Rosales, M.; Cervantes, J. An Approach to the Use of Glycol Alkoxysilane–Polysaccharide Hybrids in the Conservation of Historical Building Stones. *Molecules* **2021**, *26*, 938. https://doi.org/10.3390/molecules26040938

Academic Editors: Sławomir Rubinsztajn, Marek Cypryk and Wlodzimierz Stanczyk

Received: 16 December 2020
Accepted: 5 February 2021
Published: 10 February 2021

1. Introduction

The conservation of stones in historical buildings around the world is receiving a growing amount of interest due to the importance of preserving historical memory for new generations. Different authors have highlighted the importance of stone conservation using varied scientific approaches due to the complexity and diversity of problems that require solutions. Stone conservation is an area where chemistry, physics, material science, biology, history, architecture, archeology, restoration, geology, etc. find a common point of confluence [1–4].

In an overview of research on stone conservation, the process referred to as consolidation is considered to be an active conservation process "where stone is severely weakened by decay, and some form of consolidation may be necessary to restore some strength. Ideally, one might hope to make the stone at least as strong as it was originally, so it might resist further decay" [5]. The same overview points out that "some of the causes of stone decay are sudden and rapid in their effect. Those toward the latter part of the following list are slow and more insidious: earthquake, fire, flood, terrorism, vandalism, neglect, tourism, previous treatments, wind, rain, frost, temperature fluctuations, chemical attack, salt growth, pollution, biodeterioration, intrinsic factors, and so on" [5].

Stone consolidants have been widely used to protect historical monuments. Commercial products containing alkoxysilanes, such as tetraethoxysilane (TEOS), are commonly applied as consolidants for decayed natural stones. They are applied as low viscosity monomers or dimers in solutions that may include water, ethanol, or other organic solvents (generally MEK and acetone). An organometallic catalyst such as di-n-butyltin dilaurate (DBTL) is commonly used to increase the rate of polycondensation. Hypothetically, the fluid deeply infiltrates the porous network of the stone and, through a sol–gel process, forms a silica gel that works as a new cement for the matrix. This is done to re-establish the cohesion between loosened material grains as well as to restore the original mechanical resistance in the decayed material. The use of alkoxysilane-based products as stone consolidants to conserve decayed quartz-bearing rocks, such as sandstone, or siliceous natural materials in general, has been a common practice for several decades [6], although the first patent was approved much earlier in 1924 [7].

However, in the case of calcareous stones, their non-compatibility with alkoxysilanes has been established, so several approaches have been investigated [8–10].

In a recent review regarding the application of alkoxysilanes in the field of stone conservation, it is pointed out that conventional alkoxysilane-based consolidants have several drawbacks that hinder their successful application in carbonate stones. In terms of addressing these issues, the modification of alkoxysilanes has resulted in some improvements. The cracking tendency of alkoxysilanes has been solved by the introduction of elastic segments, surfactant templates, and nanoparticle loading, in addition to other elements. Nevertheless, there is still much room for improvement. The complexity of sol–gel chemistry and the conceptual incompatibility between alkoxysilane-based consolidants and carbonate minerals do not allow for successful prediction of the consolidation behavior. The proposed consolidants often render the treated stone hydrophobic, which is interesting if the objective is to obtain both a hydrophobic and consolidation treatment, although the risk of incompatibility exists. Despite the drawbacks, the study of new alkoxysilane-based consolidants that can provide multifunctional (consolidant, self-cleaning, or biocide) properties is still being conducted. The development of such properties again makes alkoxysilanes a feasible alternative for consolidating carbonate stones in important stone heritage if their compatibility and durability are effective. Of course, any consolidation action should not hinder or limit future interventions. In addition, due to the complexity of the interaction between the alkoxysilanes and the stone, a more detailed study of the mechanism is needed [11].

A recent investigation addresses the impact and importance of the carbonate medium in the sol–gel processes of stone consolidation based on alkoxysilanes and the possible detrimental effects in practical applications, highlighting the need for the design and development of new alkoxysilane-based consolidants to consider this effect [12].

The interest of our research group in the use of alkoxysilanes in the conservation of historical building stone has led to investigations on many topics, such as chemical, physical, geological, and mineralogical characterization of historical building stones; considering decay and biodecay evaluation; and the synthesis and application of consolidant and hydrophobic formulations based on alkoxysilanes in siliceous stones, with the aim of solving reported problems in the performance of some of the commercial formulations through the development of hybrid systems based on TEOS/SiO_2St/PDMS-OH. Most of our work in the field has focused on siliceous stones [13–21].

Tetraethoxysilane (TEOS) and alkylalkoxysilanes are widely used as precursors in consolidant and hydrophobic formulations. Due to their lack of solubility in water, the formulation must be applied in organic solvents (VOCS), promoting a clear disadvantage in terms of "green chemistry" in this field. In the search for "greener" silicon derivatives with potential use as consolidants, the use of tetrakis(2-hydroxyethyl)silane (THEOS) and tris(2-hydroxyethyl)methyl silane (MeTHEOS) as precursors is suggested because of their remarkable water solubility. Initial studies of the THEOS precursor were conducted by Mehrotra and Narain [22] and, subsequently, introduced by Hoffmann and his group through the transesterification reaction of TEOS and ethylene glycol. The complete characterization of properties achieved by Hoffman demonstrated that THEOS possesses high solubility and stability in water and, as a result, the use of typical organic solvents is suppressed [23].

On the other hand, in 2004, Shchipunov, Tatyana, and Karpenko reported the compatibility of THEOS with different natural polysaccharides, including chitosan. As observed by the authors, polysaccharides worked as accelerators, catalysts, and templates for the silica generated in situ by the sol–gel process; modification of the synthesis conditions led to different properties and structures suitable for obtaining monolithic hybrid materials. For the THEOS–chitosan system, a transparent hybrid material was obtained, and no phase separation or syneresis was detected [24–27].

An important aspect of the hybrid THEOS–chitosan was described when it was shown that it accelerated, catalyzed, and served as a template for silica generated in situ by sol–gel, thus manipulating its synthesis as well as the properties and structure of the produced monolithic hybrid materials [24]. The potential application in drug encapsulation is an example of a very important and actual field of investigation where these biocompatible hybrid materials can have an impact [28]. The hydrolysis of THEOS produces silicic acid. As a result of the complete compatibility with chitosan, the presence of sol–gel in solution results in gelation of the non-gellable chitosan as well as other polysaccharides. For example, the gelation by mineralization of carbohydrate macromolecules strengthened them and provided cross-linking [25]. The gelation time required for the sol–gel transition and the dynamic rheological properties of the resultant gel matrix could be modulated by the amount of added THEOS. The hybrid material has found application in electrochemical biosensors [27,29,30] and is able to obtain stable and intact thin films or monolithic hybrid gels [31].

On the other hand, it is important to consider that the hybrid has the presence of primary amines and hydroxyl groups in the chitosan structure, and such functionalities have a key effect on the biopolymer solubility. Additionally, they can act as reactive sites for the covalent interaction via condensation in the chitosan–siloxane network. In this direction, we have recently reported a detailed study to elucidate the covalent interaction between reactive silanols (from the complete hydrolysis of THEOS and MeTHEOS) and chitosan. The results suggest that the condensation site forming the silyl–ether bond is located at C6 of the chitosan structure [32].

The characteristics developed by the hybrid system THEOS–chitosan enable us to suggest a new application in the area of conservation of historical stone buildings that is based on the preparation of hybrid consolidant and hydrophobic formulations that can be applied in the conservation of stones with a siliceous and calcareous composition; THEOS–chitosan can be used in the consolidant formulation and MeTHEOS in the hydrophobic formulation. Bearing in mind that a great number of important historical monuments in different parts of the world have siliceous or calcareous stones as building materials, some advantages can be considered in the introduction of glycol alkoxysilane–chitosan in the field of stone conservation, i.e., that the "green" formulations based on water-soluble silanes do not need organic solvents for their application, these hybrid formulations can potentially be used as an alternative to solve the problem of compatibility between alkoxysilanes and calcareous stone, and the synergy originating from the alkoxysilane–chitosan interaction in terms of the film formation capacity of chitosan, as well as the antimicrobial activity, suggests the possibility to avoid the biodegradation of stones by many organisms. It is important to consider that the material aptitude to biological colonization by certain organisms is called the bioreceptivity and is dependent on different environmental factors, such as the pH, water availability, climate exposure, mineral composition, porosity, permeability, and nutrient sources [5,33].

The antimicrobial activity of chitosan has been widely studied; for example, the antibiofilm properties of chitosan-coated surfaces, where chitosan offers a flexible, biocompatible platform for designing coatings to protect surfaces from infection [34]. The state of the art of antimicrobial chitosan and chitosan derivatives and the effects of structural modifications on the activity and toxicity have been reviewed toward improving the understanding of the bioactivity and to develop more useful chitosan conjugates [35]. On the other hand, it has been pointed out that chitosan and its derivatives can be called environmental purification functional materials as they can effectively control the growth and reproduction of hazardous bacteria and also control toxic pollutants [36]. The antibacterial activity of chitosan extracted from a pen shell against both Gram-positive and Gram-negative bacteria was recently reported [37]. Furthermore, chitosan and its derivatives have been studied due to their antimicrobial properties in the context of preventing and treating denture stomatitis, which can be caused by fungi [38].

2. Results and Discussion

The results and discussion section considers two main aspects: synthesis and characterization of the formulations THEOS–chitosan and MeTHEOS–chitosan and their application in siliceous and calcareous building materials. Films of the silane–chitosan hybrids were obtained and characterized using different methods. Such characterization was useful for revealing the film behavior inside or on the stones as a result of the consolidation or hydrophobic treatment.

The precursors of the hybrid formulations—THEOS and MeTHEOS—were obtained according to the most reported and used method (the transesterification reaction of TEOS or MeTEOS with ethylene glycol) [22,23]. At the reaction conditions used (140 °C and 15 h of reaction time), high yields are obtained (83% to 92%). Nevertheless, the morphology of the product obtained from different syntheses under the same conditions is diverse (liquid, translucid viscous, or gel), without altering the water solubility. The analysis of the transesterification reaction products (THEOS and MeTHEOS) was conducted by ^{29}Si-NMR in solution (DMSO-d6), in respect of the TEOS and MeTEOS spectra as the reference (singlet at −82.5 ppm and −43 ppm, respectively). As demonstrated in previous studies and by recent observations, THEOS does not exist as a single molecule, as can be seen from the diverse morphology that the isolated product exhibits, and various silicon species appear in the Q unit region (−81.7, −82.3, −83.6, and −88.6 ppm) for THEOS; for MeTHEOS (−41.7, −44.1, −49.8, −52.3, and −58.2 ppm), different peaks appear in the T unit region [32,39,40].

2.1. Synthesis and Characterization of Silane–Chitosan Hybrid Films

A very wide range of tests of THEOS–chitosan and MeTHEOS–chitosan solutions were prepared, using different proportions of the reagents, in order to find out the most appropriate formulations to be applied in stone treatment. The selection of formulations to be used was carried out through observation of the film characteristics obtained in terms of the flexibility, homogeneity, transparency, and resistance to syneresis, where the excellent capability of chitosan as a film formulation was a key aspect in terms of the concentration of chitosan used. No phase separation was observed. The extensive testing revealed that a selection of formulations with similar physical characteristics were obtained, offering the possibility to apply them in experiments with different goals (see Materials and Methods). For example, the selected formulation for the films characterized by Fourier-Transform Infrared Spectroscopy-Attenuated Total Reflectance (FTIR-ATR), Scanning Electron Microscopy-Energy Dispersed X-ray Spectroscopy (SEM)–EDX, thermal stability, and solid-state Nuclear Magnetic Resonance (NMR) analysis was based on 10 mL of an aqueous solution of chitosan (0.5% in acetic acid at 1% with 72% deacetylation) and 0.5 g of THEOS. Some variations in the formulation composition were used in several analyses, such as hardness and contact angle determinations, in order to study the effect of the silane–chitosan ratio (formulations referred to as 1, 2, and 3 in Materials and Methods).

2.1.1. FTIR

The FTIR spectra of THEOS–chitosan and MeTHEOS–chitosan films present similar characteristics in terms of their band frequencies for –OH (3500–3200 cm^{-1}), –CH, –CH_2, and –CH_3 (3000–2840 and 1460–1350 cm^{-1}), amide N–C=O (1655 cm^{-1} and 1580 cm^{-1}), amine –NH_2 (1320 cm^{-1}), and the Si–O–Si network (1110–1000 cm^{-1}). The main difference is due the presence of the bands at 1270 and 770 cm^{-1} (Si–CH_3 and –C–Si–O, respectively) in the MeTHEOS–chitosan film (Figure S1).

2.1.2. SEM–EDX

SEM–EDX of THEOS–chitosan and MeTHEOS–chitosan films are presented in Figure 1. The 10,000× amplification illustrates the films' characteristics, showing that they are flexible, thin, and transparent and have no evident imperfections. The elements observed according to EDX analysis are carbon, nitrogen, oxygen, and silicon, in accordance with the hybrid composition.

2.1.3. Thermal Stability of the Hybrids

The thermal stability of the hybrid films was studied. The films were exposed to different temperatures, from room temperature to 700 °C, and FTIR–ATR spectra were collected to determine any structural changes. Comparative spectra obtained at 25 °C and 350 °C are presented (Figure 2). As can be observed, the films are thermally stable until 350 °C. In the case of MeTHEOS–chitosan, the fragment –SiCH_3 is removed around 500 °C.

2.1.4. Structural Characterization by Solid State NMR

A more detailed structural characterization of the hybrid THEOS–chitosan and MeTHEOS–chitosan films was conducted by solid state ^{13}C-NMR (CPMAS) and ^{29}Si-NMR (MAS) and reported recently [32]. The structural analysis of chitosan by ^{13}C-NMR was taken as a reference to point out that the C6, bonded to a terminal hydroxy group, is suggested to be the condensation site of THEOS and MeTHEOS, once the C6 is the most sterically favored for this purpose. The most evident change in the chemical environment corresponds to the region of the chemical shift of C6 (60.71 ppm), exhibiting 2.24 ppm of difference with respect to the C6 of the chitosan film (58.47 ppm). The ^{29}Si MAS and CPMAS spectra of the THEOS and MeTHEOS–chitosan films were collected for complementary structural analysis (Figures S2–S5).

2.2. THEOS–Chitosan and MeTHEOS–Chitosan Formulations Applied to Siliceous and Calcareous Historical Building Materials

In consideration of the expected compatibility with both siliceous and calcareous materials, the application of the hybrid silane–chitosan in the field of the conservation of historical building stones is suggested. A key aspect is the water base application, which forgoes the use of organic solvents. In order to obtain data regarding the performance of the consolidant and hydrophobic formulations, different characterization methods were used, including FTIR, SEM–EDX, hardness determination, water absorption, and measurement of the contact angle (dynamic and static) and surface free energy. Different formulations in terms of the concentration of silane–chitosan were applied in the three selected materials (caliche, Sostenes, and Compañía). The different percentages of chitosan deacetylation were considered a variable to take into consideration in different determinations (i.e., hardness and contact angle measurements).

The FTIR–ATR spectra of the three studied materials were obtained. Figure 3 illustrates the caliche without treatment (Figure 3a) and after the consolidation (Figure 3b) and hydrophobic treatment (Figure 3c). The IR spectrum shown is typical of a calcite (1550–1350 and 872 cm^{-1} corresponding to stretching and bending vibrations, respectively). The bands at 1052 cm^{-1} and 715 cm^{-1} can be assigned to Si–O stretching due to the presence of a small concentration of silicates. After the consolidation treatment, the most important modifications in the spectrum are an increase in the intensity and broadness of the band at 1096 (Si–O–Si) and 722 cm^{-1}, associated with the siloxane network; in the case of the sample treated with the hydrophobic formulation, a new small band at 1270 cm^{-1} appears, corresponding to the Si–CH_3 fragment.

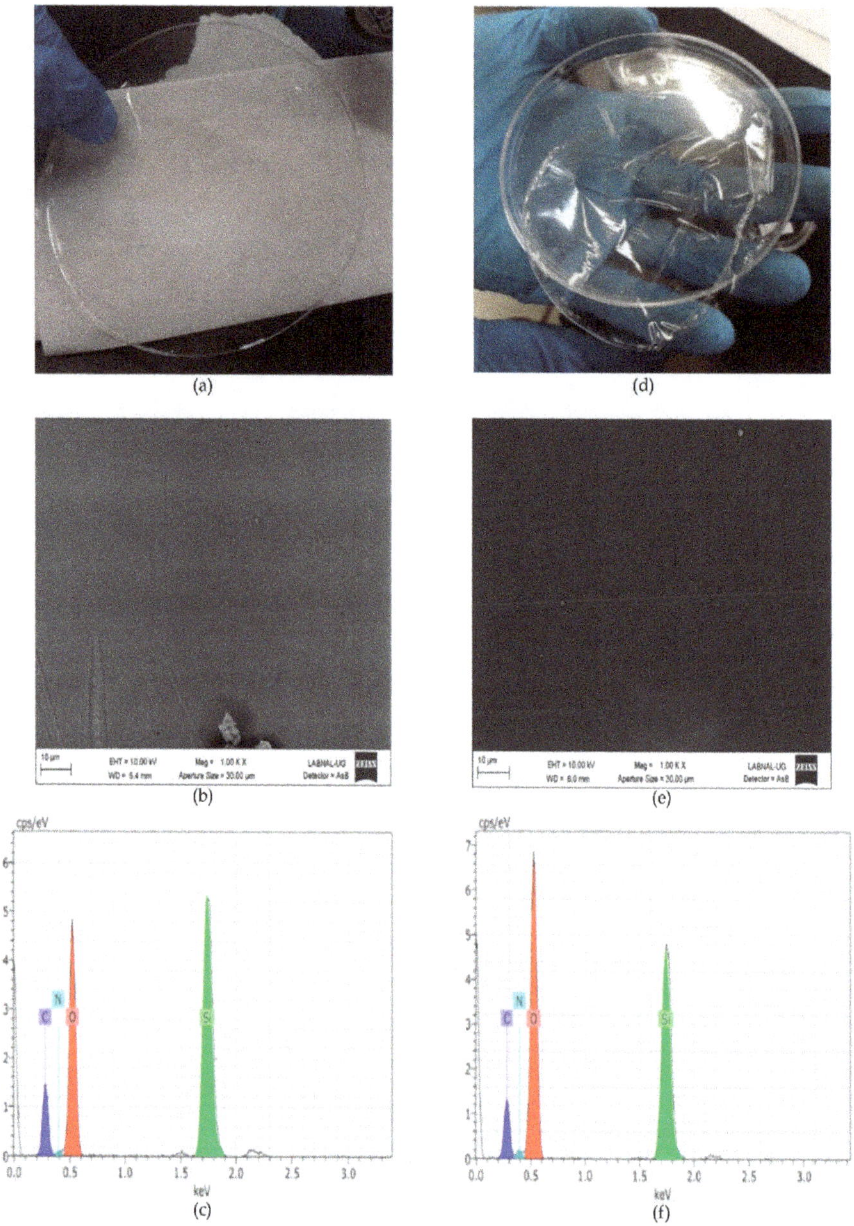

Figure 1. Characterization of the hybrid films. (**a**) MeTHEOS–chitosan film and Scanning electron microscopy (SEM) of the (**b**) tris(2-hydroxyethyl)methyl silane (MeTHEOS)–chitosan film; (**c**) EDX of the MeTHEOS–chitosan film; (**d**) THEOS–chitosan film and SEM of the THEOS–chitosan film (**e**); and (**f**) EDX of the THEOS–chitosan film.

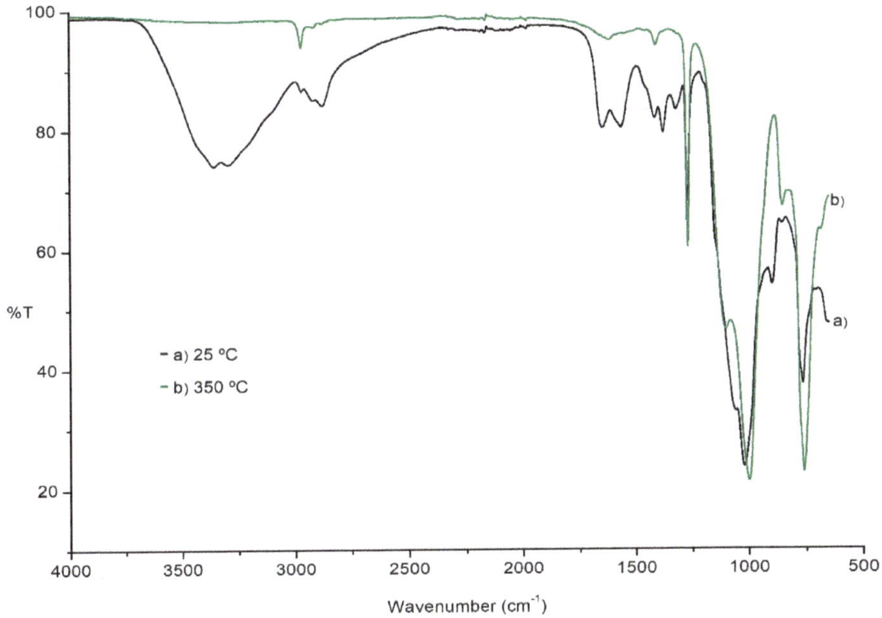

Figure 2. FTIR–ATR for a film of MeTHEOS–chitosan at (**a**) 25 °C and (**b**) 350 °C.

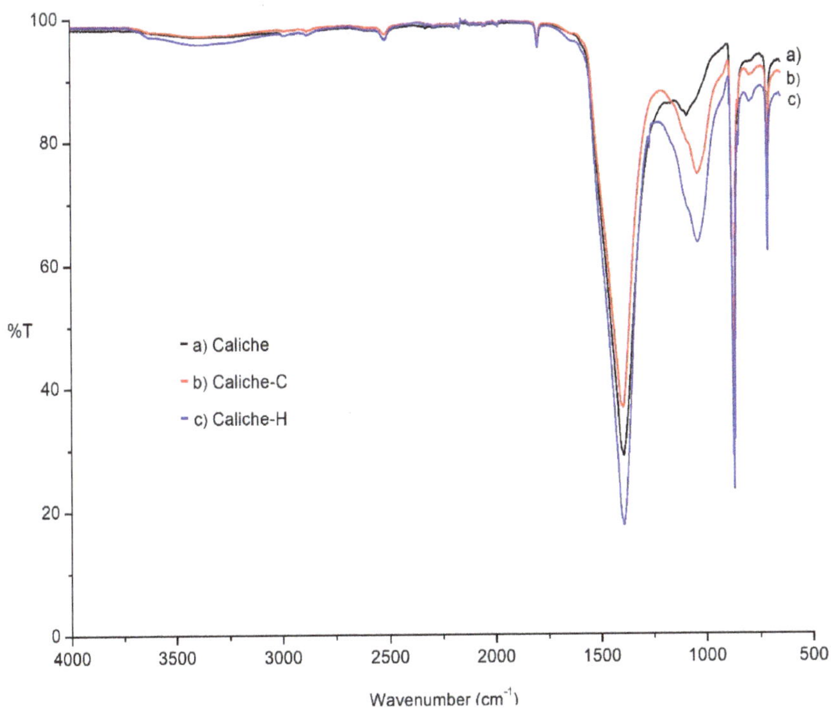

Figure 3. The FTIR–ATR spectrum of the calcite (**a**) without treatment, (**b**) consolidated, and (**c**) hydrofugated.

In the FTIR–ATR spectra of the Compañía stone, in accordance with its mineralogical composition (see Materials and Methods), bands are displayed at 1000, 1096, and 790 cm^{-1}, characteristic of cristobalite, while those at 1000, 790, and 742 cm^{-1} correspond to feldspars and quartz. The most significant modifications after the consolidation and hydrophobic treatment, as in the previous caliche sample, occurred in the region associated with the Si–O–Si network, where the band is more intense and broader (1000 cm^{-1}) and the small band at 1272 cm^{-1} (–$SiCH_3$ fragment) appears. Due the siliceous composition, Sostenes and Compañía stones present similar spectra as a result of the application of the formulations (Figures S6 and S7).

Figure 4 (Tables S1–S3), Figure 5 (Tables S4–S6), and Figure 6 (Tables S7–S9) present the results of SEM–EDX analysis of the stones before and after treatment. The order is caliche, Compañía, and then Sostenes.

In terms of the consolidation process, the aggregation of particles is evident because of the effect of the consolidant treatment. Nevertheless, the most important morphological changes are shown in caliche. In Compañía's sample, which is the least compact stone according to SEM and the one with a higher percentage of water absorption, the consolidation effect is not so evident likely due to the low quantity of added consolidant. However, as is discussed later on, the increment in hardness indicated a positive consolidation effect. On the other hand, in the Sostenes stone, the morphological change is evident; regarding the hydrophobic treatment, a coating is observed in the three stones, with an important reduction in the porosity compared with the untreated materials, though leaving the stone with enough porosity to "breathe", which is the final purpose of hydrophobic treatments in the stone conservation field.

Regarding EDX analysis, Sostenes and caliche stones display an increment in the carbon and nitrogen atomic percentage following treatment. A plausible interpretation is that the chitosan chains are exposed to the surface, not just in the case of the consolidant, but in the hydrophobic treatment (the methyl groups are surface oriented). Additionally, it is interesting to observe that the nitrogen atomic percentage is higher in consolidated Sostenes stone than in caliche, suggesting that the interaction between the consolidant and caliche possibly occurs via the free amine group. On the other hand, in Compañía stone, silicon is the element with a major atomic concentration on surface, probably suggesting, in accordance with SEM, that not enough consolidant formulation was added.

2.2.1. Hardness Determination

The effectiveness of treatment in terms of the mechanical properties was determined by hardness measurements in stones consolidated using the THEOS–chitosan formulation and was performed by indentation with a Shore D durometer. Three variables that influence the hardness increase were considered: the applied formulation (as a function of the silane/chitosan ratio); the nature of the stone; and the percentage of deacetylation of the chitosan used in the formulation (%DDA). A statistical analysis was conducted to evaluate the effect of each variable (not included here). In a next step, the Shore D hardness values were transformed to the most common hardness scale, such as Vickers, Brinell, and finally Mohs, in order to compare the hardness data obtained with respect to reference values of well-studied materials based on the Mohs scale.

The formulations named 1, 2, and 3 (see Materials and Methods) were used in hardness determination. The hardness was measured at four points of the samples before and after treatment to characterize the hardness percentage increase. Interesting results were obtained for every formulation; however, the treatment that remarkably increased the mechanical properties of the stones was formulation 2, which contains chitosan with 66% DDA (Table 1).

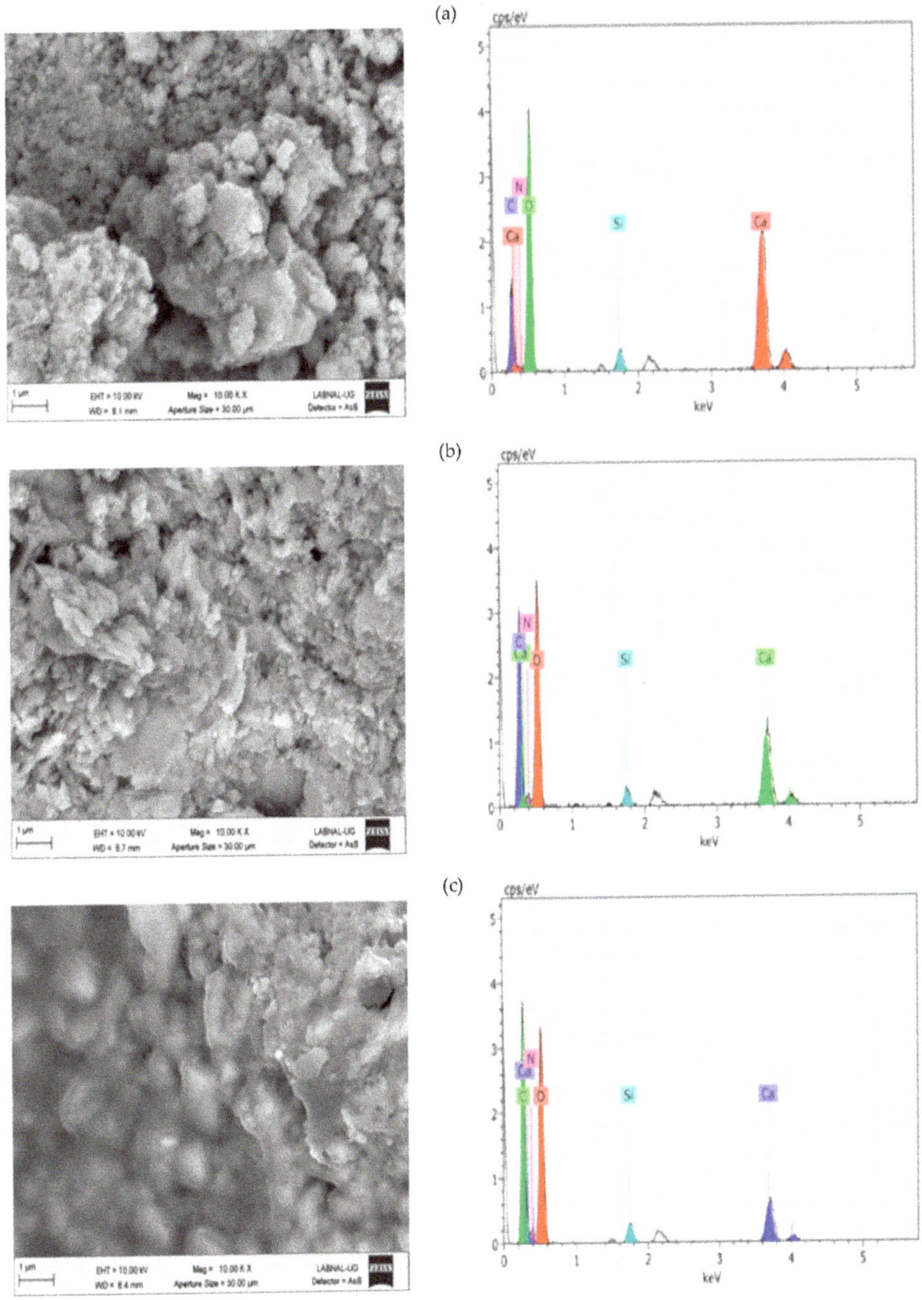

Figure 4. SEM–EDX analysis for the (**a**) caliche sample without treatment, (**b**) consolidated caliche sample, and (**c**) hydrophobic treated caliche sample.

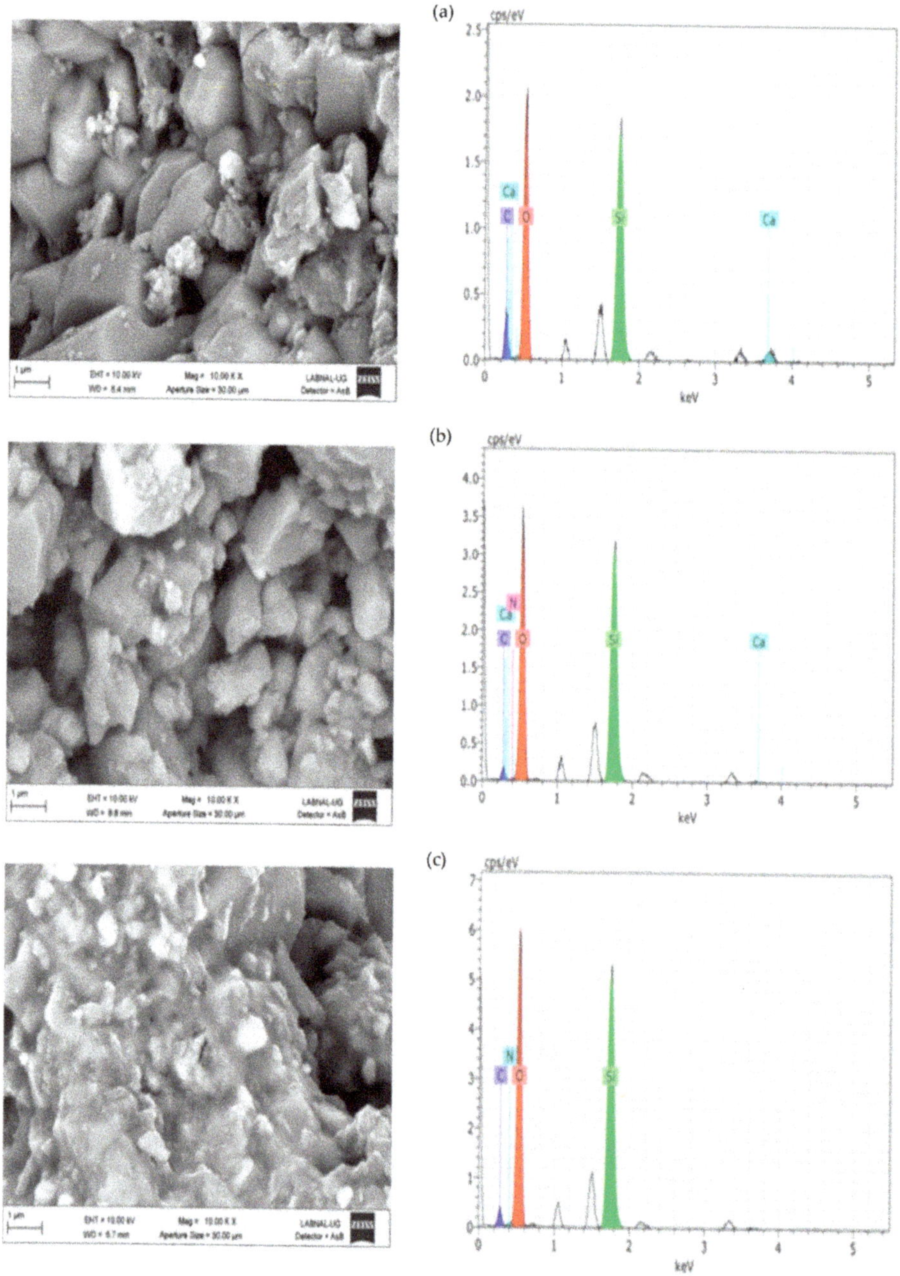

Figure 5. SEM–EDX analysis for the (**a**) Compañía sample without treatment, (**b**) consolidated Compañía sample, and (**c**) hydrophobic treated Compañía sample.

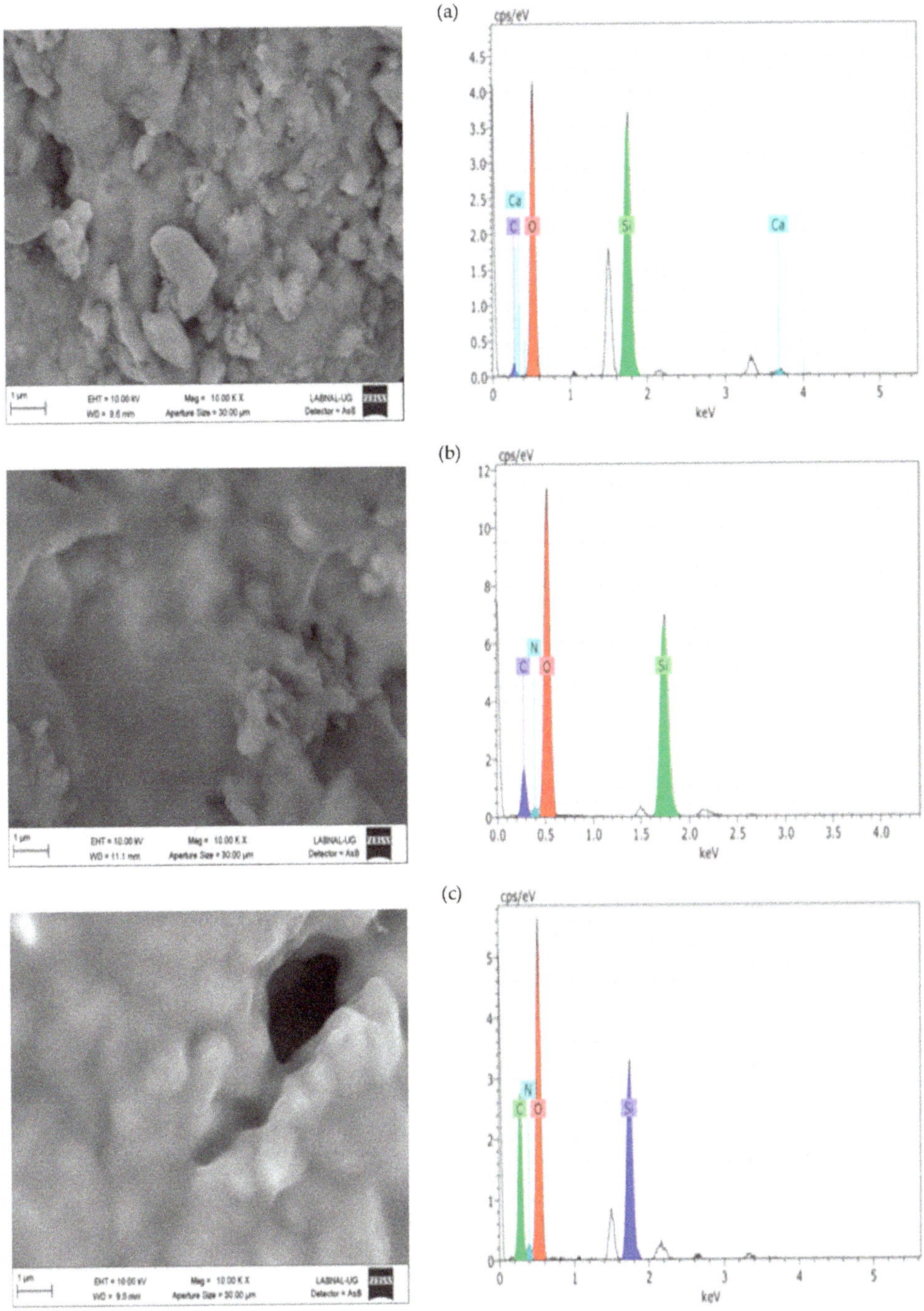

Figure 6. SEM–EDX analysis for the (**a**) Sostenes sample without treatment, (**b**) consolidated Sostenes sample, and (**c**) hydrophobic treated Sostenes sample.

Table 1. Shore D hardness and hardness increase (%) and the transformation to Vickers and Brinell scales (percentage of deacetylation of chitosan used in the formulation (%DDA) of 66).

%DDA	Stone	Formulation	Shore Hardness before	Vickers Hardness	Brinell Hardness	Shore Hardness after	Vickers Hardness	Brinell Hardness
66%	Caliche	1	71	581	511	81 (14%)	670	583
		2	70	564	497	81 (16%)	675	587
		3	72	585	515	81 (13%)	675	587
	Compañía	1	66	531	471	74 (12%)	604	530
		2	51	390	357	66 (29%)	531	471
		3	55	423	383	68 (23%)	550	486
	Sostenes	1	60	475	425	80 (33%)	663	577
		2	70	567	499	80 (14%)	665	579
		3	74	609	534	83 (12%)	687	597

Tables 1 and 2 present illustrative data on the Shore D hardness determination and transformation, first to Vickers and Brinell scales, and then (Table 2) from the Brinell to Mohs scale. In any case, the hardness increase is evident, with some important variations, where the influence of the formulation (silane/chitosan ratio) seems to have a certain effect. However, it is also important to bear in mind the different stone compositions. In the case of caliche, the increase in hardness is quite similar, having a major effect on the siliceous materials.

Table 2. Hardness transformation from the Brinell to Mohs scale and % of hardness increase (66% DDA).

%DDA	Stone	Formulation	Brinell	Mohs	Brinell [f,1]	Mohs [f,1]
66%	Caliche	1	511	6	583 (14%)	8 (33%)
		2	497	6	587 (18.1%)	8 (33%)
		3	515	6	587 (13.9%)	7 (16.6%)
	Compañía	1	471	6	530 (12%)	7 (16.6%)
		2	357	5	471 (31.9%)	6 (20%)
		3	383	5	486 (26%)	6 (20%)
	Sostenes	1	425	6	577 (35.7%)	7 (16.6%)
		2	499	6	579 (16%)	7 (16.6%)
		3	534	6	597(11.7%)	7(16.6%)

[f,1] = hardness measured after the consolidation treatment and increase percentage.

The hardness values transformed to the Mohs scale and reported in Table 2 indicate a hardness increment of one unit in siliceous materials, and in the case of caliche (formulations 1 and 2), even 2 units. In general, the most important increase in hardness occurred in caliche. In terms of the Mohs scale, the hardness values from 5 to 7 obtained for the samples range between apatite to orthoclase and quartz. The hardness studies indicate that the samples treated with THEOS–chitosan displayed an important increase in the mechanical properties of the three materials.

2.2.2. Water Absorption

Water absorption was tested using the Karsten tube technique, and measurements were taken before and after the application of the hydrophobic treatment (MeTHEOS–chitosan) on the stones (Table 3). The stone samples subjected to treatment present different mineral composition and water absorption values.

The penetration of water in the Compañía stone was quite high (51%) and was reduced to 7% with the hydrophobic treatment; such behavior makes sense due to its high pore diameter (macropores) in comparison with the other stones. Sostenes samples, that also possess a siliceous composition, had a water absorption value of 29% before treatment with a reduction to 10% as a result of the hydrophobic treatment. The calcareous stone (caliche) from the archeological site with an initial water absorption value of 46% exhibited remarkable reduction to 23%.

Table 3. Water absorption percentage on untreated and hydrophobic formulation (MeTHEOS)-treated stones.

Stone	Caliche	Caliche–MeTHEOS	Compañía	Compañía–MeTHEOS	Sostenes	Sostenes–MeTHEOS
Dry weight (g)	48.0	29.9	27.6	31.9	26.2	31.5
Wet weight (g)	70.1	33.1	41.7	34.2	33.9	34.7
Water absorption (%)	45.8	22.9	50.9	7.4	29.4	10.3

2.2.3. Contact Angle Measurements

The evaluation of the hydrophobic formulation MeTHEOS–chitosan was studied by static and dynamic contact angle measurements using the same formulations 1, 2, and 3 and the % of DDA of 66. Because of the natural existence of defects on certain materials, as is the case of the stones studied in the current investigation, it has been suggested that a static water contact angle does not necessarily characterize the intrinsic water wettability [41]. Dynamic contact angle determination in the three stones is presented. The dynamic contact angle was obtained by the degree of hydrophobicity calculated by the hysteresis, representing the difference as θ_R (receding angle) − θ_A (advancing angle). The hysteresis values and the average of three measurements in different surface sections of the three stones are reported in Table 4. The dynamic angle measurements indicate that the surfaces of the three stones studied display water repellency.

Table 4. Dynamic contact angle. Three contact angle measurements were performed at different points of the stones.

Stone		θ_A [1] (°)	θ_R [2] (°)	$\overline{\theta_A}$ (°)	$\overline{\theta_R}$ (°)	$\overline{\theta_R}-\overline{\theta_A}$ (°)
Caliche		67.2	95.2	73.4 ± 5.5	109.5 ± 12.5	36.1
		75.6	114.9			
		77.5	118.3			
Compañía		80.1	131.0	77.7 ± 1.2	113.6 ± 0.2	35.9
		79.1	132.4			
	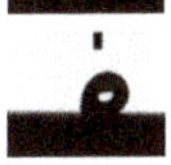	85.0	132.7			
Sostenes	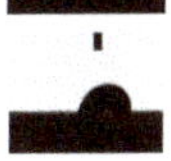	87.0	104.9	82.7 ± 3.7	107.5 ± 2.7	24.8
		80.6	110.3			
	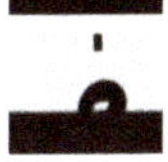	80.6	107.2			

[1] θ_A = advancing angle. [2] θ_R = receding angle.

The static contact angle was measured in three mediums (water, diiodomethane, and formamide) to take into consideration the different contributions of polar and non-polar mediums. The information obtained in the three mediums was useful for calculating the surface free energy or free energy of hydrophobicity by using the Owens and Van Oss (acid–base) methods [42,43].

The results are presented in Table 5 (1, 2, and 3 correspond to the formulation applied).

Table 5. Static angle determinations for the treated stones.

Sample	Formulation	Medium		
		Water	Formamide	Diiodomethane
Caliche	1	89.1	83.3	54.2
	2	108.1	88.1	79.0
	3	96.2	92.1	70.6
Compañía	1	139.7	134.2	103.6
	2	135.4	132.1	109.3
	3	134.8	131.5	110.6
Sostenes	1	110.9	102.7	74.2
	2	116.3	89.0	84.8
	3	105.1	103.1	77.8

The value obtained for static contact angle showed that hydrophobic properties were achieved after the application of the MeTHEOS–chitosan formulation. In terms of the static contact angle in water, all values are over 90°, with caliche as an exception (formulation 1, 89.1°). Some authors consider that 90° demarcation can generally be applied to classify hydrophilic and hydrophobic behaviors; however, they consider contact angles closer to 90° to be relatively hydrophobic and lower contact angles to be relatively hydrophilic [41].

2.2.4. Determination of the Surface Free Energy

In general terms, a sample with a low surface energy will cause poor wetting (a high contact angle). The reason for this is that the surface is not capable of forming strong bonds, so there is little energetic reward for the liquid to break bulk bonding in favor of interacting with the surface. On the contrary, a high surface energy will generally cause good wetting with a low contact angle. A surface will always try to minimize its energy. This can be done by adsorbing a material with a lower energy onto its surface [42,43]. The energy surface data are presented in Table 6.

Table 6. Surface free energy of the stones treated with MeTHEOS–chitosan using the Owens and Van Oss (acid–base) methods [1].

Stone	Formulations	%DDA	Owens	Van Oss
Caliche	1		46.9	48.6
	2		1.2	18.2
	3		33.6	19.5
Compañía	1		10.2	5.8
	2	66%	8.9	4.2
	3		1.6	3.9
Sostenes	1		37.0	16.3
	2		0.00	17.1
	3		21.4	14.6

[1] Surface energy units, mN/m.

The data interpretation reported in Table 6 is based on the criterium of a low surface energy value of 40 mN/m as a reference to consider a hydrophobic surface, although it is dependent on the model used; in the Owens model, the interval ranges from 49 to 3.16 mN/m, while in the Van Oss (acid–base), it ranges from 47 to 0 mN/m, so a lower value means a more hydrophobic surface [42,43]. According to the 40 mN/m reference value, or either the Owens or Van Oss model, caliche (formulation 1) is the only sample considered not to be hydrophobic. Some stones have a very low surface energy value, in agreement with the static contact angle obtained; Compañía is the most hydrophobic, followed by Sostenes and, finally, caliche. Moreover, the energy surface data indicate that formulation 2, in some way, is the most appropriate in terms of the silane/chitosan ratio (1 g of THEOS and 10 mL of a 0.5% aqueous solution of chitosan, and 66% DDA).

3. Materials and Methods

3.1. Chemicals

Tetraethoxysilane (98.5%), triethoxymethylsilane (98.5%), $CDCl_3$, DMSO-d6, CH3COOD, D_2O, and chitosan (72% deacetylation) were obtained from Sigma Aldrich. The ethylene glycol (JT. Backer, 99%) was distilled prior to its use in each reaction.

3.2. Synthesis of THEOS and Preparation of THEOS/MeTHEOS–Chitosan Formulations

The synthesis of THEOS and MeTHEOS by transesterification reactions was performed in a dry nitrogen atmosphere using Schlenk techniques. The reaction system used a three neck round bottom flask, a Vigreux column, a condenser, and a collector flask.

3.2.1. Synthesis of THEOS by the Transesterification of TEOS

Into a three neck round bottom flask purged under N_2 flow, 4.3 mL of freshly distilled ethylene glycol (4.76 g, 0.0768 moles) was added. After 30 min under magnetic stirring, 4.3 mL of TEOS (4 g, 0192 moles) was added drop by drop. The first drop of ethanol collected was considered the reaction initiation. The reaction temperature was 140 °C and was stable until the reaction ended (15 h). The reaction crude was concentrated under the vacuo line.

3.2.2. Synthesis of THEOS–Chitosan and MeTHEOS–Chitosan Solutions

Formulation solutions were prepared by the addition of 0.5 g of THEOS or MeTHEOS to 10 mL of an aqueous solution of chitosan (0.5%) in acetic acid (1%) under magnetic stirring until complete dissolution. The % of DDA of chitosan was 72% (Sigma Aldrich Química, S.L., Toluca, Mexico). The described solutions were used to obtain hybrid films that were characterized by FTIR, SEM–EDX, and solid state NMR and regarding their thermal stability, and then used in consolidation and hydrophobic treatments of the stones. The formulations named 1, 2, and 3 were prepared using chitosan obtained from the extraction of shrimp exoskeleton with three different degrees of deacetylation (%DDA): 62%, 66%, and 70%. Formulation 1 (0.5 g of THEOS or MeTHEOS and 10 mL of 0.5% aqueous solution of chitosan), formulation 2 (1 g of THEOS or MeTHEOS and 10 mL of a 0.5% aqueous solution of chitosan), and formulation 3 (0.5 g of THEOS or MeTHEOS and 10 mL of a 1% aqueous solution of chitosan) were applied to the stones and used in the hardness and contact angle determinations.

3.2.3. Synthesis of THEOS–Chitosan and MeTHEOS–Chitosan Films

The formulation solution obtained in each case was dispersed on plastic Petri dishes and left to dry at room temperature. Further characterization was performed by FTIR and solid state NMR (^{29}Si and ^{13}C).

3.3. Analytical Methods

3.3.1. NMR

^{29}Si, ^{13}C, and ^{1}H-NMR spectra in solution were collected using a Bruker AVANCE III 500 MHz spectrometer (Probe BBO-S2 5 mm). The internal references used were TMS (0 ppm) and hexamethyldisiloxane (chemical shifts at 6.54, 2.90, and 0.04 ppm in D_2O, and 6.54, 1.96, and 0.06 ppm in DMSO-d6). In the case of ^{29}Si, the one-dimensional sequence and inverse decoupling with a 90° pulse was used (d1 from 2 to 5 s, dt = 30 ms, ds = 4, and ns = 512). The ^{13}C-NMR spectra were recorded using the one-dimensional sequence and proton decoupling with a 30° pulse (d1 = 1 to 2 s, dt = 30 ms, ds = 4, and ns = 128), and ^{1}H spectra were obtained with the one-dimensional sequence with d1 = 1 s, ds = 2, and ns = 16.

The ^{29}Si MAS and CPMAS and ^{13}C CPMAS NMR spectra were collected using a Bruker AVANCE III 400 MHz spectrometer (probe: HRMAS 4 mm) using talc (−90 ppm as the reference for ^{29}Si) and adamantane (28.46 and 37.52 ppm for CH and CH_2, respectively). The parameters for the chitosan ^{13}C CPMAS experiment were ns = 4096 and d1 = 4 s; for the THEOS/chitosan hybrid ^{13}C CPMAS experiment, ns = 8192 and d1 = 4 s; for the THEOS/chitosan hybrid ^{29}Si CPMAS experiment, ns = 8192 and d1 = 4 s; and for the THEOS/chitosan hybrid ^{29}Si MAS experiment, ns = 14,336 and d1 = 6 s. The samples were placed in 4.0 mm zirconia rotors with a spinning rate of 8 kHz.

3.3.2. FTIR Analysis

The spectra of hybrid films from 4000 cm^{-1} to 650 cm^{-1} were collected using a Perkin Elmer Spectrum FTIR 1600 coupled with an ATR accessory (germanium point, 100 μm in diameter). An average of 16 scans was obtained, with a resolution of 4 cm^{-1}. Similar experimental conditions were used in the case of treated and untreated stone samples. The FTIR spectra were obtained from powders (−100 mesh) of each stone.

3.3.3. Scanning Electron Microscopy (SEM)

A palladium–gold alloy was vacuum evaporated on the dried samples. The outer surfaces of the treated stones were then studied using a EVO15-HD ZEISS scanning electron microscope at a 15 kV accelerating voltage under various magnifications (1000, 5000, and 10,000).

3.3.4. Stone Materials and Treatment

Samples of partially decayed siliceous stone (pink tuff) with the measurements 5 cm × 3 cm × 1 cm were collected from three different monuments or locations; two of them correspond to a tuff with a siliceous composition and are from two different historical monuments located in the city of Guanajuato, México (UNESCO World heritage City since 1988). The first is from the basement of a middle 20th century monument, called the statue of General Sostenes Rocha (who fought against the French army in México in the second half of the XIX century). Mineralogical analysis and XRD showed that the composition of the stone is mainly alkaline feldspars (46%), quartz (27%), mica (10%), kaolinite (9%), and smectite (3%), with traces of hematite. The second is from the church known as Oratorio de San Felipe Neri (traditionally called Compañía), which is a religious historical monument from the middle of the XVIII century [33]. The reported mineralogical analysis indicated alkaline feldspar (65%), quartz (29%), calcium silicate (3%), and hematite (4%) contents [44]. Additionally, a third sample from an archeological site called "Cerro de Los Remedios" (located in Comonfort county, Guanajuato state, México) was studied. As part of the basement of a pyramid, this stone possesses a calcareous nature (caliche), with $CaCO_3$ (93%), CaO (1.5%), halloysite (5%), and traces of hematite. The relatively high amount of kaolinite present in the stone might be taken as evidence that some of the original feldspars have been hydrolyzed to clays by the weathering process known as kaolinization [45]. A comparative composition of the samples is presented in Table 7.

To evaluate the consolidation effect of the hybrid formulations, some samples (already cleaned and dried) were treated with the THEOS–chitosan formula, and to test the hydrophobic properties, others were treated with the MeTHEOS–chitosan formulation. The formulations were applied on stone samples by brushing in one phase under laboratory conditions until saturation.

Then, all samples were carefully wrapped in black plastic polypropylene sheets (as used in practical conservation) to permit gelling and aging for 2 weeks.

Table 7. Mineralogical composition of the siliceous and calcareous stones.

Mineral	Caliche	Compañía	Sostenes
Alkaline feldspar	-	65%	46%
Quartz	-	29%	27%
Calcite	93%	-	-
CaO	1.5%	-	-
Mica	-	-	10%
Kaolinite	-	-	9%
Calcium silicate	-	3%	-
Halloysite	5%	-	-
Smectite	-	-	3%
Hematite	traces	4%	traces

3.3.5. Characterization of the Samples Treated with the Consolidant and the Hydrophobic Formulations

The hardness changes of the stones untreated and after treatment were measured as the Shore hardness using an REX 2000D indentation durometer. The hardness values obtained were transformed to Vickers, Brinell, and Mohs scales.

Water absorption measurements were carried out using the Karsten (Rilem) pipe [46]. The graduated pipe was fixed onto the sample and filled with water. Water absorption for each sample was measured as the difference between the quantities of water (mL) absorbed after five and thirty minutes.

Static and dynamic contact angles were measured using the OCA 15 Dataphysics system. Contact angle data obtained in the three classical liquids of different polarities (water, formamide, and diiodomethane) were used to calculate the surface free energy or free energy of hydrophobicity. The surface free energy was calculated from the OCA 15 Dataphysics software using the Owens and Van Oss (acid–base) methods [42,43].

4. Conclusions

As a result of the present investigation, a new approach to the application of glycol alkoxysilane–chitosan hybrids, including THEOS–chitosan and MeTHEOS–chitosan, in the area of stone conservation of historic stone buildings is presented and suggested, with THEOS–chitosan used as a consolidant and MeTHEOS–chitosan as a water repellent. Several aspects have been covered. First, a more detailed characterization of the hybrids has been described and discussed, as we believe that it was important to address the lack of such information. Once the hybrids had been obtained by a very simple synthetic procedure, they were applied to three different historic building stones and their performance was evaluated in detail. The application of the formulations to the stones is water-based, which implies the elimination of organic solvents, as an important contribution, but also suggestive of their use in the conservation of either siliceous or calcareous natural stones. By this means, synergetic benefits arise from the interaction of alkoxysilanes and chitosan in the performance of the formulations. The effectiveness of the consolidation and hydrophobic treatments was evaluated by different spectroscopic methods, such as FTIR–ATR and SEM–EDX, and physical analysis, such as hardness measurements, in the case of the consolidation, water absorption, characterization of the dynamic and static contact angle, and energy surface determination. The evaluation of the effectiveness of treatments is considered positive in terms of the consolidation and water repellency.

Several perspectives will now be presented for further study of the interaction of the formulations with the calcareous material, which in the present case, is caliche. A primary suggestion is that the interaction between the consolidant and caliche occurred via the free amino group of the chitosan. It is very important to point out that data on colorimetric changes after treatments have not been obtained, bearing in mind that the assessment of such analyses is quite important when stones of historical buildings are treated. No apparent colorimetric changes were observed after treatment in any sample, although this was recorded by simple observations. However, corresponding analyses must be performed and considered in perspectives; for example, the use of the Munsell method, which is commonly applied to observe chromatic variations as a result of sample treatment. Another interesting aspect that is currently under study, with preliminary results availables, is the use of the intrinsic fluorescence emission of chitosan. We have observed that such a property is maintained in the hybrid and might be a useful tool for ascertaining the effectiveness of the dispersion or penetration of the formulations on or inside the stone. Additionally, the antimicrobial property of chitosan is under study, which could lead to a formulation that will also prevent or solve cases of biodecay.

Supplementary Materials: The following are available online: Figure S1: FTIR–ATR for films of (a) MeTHEOS–chitosan and (b) THEOS–chitosan, Figure S2: ^{29}Si CPMAS NMR of THEOS–chitosan film, Figure S3: ^{13}C CPMAS NMR of MeTHEOS–chitosan film, Figure S4: ^{29}Si MAS NMR of MeTHEOS–chitosan film, Figure S5: ^{29}Si MAS NMR of MeTHEOS–chitosan film, Figure S6: FTIR–ATR spectrum of a Compañía sample without treatment, consolidated, and hydrofugated, Figure S7: FTIR–ATR spectrum of a Sostenes sample without treatment, consolidated, and hydrofugated, Table S1: SEM–EDX elemental composition for the MeTHEOS–chitosan hybrid, Table S2: SEM–EDX elemental composition for the THEOS–chitosan hybrid, Table S3: SEM–EDX elemental composition for the caliche sample without treatment, Table S4: SEM–EDX elemental composition for the consolidated caliche sample, Table S5: SEM–EDX elemental composition for the hydrophobic treated caliche sample, Table S6: SEM–EDX elemental composition for the Compañía sample without treatment, Table S7: SEM–EDX elemental composition for the consolidated Compañía sample, Table S8: SEM–EDX elemental composition for the hydrophobic treated Compañía sample, Table S9: SEM–EDX elemental composition for the Sostenes sample without treatment, Table S10: SEM–EDX elemental composition for the consolidated Sostenes sample, and Table S11: SEM–EDX elemental composition for the hydrophobic treated Sostenes sample.

Author Contributions: Conceptualization, J.C., M.M.-Z. and R.Z.-N.; methodology, M.M.-Z., I.B.-F., E.R.-O., G.Á.-G., A.G.-C. and A.V.; validation, J.C., M.M.-Z., R.Z.-N. and I.B.-F.; formal analysis, I.B.-F., M.M.-Z., G.Á.-G., E.R.-O., A.G.-C., R.Z.-N., M.M.-R. and A.V.; investigation, I.B.-F., M.M.-Z., E.R.-O., G.Á.-G. and A.G.-C.; resources, J.C., R.Z.-N. and G.Á.-G.; data curation, M.M.-Z., J.C., E.R.-O., A.G.-C., G.Á.-G., A.V. and M.M.-R.; writing—original draft preparation, J.C., M.M.-Z. and I.B.-F.; writing—review and editing, J.C. and M.M.-Z.; visualization, J.C., M.M.-Z. and I.B.-F.; supervision, J.C., M.M.-Z., G.Á.-G., A.G.-C. and E.R.-O.; project administration, J.C., G.Á.-G. and A.G.-C.; funding acquisition, J.C., A.G.-C., E.R.-O., A.V. and M.M.-R. All authors have read and agreed to the published version of the manuscript.

Funding: This research was funded by CONACYT-México award 284510 supported by the Fondo Sectorial de Investigación para la Educación and the University of Guanajuato (Guanajuato-México).

Institutional Review Board Statement: Not applicable.

Informed Consent Statement: Not applicable.

Data Availability Statement: Not applicable.

Acknowledgments: The authors wish to thank the support of the following laboratories located at the University of Guanajuato: Laboratorio Nacional de Caracterización de Propiedades Fisicoquímicas y Estructura Molecular University of Guanajuato-CONACYT-México; laboratorio de Análisis Instrumental, "Q. Fernando de Jesús Amezquita López", laboratory of Analytical Chemistry Services, Chemistry Department; laboratory in charge of Mario Avila (contact angle measurements). Finally, I.B.-F. wishes to thank CONACYT-México for the fellowship received (CVU 919318).

Conflicts of Interest: The authors declare no conflict of interest. The funders had no role in the design of the study; in the collection, analyses, or interpretation of data; in the writing of the manuscript; or in the decision to publish the results.

Sample Availability: Samples of the compounds are not available from the authors.

References

1. Prikryl, R.; Smith, B.J. *Building Stone Decay: From Diagnosis to Conservation*; Special Publication No. 271; Geological Society: London, UK, 2007.

2. Smith, B.J.; Gomez-Heras, M.S.; McCabe, S. Understanding the Decay of Stone-Built Cultural Heritage. *Prog. Phys. Geog.* **2008**, *32*, 439. [CrossRef]
3. Torraca, G. *Lectures on Materials Science for Architectural Conservation*; The Getty Conservation Institute: Los Angeles, CA, USA, 2009.
4. Siegesmund, S.; Snethlage, R. *Stone in Architecture*; Springer: Berlin/Heidelberg, Germany, 2014; pp. 1–415.
5. Price, C.A.; Doehne, S. *Stone Conservation: An Overview of Current Research*; Getty Conservation Institute: Los Angeles, CA, USA, 2010.
6. Wheeler, G. *Alkoxysilanes and the Consolidation of Stone*; The Getty Conservation Institute: Los Angeles, CA, USA, 2005.
7. Laurie, A.P. Preservation of Stone. U.S. Patent 1561988, 17 November 1924.
8. Da Fonseca, B.S.; Picarra, S.; Pinto, A.P.F.; Ferreira, M.J.; Montemor, M.F. TEOS-based consolidants for carbonate stones: The role of N1-(3-trimethoxysilylpropyl)diethylenetriamine. *New J. Chem.* **2017**, *41*, 2458–2467. [CrossRef]
9. Pinto, A.P.F.; Rodrigues, J.D. Impacts of consolidation procedures on color and absorption kinetics of carbonate stones. *Stud. Conserv.* **2014**, *59*, 79–90. [CrossRef]
10. Sassoni, E.; Franzoni, E.; Pigino, B.; Scherer, G.W.; Naidu, S. Consolidation of calcareous and siliceous sandstones by hydroxyapatite: Comparison with a TEOS-based consolidant. *J. Cult. Herit.* **2013**, *14*, 103–108. [CrossRef]
11. Feigao, X.; Weiping, Z.; Dan, L. Recent advance in alkoxysilane-based consolidants for stone. *Prog. Org. Coat.* **2019**, *127*, 45–54.
12. Pinto, A.P.F.; Montemor, M.F. Alkoxysilane-based sols for consolidation of carbonate stones: Impact of the carbonate medium in the sol-gel processes. *J. Cult. Herit.* **2019**, *37*, 63–72.
13. Cervantes, J.; Mendoza-Díaz, G.; Alvarez-Gasca, D.E.; Martínez-Richa, A. Application of ^{29}Si and ^{27}Al magic angle spinning nuclear magnetic resonance to studies of the building materials of historical monuments. *Solid State NMR* **1999**, *13*, 263–269. [CrossRef]
14. Zárraga, R.; Alvarez-Gasca, D.E.; Cervantes, J. Solvent effect on TEOS film formation in the sandstone consolidation process. *Silicon Chem.* **2002**, *1*, 397–402. [CrossRef]
15. Salazar-Hernández, C.; Zárraga, R.; Alonso, S.; Sugita, S.; Calixto, S.; Cervantes, J. Effect of solvent type on polycondensation of TEOS catalyzed by DBTL as used for stone consolidation. *J. Sol-Gel Sci. Technol.* **2009**, *49*, 301–310. [CrossRef]
16. Zarraga, R.; Cervantes, J.; Salazar-Hernández, C.; Wheeler, G. Effect of the addition of hydroxyl-terminated polydimethylsiloxane to TEOS-based stone consolidants. *J. Cult. Herit.* **2010**, *11*, 138–144. [CrossRef]
17. Salazar-Hernández, C.; Cervantes, J.; Alonso, S. Viscoelastic characterization of TEOS sols in three different solvents when DBTL is used as polycondensation catalyst. *J. Sol-Gel Sci. Technol.* **2010**, *54*, 77–82. [CrossRef]
18. Salazar-Hernández, C.; Puy-Alquiza, M.-J.; Salgado, P.; Cervantes, J. TEOS—colloidal silica—PDMS-OH hybrid formulation used for stone consolidation. *Appl. Organomet. Chem.* **2010**, *24*, 481–488. [CrossRef]
19. Reyes-Zamudio, V.; Angeles-Chávez, C.; Cervantes, J. Clay minerals in historic buildings. *Therm. Anal. Calor.* **2011**, *104*, 405–413. [CrossRef]
20. Puy-Alquiza, M.J.; Miranda-Aviles, R.; Salazar-Hernández, C.; Vega-González, M.; Cervantes, J. Characterization petrophysical of the Losero formation in the historical architecture of the Guanajuato city Mexico. *Ing. Investig. Tecnol.* **2013**, *15*, 191–205.
21. Salazar-Hernández, C.; Cervantes, J.; Puy-Alquiza, M.J.; Miranda, R. Conservation of building materials of historic monuments using a hybrid formulation. *J. Cult. Herit.* **2015**, *16*, 185–191. [CrossRef]
22. Mehrotra, R.C.; Narain, R.P. Reactions of tetramethoxy- and triethoxysilanes with glycols. *Indian J. Chem.* **1967**, *5*, 444–448.
23. Meyer, M.; Fischer, A.; Hoffmann, H. Novel ringing silica gels that do not shrink. *J. Phys. Chem. B* **2002**, *106*, 1528–1533. [CrossRef]
24. Shchipunov, Y.A.; Karpenko, T.Y. Hybrid polysaccharide-silica nanocomposites prepared by the solgel technique. *Langmuir* **2004**, *20*, 3882–3887. [CrossRef]
25. Shchipunov, Y.A.; Karpenko, T.Y.; Krekoten, A.V. Hybrid organic-inorganic nanocomposites fabricated with a novel biocompatible precursor using sol-gel processing. *Comp. Int.* **2005**, *11*, 587–607. [CrossRef]
26. Shchipunov, Y.A.; Kojima, A.; Imae, T. Polysaccharides as a template for silicate generated by sol–gel processes. *J. Colloid Interface Sci.* **2005**, *285*, 374–380. [CrossRef]
27. Shchipunov, Y.A.; Karpenko, T.Y.; Krekoten, A.V.; Postnova, I.V. Gelling of otherwise non gelable polysaccharides. *J. Colloid Interface Sci.* **2005**, *287*, 373–378. [CrossRef] [PubMed]
28. Kim, J.M.; Chang, S.M.; Kong, S.M.; Kim, K.S.; Kim, J.; Kim, W.S. Control of hydroxyl group content in silica particle synthesized by the sol-precipitation process. *Ceram. Int.* **2009**, *35*, 1015–1019. [CrossRef]
29. Wang, G.; Zhang, L. Using Novel Polysaccharide-Silica Hybrid Material to Construct an Amperometric Biosensor for Hydrogen Peroxide. *J. Phys. Chem. B* **2006**, *110*, 24864–24868. [CrossRef] [PubMed]
30. Zhang, L.; Wang, G.; Xing, Z. Polysaccharide-assisted incorporation of multiwalled carbon nanotubes into sol–gel silica matrix for electrochemical sensing. *J. Mater. Chem.* **2011**, *21*, 4650–4659. [CrossRef]
31. Wang, G.; Zhang, L. Electroactive polyaniline/silica hybrid gels: Controllable sol-gel transition adjusted by chitosan derivatives. *Carbohydr. Polym.* **2018**, *202*, 523. [CrossRef]
32. Bravo-Flores, I.; Meléndez-Zamudio, M.; Guerra-Contreras, A.; Ramírez-Oliva, E.; Álvarez-Guzmán, G.; Zárraga-Núñez, R.; Villegas, A.; Cervantes, J. Revisiting the system silanes-polysaccharides: The case THEOS-chitosan and MeTHEOS-chitosan. *Macromol. Rapid. Commun.* **2021**. [CrossRef] [PubMed]
33. Warscheid, T.; Braams, J. Biodeterioration of stone: A review. *Int. Biodeterior. Biodegrad.* **2000**, *46*, 343–368. [CrossRef]

34. Carlson, R.P.; Taffs, R.; Davison, W.M.; Stewart, P.S. Anti-biofilm properties of chitosan-coated surfaces. *J. Biomater. Sci. Polym. Ed.* **2008**, *19*, 1035–1046. [CrossRef]
35. Sahariah, P.; Másson, M. Antimicrobial Chitosan and Chitosan Derivatives: A Review of the Structure-Activity Relationship. *Biomacromolecules* **2017**, *18*, 3846–3868. [CrossRef]
36. Atay, H.Y. Antibacterial Activity of Chitosan-Based Systems. In *Functional Chitosan: Drug Delivery and Medical Applications*, 1st ed.; Jana, S., Jana, S., Eds.; Springer: Singapore, 2019; pp. 457–489.
37. Sudatta, B.P.; Sugumar, V.; Varma, R. Extraction, characterization and antimicrobial activity of chitosan from pen shell, Pinna bicolor. *Int. J. Biol. Macromol.* **2020**, *163*, 423–430. [CrossRef]
38. Walczak, K.; Schierz, G.; Basche, S.; Petto, C.; Boening, K.; Wieckiewicz, M. Antifungal and Surface Properties of Chitosan-Salts Modified PMMA Denture Base Material. *Molecules* **2020**, *25*, 5899. [CrossRef]
39. Hartmann,.S. Hierarchically Organized (Hybrid) Silica Monoliths for the Application as Stationary Phases in HPLC. Ph.D. Thesis, Ulm University, Ulm, Germany, December 2009.
40. Brandhuber, D.; Torma, V.; Raab, C.; Peterlik, H.; Kulak, A.; Husing, N. Glycol-modified silanes in the synthesis of mesoscopically organized silica monoliths with hierarchical porosity. *Chem. Mater.* **2005**, *17*, 4262–4271. [CrossRef]
41. Kozbial, A.; Trouba, C.; Haitao, L.; Li, L. Characterization of the Intrinsic Water Wettability of Graphite Using Contact Angle Measurements: Effect of Defects on Static and Dynamic Contact Angles. *Langmuir* **2017**, *33*, 959–967. [CrossRef] [PubMed]
42. Della Volpe, C.; Maniglio, D.; Brugnara, M.; Siboni, S.; Morra, M. The solid surface free energy calculation I. In defense of the multicomponent approach. *J. Colloid Interface Sci.* **2004**, *271*, 434–453. [CrossRef] [PubMed]
43. Kwok, D.Y.; Neuman, A.W. Contact angle measurement and contact angle interpretation. *Adv. Colloid Int. Sci.* **1999**, *81*, 167–249. [CrossRef]
44. Zarraga-Núñez, R.; Cervantes, J.; Álvarez-Gasca, D.A.; Zamudio, V.R.; Salazar-Hernández, C. La Investigación Científica en la Conservación de Monumentos de Cantera. *Acta Univ.* **2006**, *16*, 38–50. [CrossRef]
45. Prada-Pérez, J.; Álvarez-Pérez, A. *Mineralogy of Decay Processes in Historic Buildings*; Melgranejo: Barcelona, Spain, 1997.
46. RILEM, II. 4-Water absorption under low pressure (pipe method). In *RILEM Technical Commision 25-PEM*; RILEM, II: Paris, France, 1980.

molecules

MDPI

Article

Controlled Synthesis of Polyphosphazenes with Chain-Capping Agents

Robert A. Montague † **and Krzysztof Matyjaszewski** *

Department of Chemistry, Carnegie Mellon University, 4400 Fifth Avenue, Pittsburgh, PA 15213, USA; rmontague.ky@roadrunner.com
* Correspondence: matyjaszewski@cmu.edu or km3b@andrew.cmu.edu
† Current address: 2217 Forest Avenue, Ashland, KY 41101, USA.

Abstract: *N*-alkyl phosphoranimines were synthesized via the Staudinger reaction of four different alkyl azides with tris(2,2,2-trifluoroethyl) phosphite. *N*-adamantyl, *N*-benzyl, *N*-*t*-butyl, and *N*-trityl phosphoranimines were thoroughly characterized and evaluated as chain-capping compounds in the anionic polymerization of *P*-tris(2,2,2-trifluoroethoxy)-*N*-trimethylsilyl phosphoranimine monomer. All four compounds reacted with the active chain ends in a bulk polymerization, and the alkyl end groups were identified by ^{1}H-NMR spectroscopy. These compounds effectively controlled the molecular weight of the resulting polyphosphazenes. The chain transfer constants for the monomer and *N*-benzyl phosphoranimine were determined using Mayo equation.

Keywords: *N*-alkyl phosphoranimines; polyphosphazenes; phosphoramidate esters; fluoride; trifluoroethoxide; *N*-methylimidazole initiators; anionic; chain-capping agents

Citation: Montague, R.A.; Matyjaszewski, K. Controlled Synthesis of Polyphosphazenes with Chain-Capping Agents. *Molecules* **2021**, *26*, 322. https://doi.org/10.3390/molecules26020322

Academic Editors: Sławomir Rubinsztajn, Marek Cypryk and Wlodzimierz Stanczyk

Received: 20 December 2020
Accepted: 7 January 2021
Published: 10 January 2021

1. Introduction

The most important examples of inorganic polymers [1–3] are the polysiloxanes—$(R_2Si\text{-}O)_n$ [4–8], polysilanes—$(R_2Si)_n$ [9–11], and the polyphosphazenes—$(R_2P{=}N)_{n-}$ [12–14]. They have been the subject of significant research due to desirable properties, such as biocompatibility, high thermal resistance, oxidative stability, UV resistance, interesting photoelectronic behavior, and high flexibility that are often difficult or impossible to achieve in carbon-based organic polymers. Much of the current research on polyphosphazenes is focused on applications in the life sciences as drug delivery vehicles and other biological applications [15–18].

The field of polyphosphazene synthesis has expanded significantly with the advent of more efficient and faster synthetic routes such as the ambient temperature preparation of poly(dichlorophosphazene) by the PCl_5-initiated polymerization of *N*-trimethylsilyl-*P*-trichlorophosphoranimine, [13,19] the polymerization of *N*-trimethylsilyl-*P*-tris(2,2,2-trifluoroethoxy)phosphoranimine by antimony pentachloride initiator, [20] and poly (organophosphazenes) from partially halogenated phosphoranimines [21]. These reactions are examples of cationic polymerization of phosphoranimines. The cationic route has also led to polyphosphazenes with functional and terminal groups at the chain ends [22–27] as well as many other derivative copolymers with more complex architecture [28–32].

The first catalyzed/initiated polymerization of a phosphoranimine monomer by fluoride ion was previously reported (Scheme 1a) and then extensively investigated [33]. The successful conversion of these monomers to polyphosphazenes initiated/catalyzed by fluoride anion, trifluoroethoxide anion, and *N*-methylimidazole was described [34–36]. Polyphosphazene random and block copolymers with mixed alkoxy/alkoxyether substituents were subsequently prepared by the fluoride/anionic method and characterized [37]. Other examples include polyphosphazenes with alkyl or aryl substituents [38], polyphosphazenes with electronegative nitropropoxy groups bound to the phosphorus atom of the *N*-silylated phosphoranimine [39], polymerization of *N*-silyl-*P*-diethyl phosphoranimine with fluoride and phenoxide initiators [40], and *P*-tris(trifluoroethoxy)-*N*-trimethylsilyl phosphoranimine polymerized with water in the presence of *N*-methylimidazole initiator in diglyme solution polymerization [41].

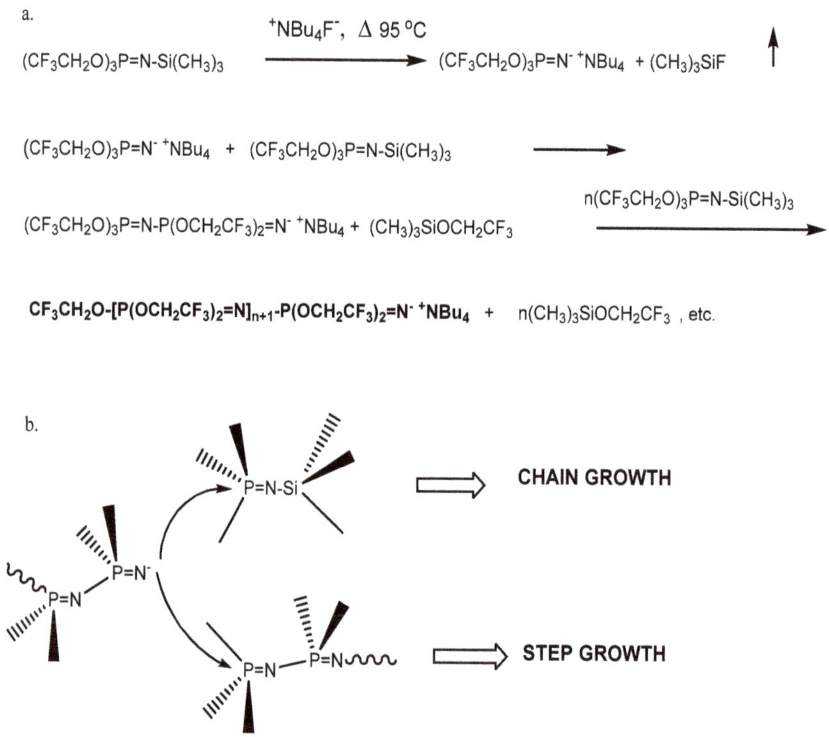

Scheme 1. (**a**) Anionic chain growth condensation polymerization. (**b**) Chain growth vs. step growth macrocondensation.

The use of chain capping agents is a technique to prepare well-defined polyphosphazenes produced by anionic polymerization of phosphoranimines, since active chain segments are subject to late-stage condensation reactions that can increase molecular weight with diminished control of chain length. In Scheme 1b, active anionic chain ends of the forming polymer can react with a monomer molecule via chain growth condensation (with elimination of trimethylsilyl trifluoroethoxide), or it may react with a partially polymerized chain segment to advance molecular weight via a macrocondensation. The preparation of four *N*-alkyl phosphoranimine compounds and their efficiency as chain end-capping agents in polyphosphazene synthesis was briefly mentioned before [42].

In this paper, we report more detailed experimental data obtained using these compounds, and present evidence of their incorporation as unique chain end groups during the formation of polyphosphazenes by the anionic route.

2. Results and Discussion

2.1. Synthesis and Characterization of N-Alkyl Phosphoranimines

The synthetic route to the four *N*-alkyl phosphoranimines prepared in this study was via the Staudinger reaction [43] is shown in Scheme 2, in which tris(2,2,2-trifluoroethyl) phosphite reacted with an alkyl azide to form the *P*-tris(2,2,2-trifluoroethoxy)-*N*-alkyl phosphoranimine with the evolution of nitrogen gas.

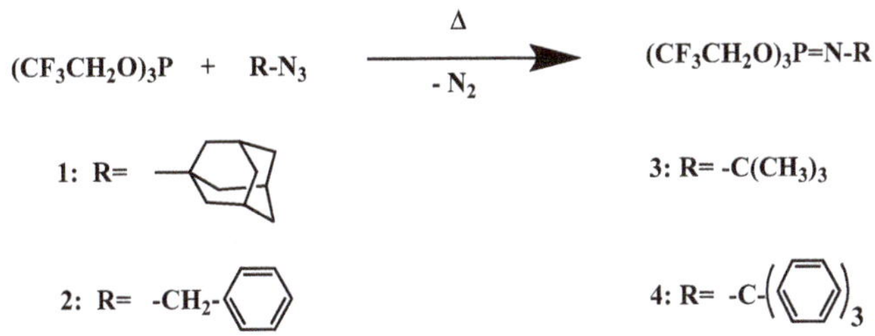

Scheme 2. Staudinger synthesis of *P*-tris(trifluoroethoxy)-*N*-alkylphosphoranimines.

The *N*-alkyl compounds formed consisted of the *N*-adamantyl, *N*-benzyl, *N*-*t*-butyl, and *N*-trityl phosphoranimines. Yields, boiling or melting points, densities, and refractive indices are shown in Table 1.

Table 1. Physical data for *N*-alkyl phosphoranimines.

Compound	% Yield	B.P.(M.P.)	Density	Refractive Index, n_D^{20}
1 (Adamantyl)	46.3	93 °C/2.5 torr, (0–2 °C)	1.606 g/mL	1.4076
2 (Benzyl)	98.9	95 °C/2.5 torr	1.664 g/mL	1.4022
3 (*t*-Butyl)	34.4	93 °C/52 torr	1.568 g/mL	1.3445
4 (Trityl)	56.5	(80–85 °C)		

The new compounds were characterized by NMR, fast atom bombardment, and GC-mass spectrometry, with the resulting data presented in Table 2. The data obtained from the NMR spectra are consistent with the proposed structures of the four compounds, as are the mass spectrometry measurements for molecular weight of the parent ions.

Table 2. NMR and mass spectrometry data for *N*-alkyl phosphoranimines.

Compound.	^{31}P-NMR, ppm	^{1}H-NMR ppm	Mass Spectrometry	Theor. MW
1 (Adamantyl)	−27.3 (m)	4.28 (p): 6H $^3J_{(POCH)}$ = 7.6 Hz 1.73(d): 6H 1.65(br. S.):6H 2.05(br. s.):3H	477 **	477
2 (Benzyl)	−13.19 (m)	4.20 (p): 6H $^3J_{(POCH)}$ = 7.2 Hz 4.32 (d): 2H$^3J_{(PNCH)}$ = 22 Hz 7.30 (m): 5H	433 ^	433
3 (*t*-Butyl)	−28.95 (m)	4.28 (p): 6H$^3J_{(POCH)}$ = 8.2 Hz 1.23 (s): 9H	399 *	399
4 (Trityl)	−29.01 (p)	4.05 (p): 6H$^3J_{(POCH)}$ = 7.8 Hz 7.31(m): 15H	585 **	585

* Molecular ion; ^ Partially decomposes on silica GC column; ** Molecular ion, FAB mass spec.

Elemental analyses of these compounds for found and theoretical percentages are displayed in Table 3. The determined percentages of elements from the analyses are in good agreement with the assigned structures.

Table 3. Elemental analyses of N-alkyl phosphoranimines.

Compound	% C	% F	% H	% N	%O	%P	Theory/Found
1 (Adamantyl)	40.28 40.58	35.83 35.49	4.45 4.54	2.94 3.22	10.06 -	6.47 6.33	Theory Found
2 (Benzyl)	36.06 35.73	39.47 38.12	3.03 2.98	3.24 3.16	11.8 -	7.15 6.77	Theory Found
3 (*t*-Butyl)	30.10 29.36	42.83 41.88	3.80 3.65	3.51 3.27	12.03 -	7.76 7.75	Theory Found
4 (Trityl)	51.32 48.14	29.22 23.89	3.63 3.60	2.39 2.25	8.21 -	5.29 *	Theory Found

Notes: Fluorine interferes with oxygen determination. * Sample size insufficient for P-analysis.

FTIR spectra further confirmed the structures with the appearance of the broad absorption bands of the P=N moieties between 1270–1310 cm^{-1}, and the characteristic absorptions of aliphatic or aromatic protons as expected for the particular structure. The data collected are shown in Table 4, and an example IR spectrum of the N-adamantyl phosphoranimine is shown in Figure 1.

Table 4. FTIR band assignments of N-alkyl phosphoranimines.

Compound	Absorbance Wavenumbers cm^{-1}	Functional Group Assignment
1 (Adamantyl)	2870–2950 1435 1270 1180, 960 1090 660	CH aliph. P-O-C P=N P-O C-O CF_3
2 (Benzyl)	3020–3120 1305 1150, 970 1090 700–750 670	CH arom. P=N P-O C-O CH arom. CF_3
3 (*t*-Butyl)	2970 1420 1310 1170, 970 1085 675	CH aliph. P-O-C P=N P-O C-O CF_3
4 (Trityl)	3000–3110 1420–1440 1300 1180 1090 720–750 660–670	CH arom. P-O-C P=N P-O C-O CH arom. CF_3

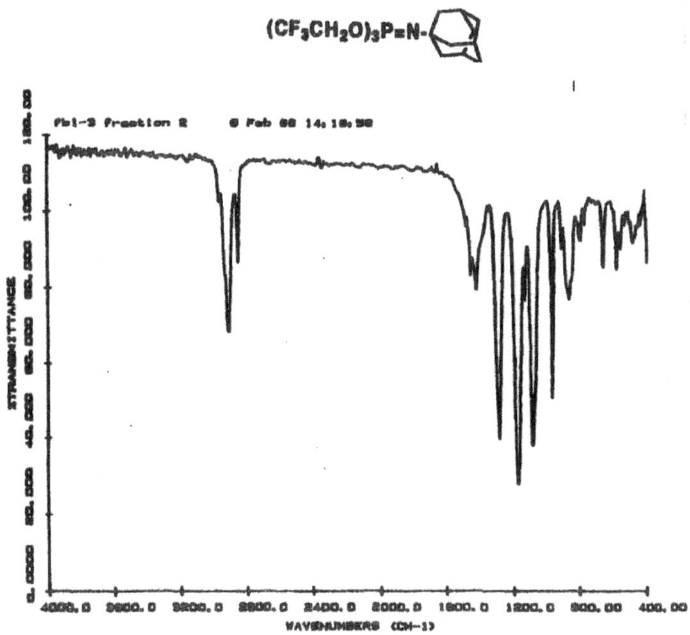

Figure 1. FT-IR spectrum of *P*-tris(trifluoroethoxy)-*N*-adamantyl phosphoranimine.

2.2. Reactivity of N-Alkyl Phosphoranimines

The phosphoranimines are air-sensitive and hydrolyze easily to the corresponding phosphoramidite esters, $(CF_3CH_2O)_2P(=O)$-NHR, as confirmed by mass spectrometry and NMR. For example, the GC-mass spectrum of the *N*-adamantyl phosphoranimine in (wet) diethyl ether showed two major peaks with fragments that are consistent for the phosphoranimine (molecular ion 477), and the corresponding *N*-adamantyl phosphoramidate ester (molecular ion 395). A third peak in trace amount is in good agreement with the 1-azidoadamantane starting reagent, and all three species contain the adamantyl fragment (135).

The proton-coupled ^{31}P-NMR spectrum of the GC-MS sample showed the phosphoranimine signal at −27.4 ppm, and the new species at +7.0 ppm. The new signal is a sextet with a coupling constant of 7.3 Hz, as expected for $^3J_{POCH}$ coupling.

The ^{1}H-NMR spectrum of this sample showed the ethyl ether proton signals at 1.18 ppm (t) and 3.45 ppm (q); two distinct sets of the different adamantyl protons from 1.58 to 2.10 ppm; amidate proton at 2.80 ppm (d); and trifluoroethoxy protons at 4.27 ppm (p). The amidate doublet has a coupling constant of $^2J_{PNH}$ = 9 Hz. This data supports the structure assignment for the hydrolysate.

The phosphoramidate ester can arise from nucleophilic attack by water (or OH^-) on the electrophilic phosphorus atom of the phosphoranimine, as shown in Scheme 3.

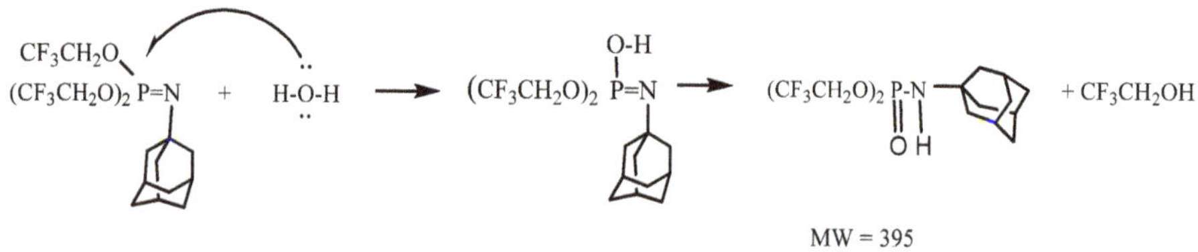

Scheme 3. Reaction of *P*-tris(trifluoroethoxy)-*N*-adamantylphosphoranimine with H_2O.

Similar behavior was reported in the study of an *N*-silylated phosphoranimine [44]. This could be a general reaction of phosphoranimines bearing trifluoroethoxy (or other leaving) groups bound to the phosphorus center.

A sample of the *N*-benzyl phosphoranimine was exposed to air and was converted to a colorless crystalline solid with a $\delta p = +8.3$ ppm, which is consistent with the corresponding *N*-benzyl phosphoramidate ester, $(CF_3CH_2O)_2P = (O)\text{-}NHCH_2Ph$ [45].

In addition to the phosphoramidate ester major product, a trace signal at −152.5 ppm was also observed. This high field resonance arises from the hexacoordinate species $(CF_3CH_2O)_6P^-$, in which the phosphorus center carries a formal negative charge. This compound was previously prepared and its ^{31}P-NMR chemical shift was reported as −154.6 ppm [46]. Electronegative groups stabilize the phosphorus hexacoordinated state [47].

This species has also been observed as a trace component of polymerizing mixtures of the *P*- tris(2,2,2-trifluoroethoxy)-*N*-trimethylsilyl phosphoranimine monomer, and in unreacted fractions of tris(2,2,2-trifluoroethyl) phosphite that were distilled from reaction mixtures. It is probably formed by a thermal rearrangement of phosphorus compounds bearing trifluoroethoxy ligands.

A pentacoordinated phosphorus compound was observed in the ^{31}P-NMR spectrum of a sample of the clear colorless *N*-benzyl phosphoranimine that had crystallized after several weeks in a desiccator. Its chemical shift of −72 ppm is typical of five-coordinate phosphorus compounds as previously reported [45], and sufficiently electronegative ligands tend to stabilize such species [45]. The penta-(2,2,2-trifluoroethoxy) phosphorane, $P(OCH_2CF_3)_5$, was previously prepared and its ^{31}P-NMR chemical shift reported as −76.6 ppm [46]. Another example of a penta-coordinated phosphorus compound with an *N*-benzyl ligand is $(CF_3)_3$ (F)P[N(CH_3) (CH_2Ph)] with a $\delta p = -68.8$ ppm [48].

The mass spectrum of the crystallized sample of the *N*-benzyl phosphoranimine showed a high mass peak of 351. The pentacoordinate phosphorus compounds shown in Scheme 4 are potential isomers with MW = 353 and are reasonable structural assignments for the species observed. They may exist in equilibrium with their phosphonium salts of the general formula, $R_4P^+R^-$ [45].

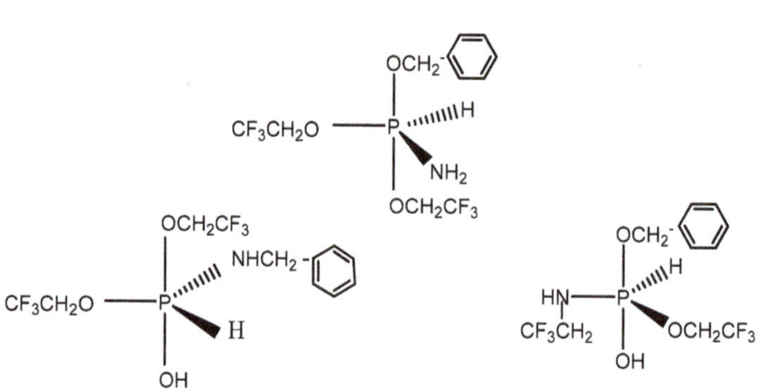

Scheme 4. Pentacoordinated phosphorus structures.

2.3. *Polymerization Studies*

Bulk Polymerization with Addition of *N*-alkyl Phosphoranimines

Experiments were conducted in which small samples of these compounds were added to polymerizing mixtures of the tris(2,2,2-trifluoroethoxy) phosphoranimine monomer and corresponding polyphosphazene. The initial experiment probed the reactivity of the *N*-alkyl phosphoranimine in bulk polymerization. *P*-tris(2,2,2-trifluoroethoxy)-*N*-trimethylsilyl phosphoranimine monomer, 1.5 mmol $(CF_3CH_2O)_3P = N\text{-}Si(CH_3)_3$, was treated with 1 mol% tetrabutylammonium fluoride (TBAF) in a dry NMR tube and heated at 150 °C for 15 min. Then, 0.6 mmol of *N*-benzyl phosphoranimine was injected by syringe, mixed, and heated for another 15 min.

The reaction was cooled to room temperature, and ^{31}P-NMR spectra confirmed polymer formation, along with a few percent oligomers. There were two very small doublets (+8.5, −2.5 ppm, J_{PNP} = 73 Hz) in the spectrum of the reaction mixture (less than 2%) which indicate formation of a phosphazene dimer as a minor side product, probably formed by a coupling reaction of monomer and the *N*-benzyl compound. A small signal at −72 ppm indicated the presence of ca. 2% of the pentacoordinate phosphorane discussed above.

The sample was then heated for an additional 15 min and final spectra recorded. A control reaction without *N*-benzyl phosphoranimine was run concurrently. The polymer samples were dissolved in diglyme and precipitated from excess cold chloroform, then re-dissolved in diglyme and precipitated again from 90/10 $CHCl_3$/MeOH with thorough washing of the white polymer solids.

The ^{1}H-NMR spectra of the isolated polymer in d_6-acetone solution showed the polymer signal at 4.55 ppm, *n*-butyl protons from the TBAF initiator, and new signals at 7.35 ppm indicating the aromatic protons of the benzyl group (Figure 2).

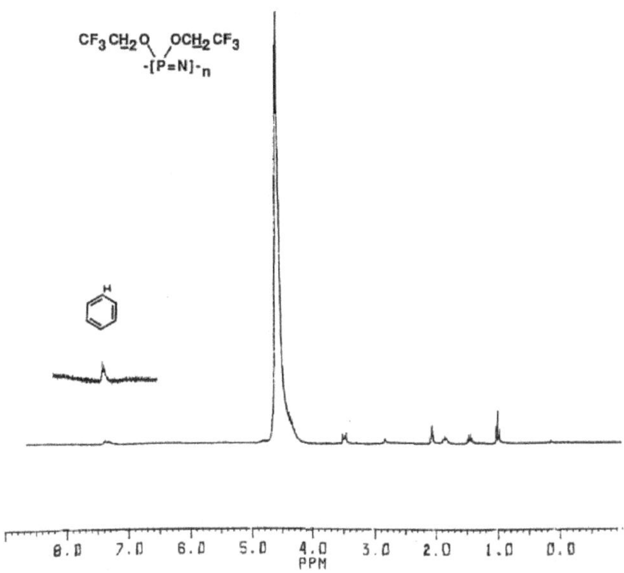

Figure 2. ^{1}H-NMR spectrum of *N*-benzyl-capped poly(bis(trifluoroethoxy)phosphazene).

The ^{31}P-NMR spectrum of this sample showed the polymer signal at −7.05 ppm, but no signals for the *N*-benzyl phosphoranimine, dimer, or phosphorane. The aromatic proton signals were thus ascribed to reaction of the *N*-benzyl compound with the polymer chain. Integration of the polymer and benzyl proton signal ratio correlated with incorporation of one *N*-benzyl phosphoranimine residue per chain, indicative of a functionalized chain-end. Similar results were obtained using the other *N*-alkyl phosphoranimines. After heating

6 mmol of monomer and 1 mol% TBAF initiator for 15 min at 150 °C, 3 mmol of the *N*-alkyl compound (neat or in solution) was added by syringe and heating continued for a total time of 1 h.

The polymer samples were precipitated twice as described before, and the ^{1}H-NMR spectra showed signals for polymer, the *n*-butyl groups of the TBAF, and the corresponding alkyl group. The presence of the trityl-capped chains is particularly distinctive by the multiple aromatic proton signals of the formed polymer product, as seen in Figure 3. The proton signals for the adamantyl and *t*-butyl chain-end groups are partially obscured by the TBAF initiator residue, but they can be seen and identified. The ^{31}P-NMR spectra of the experimental polymer samples showed no signals for unreacted *N*-alkyl phosphoranimines. The control reaction of monomer and TBAF showed proton signals for polymer and *n*-butyl groups, and GPC confirmed higher molecular weight. The yield of the polymer was significantly higher, indicating a more complete monomer conversion.

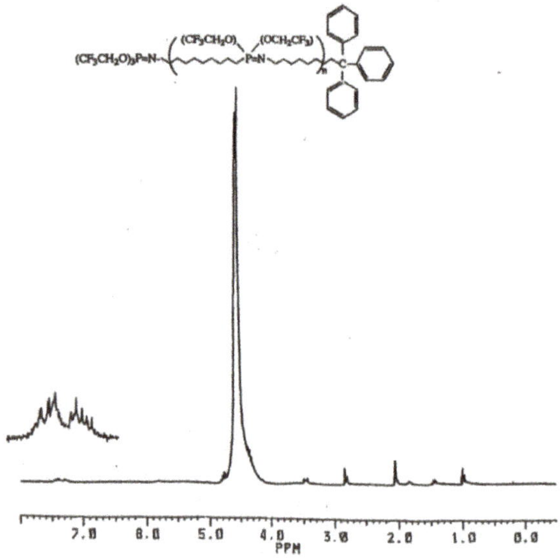

Figure 3. ^{1}H-NMR spectrum of *N*-trityl-capped poly(bis(trifluoroethoxy)phosphazene).

The molecular weights of the polymer samples obtained show the effect of the addition of the *N*-alkyl compounds. As seen in Table 5, the molecular weight and yield of polymer were significantly reduced by the addition of the *N*-alkyl compound compared to the control.

Table 5. Molecular weights, dispersity, yields of *N*-capped polyphosphazenes (bulk polymerization at 150 °C 1% TBAF, 3 mmol *N*-alkyl phosphoranimine addition, heated for 1 h).

Polymer End Group	M_n	M_w/M_n	% Grav. Yield
Adamantyl	21,960	1.40	22
Benzyl	19,867	1.34	25
t-Butyl	21,297	1.49	34
Trityl	17,362	1.51	33
Control (no *N*-alkyl)	29,012	1.70	71

In related experiments, bulk polymerizations with monomer-equivalent additions of *N*-benzyl phosphoranimine (NBP) at specific intervals during 150 °C polymerization using 1 mol% TBAF initiator showed that the polymer molecular weight was terminated upon addition compared to controls (*) with no *N*-benzyl compound (Table 6). Neither conversion nor molecular weight increased after addition of the NBP.

Table 6. Effect of addition time of *P*-Tris(2,2,2-trifluoroethoxy)-*N*-trimethylsilyl phosphoranimine on conversion and molecular weight (bulk at 150 °C, 1% TBAF).

NBP Sample	Minutes before Add.	Time, Min.	M_n	M_w/M_n	% Conversion
1	2	20	7123	1.08	9
2 *	–	2	9757	1.29	35
3	5	20	11,472	1.45	29
4 *	–	5	15,646	1.40	54
5 *	–	20	18,188	1.72	81

* controls, no *N*-benzyl phosphoranimine.

The effect of various amounts of *N*-benzyl phosphoranimine was examined by preparing samples containing monomer, 1 mol% TBAF initiator, and *N*-benzyl compound in the ratios shown in Table 7. After heating at 150 °C for 45 min, the molecular weights of the polymers obtained were measured by GPC and conversion determined by ^{31}P-NMR spectra integration. These data show that at least 2% *N*-benzyl phosphoranimine was needed to effectively limit polymer molecular weight and conversion of monomer, with 20% inhibiting polymerization.

Table 7. Effect of various monomer: *N*-benzyl phosphoranimine ratios on conversion and molecular weight.

Monomer: *N*-Benzyl mol. Ratio	M_n	M_w/M_n	% Conversion
5:1	Polymer not detected	–	–
20:1	4849	1.02	7
50:1	10,183	1.50	52
100:1	14,952	1.73	86
Control	15,195	1.84	86

The polymer molecular weight, dispersity, and yield are decreased by the *N*-alkyl phosphoranimine addition. However, the data also show that perfect stoichiometric agreement between the amount of the capping agent and degree of polymerization (DP) was not always observed. For example, the 20:1 and 50:1 samples have DPs of about 20 and 50, respectively, but the other ratios exhibit some variability. The lower than expected DP of the 100:1 sample may arise from other transfer reactions in the system.

Using the data from Table 7, a graph was constructed, as shown in Figure 4. The following Mayo equation was used to determine the chain transfer to the capping agent constant, $C_{tr,X}$, and the constant for transfer to monomer, $C_{tr,M}$:

$$1/DP_n = C_{tr,M} + C_{tr,X}\,[X]/[M]$$

From the graph, the slope of the line is the chain transfer constant to capping agent and the y-intercept is the chain transfer to monomer constant. From the equation, the values for $C_{tr,X}$ and $C_{tr,M}$ are 8.54×10^{-1} and 7.2×10^{-3}, respectively. There was a very small signal from trimethylsilyl protons in the ^{1}H-NMR spectrum of precipitated polymer at 0.12 ppm (Figure 2), which indicated that some chains could be terminated by trimethylsilyl groups, albeit in trace amount.

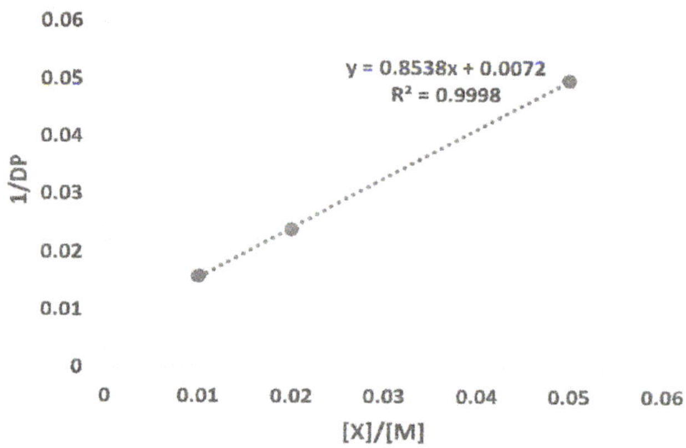

Figure 4. Mayo plot of 1/DP as function of [X]/[M] ratio.

A final experiment was conducted in which a 1:1.7 mol mixture of monomer and *N*-adamantyl phosphoranimine with 1% (mol) TBAF was heated for several hours at 150 °C, well beyond the normal time required for complete conversion of monomer to polymer in the presence of the fluoride. Analysis of the reaction mixture by ^{31}P-NMR showed that only 13% conversion of the monomer occurred, producing polymer and oligomers, with 87% of the spectrum signals arising from unreacted monomer and the *N*-adamantyl compound. Under these forcing conditions, the polymerization was retarded to a major extent, but not completely inhibited by the *N*-adamantyl phosphoranimine. The amount of polymer in the reaction mixture was less than 7% and could not be isolated for inspection of end groups.

Significantly, no phosphorus doublets were observed in the ^{31}P-NMR spectrum, thus, the monomer and *N*-adamantyl compound did not react to form a phosphazene dimer. This result and that of the end-capping experiments indicate that the *N*-adamantyl phosphoranimine reacts preferentially with active chain ends.

The active species, particularly in the early chain-growth stage of the polymerization, should be an anionic polymer chain-end [49]. The phosphorus atom of the *N*-alkyl phosphoranimine has similar reactivity as in the *N*-silylated monomer ($C_{tr,X}$ ~1) but forms an unreactive group. The stable alkyl group of the new phosphoranimine leads to termination of the chain, whereas the labile silyl group of the monomer results in continued propagation of the chain, as shown in Scheme 5.

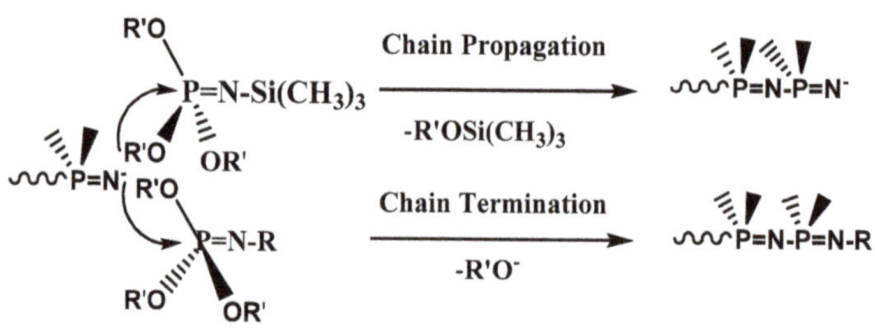

R= Adamantyl, benzyl, t-butyl, trityl groups
R'= -CH$_2$CF$_3$

Scheme 5. Chain propagation vs. chain termination.

Chain termination with these compounds is a degradative transfer reaction in which an active chain is terminated and a new active species, trifluoroethoxide anion, is produced. The number of active species in the system is unchanged, and the anion can initiate a new chain, as reported in an earlier study [49,50].

3. Conclusions

Four *N*-alkyl phosphoranimines were synthesized and found to be effective chain capping agents for polyphosphazenes prepared by anionic initiation of a phosphoranimine monomer. The NMR and GPC data support their incorporation into the chain as end groups. The use of chain-capping/terminating compounds in the anion-initiated polymerization of a phosphoranimine monomer has been demonstrated and may provide a useful technique for further functionalization of polyphosphazenes. For example, it may be possible to add alkenyl groups to short phosphazene chain ends as a route to prepare inorganic/organic graft copolymers, as well as to prepare block copolymers with controlled molecular weight.

4. Materials and Methods

4.1. Materials

Tris(2,2,2 trifluoroethyl) phosphite and trimethylsilyl azide were obtained from Aldrich Chemical Co., Milwaukee, WI, USA and purified by distillation at reduced pressure from 4 Angstrom molecular sieves to molecular sieves. Tetra-*n*-butyl ammonium fluoride TBAF in THF, 1.0M and TBAF-silica gel were obtained from Aldrich and were used as received. Benzyl azide was obtained from Johnson-Mathey Co., London, England and purified by recrystallization from dry benzene-hexane solution. 1-Azidoadamantane and sodium azide from Aldrich were used as received. Trityl bromide and *t*-butyl alcohol (Fisher Scientific-Fisher Scientific Co., LLC Hampton, NH, USA), and $BF_3.ET_2O$ from Aldrich were used as received.

4.2. Monomer Synthesis

A clean 100 mL three-neck round bottom flask with magnetic stir-bar, water jacketed condenser, glass or Teflon stopcock, and Claisen tube with 250 mL pressure-equalizing addition funnel was oven-dried at 150 °C overnight, assembled hot with natural rubber septa and gas inlet and outlet, and cooled under a dry nitrogen purge. All ground-glass joints were sealed with a light application of stopcock grease and wrapped with Teflon tape. The flask was covered with aluminum foil to prevent photolysis of the azide. In accordance with a literature procedure [51], distilled tris(2,2,2-trifluoroethyl) phosphite (0.4 mol) was combined with equimolar distilled trimethylsilyl azide and refluxed for 24 h at 120 °C, followed by two successive additions of equimolar azide under agitation at 24 h intervals, for a final mole ratio of 1:3 phosphite:azide. During each azide addition, the reaction flask was cooled to 0 °C, then the temperature was slowly raised to 120 °C. At the end of the 72-h period, an orange/amber liquid was observed in the flask. Upon distillation at reduced pressure, a first fraction of unreacted clear, colorless azide was obtained at 40–50 °C/85 mm, and a second fraction of clear, colorless liquid monomer distilled at B.P. 50–60 °C/0.5 mm. Yield of *P*-tris (2,2,2 trifluoroethoxy)-*N*-trimethylsilyl phosphoranimine monomer: 95% based on the phosphite.

4.3. Synthesis of N-Alkyl Phosphoranimines

The adamantyl azide (21 mmol) and an equimolar amount of the tris(2,2,2-trifluoroethyl) phosphite were combined in a three-neck round bottom 100 mL flask with condenser over an electrically heated and thermostated oil bath. The oil temperature was increased gradually from 50 °C to 140 °C over a 1.75 h period, in order to moderate the pressure increase in the system as the nitrogen outgassed. The *N*-adamantyl phosphoranimine distilled under reduced pressure (93 °C/2.5 torr) as a clear, colorless liquid which crystallized as colorless needles with a melting point near 0 °C as shown in Table 1.

In a similar fashion, 136 mmol benzyl azide and equimolar amount of phosphite were combined with stirring at 40–80 °C for 22 h. ***CAUTION: very vigorous reaction with rapid outgassing and pressure build-up in apparatus!*** *N*-benzyl phosphoranimine distilled as a clear colorless liquid at 95 °C/2.5 torr. The *t*-butyl azide was prepared by the reaction of 100 mmol *t*-butyl alcohol and 120 mmol trimethylsilylazide in the presence of 120 mmol $BF_3.Et_2O$, following a published procedure [52]. Due to difficulty in isolating the azide by distillation, 100 mmol phosphite was added to this reaction at room temperature (RT) for the Staudinger coupling in situ, at 90 °C for 18 h. *N*-*t*-butyl phosphoranimine distilled from the reaction as a clear, colorless liquid. Trityl azide was prepared by the reaction of sodium azide suspended in MeCN with trityl bromide in benzene at RT over several days, in a manner similar to published methods [53,54]. The crystallized azide (13.5 mmol) was combined with an equimolar amount of phosphite at 60–140 °C over a 2-h period. *N*-trityl phosphoranimine crystallized as an amber, fibrous solid in high purity.

4.4. Bulk Polymerization of Monomer and N-Alkyl Phosphoranimines

The bulk polymerizations of the monomer phosphoranimine with and without the *N*-alkyl compounds were conducted in NMR tubes as indicated in the Results and Discussion section. Monomer in the specified amount was charged via syringe and the specified amount of TBAF initiator solution was added by microliter syringe. Heat was supplied by a mineral oil bath over an RCT Tekmar hotplate unit for the specified time period. The *N*-alkyl phosphoranimine was added neat if liquid, or as a solution by syringe as indicated. After the reaction was cooled to RT, the work-up entailed dissolving the solid polymer mass in 2–5 mL THF, and adding the solution to excess cold chloroform to precipitate the polymer. The polymer mass was re-dissolved in a 90/10 blend of chloroform and methanol and re-precipitated. The re-precipitated polymer was allowed to stand under the mother liquor at −20 °C overnight to maximize precipitation, and was collected on a clean, tared glass frit with thorough washing to remove any unreacted materials. The collected white polymer solid was dried in a vacuum desiccator overnight before weighing. Yield was calculated on the basis of monomer weight minus the condensate by-product as the theoretical yield.

4.5. Characterization of N-Alkyl Phosphoranimines and Polymers

NMR spectra were recorded on an IBM NR/300 MHz FT NMR spectrometer. Trimethyl phosphite in C_6D_6 was used as an external standard for ^{31}P-NMR spectra, with a δ P of 141.0 ppm (85% phosphoric acid H_3PO_4 = 0 ppm). The ^{31}P-NMR and ^{1}H-NMR spectra were obtained in $CDCl_3$ solution or in d^6 acetone as internal standard (^{1}H-NMR = 7.24 ppm). Abbreviations for NMR signals: s = singlet; m = multiplet, d = doublet, t = triplet, q = quartet, p = pentet, br. = broad. Integration symbols, such as "6H", signify six protons of a particular type as shown in the table.

Mass spectra were obtained using a Hewlett Packard 5890 gas chromatograph with a silica column and equipped with a 5970 series mass spectrometer. Run time was 40 min with 2 min solvent delay; T (initial) = 100 °C for 10 min, heating rate of 10 °C/minute, and T (final) = 250 °C for 5 min. FTIR spectra were obtained on a Nicolet 5DXB FT-IR spectrometer. Samples were analyzed in KBr pellets, or thin translucent films between sodium chloride plates as appropriate, using polystyrene film standard. Melting points were measured with a digital melting point apparatus from Electrothermal Eng. Ltd. at a heating rate of 1 °C/minute. Refractive indices were measured on the Bausch and Lomb Abbe'−3L refractometer at 20 °C. Elemental analyses were provided by Midwest Microlab of Indianapolis, IN. Gel Permeation Chromatography (GPC) was performed on polymer samples by first dissolving the solid polymer (0.2–0.5 g) in 1 mL THF-HPLC grade, and filtering through 0.5 micron Teflon filter, 20 microliters of this solution was injected into the carrier solvent stream (THF with 0.1% tetra-*n*-butyl ammonium bromide, a literature procedure for polyphosphazenes [55]) at a flow rate of 1.0 mL/minute. Ultrastyragel columns (10,000; 1000; 100 Angstroms) and a Waters 410 differential refractometer were used at 35 °C internal temperature. Data acquisition and calculations were performed with a Nelson 900 analytical interface and Samsung 286 personal computer. Calibration was based on polystyrene standards of low to high molecular weights.

Author Contributions: Conceptualization: R.A.M. and K.M. Methodology: R.A.M. and K.M. Validation: R.A.M. and K.M. Investigation planning: R.A.M. and K.M. Data acquisition: R.A.M. Formal analysis: R.A.M. and K.M. Writing—original draft: R.A.M. Writing—review and editing: K.M. Project administration: K.M. Funding acquisition: K.M., R.A.M. Both authors have read and agreed to the published version of the manuscript. All authors have read and agreed to the published version of the manuscript.

Funding: K.M. acknowledges support from PPG Industries, Inc., Eastman Kodak, Xerox Corp., Hoechst-Celanese. R.A.M. acknowledges support from PPG Industries, Inc.

Institutional Review Board Statement: Not applicable.

Informed Consent Statement: Not applicable.

Data Availability Statement: The data presented in this study are available in this article.

Acknowledgments: This paper is dedicated to Professor Julian Chojnowski on the occasion of his 85th birthday. R.A.M. and K.M. acknowledge and thank Frank Burkus for his assistance with the experiments in the preliminary research study [56], and Kasi Somajajula of the University of Pittsburgh for the FAB-Mass Spectrometry measurements.

Conflicts of Interest: The authors declare no conflict of interest. The sponsors had no role in the design, execution, interpretation, or writing of the study.

Sample Availability: Samples of the compounds are not available from the authors.

References

1. Allcock, H.R. Developments at the Interface of Inorganic, Organic, and Polymer Chemistry. *Chem. Eng. News* **1985**, *63*, 22–36. [CrossRef]
2. Allcock, H.R. Inorganic—Organic Polymers. *Adv. Mater.* **1994**, *6*, 106–115. [CrossRef]
3. Manners, I. Polymers and the Periodic Table: Recent Developments in Inorganic Polymer Science. *Angew. Chem.* **1996**, *35*, 1602–1621. [CrossRef]
4. Jutzi, P.; Schubert, U. *Silicon Chemistry: From the Atom to Extended Systems*; Wiley Publishing Co.: Hoboken, NJ, USA, 2007.
5. Chojnowski, J. Kinetically Controlled Siloxane Ring-Opening Polymerization. *J. Inorg. Organomet. Polym.* **1991**, *1*, 299–323. [CrossRef]
6. Chojnowski, J.; Cypryk, M.; Fortuniak, W.; Ścibiorek, M.; Rózga-Wijas, K. Synthesis of Branched Polysiloxanes with Controlled Branching and Functionalization by Anionic Ring-Opening Polymerization. *Macromolecules* **2003**, *36*, 3890–3897. [CrossRef]
7. Kazmierski, K.; Hurduc, N.; Sauvet, G.; Chojnowski, J. Polysiloxanes with Chlorobenzyl Groups as Precursors of New Organic-Silicone Materials. *J. Polym. Sci. Part A Polym. Chem.* **2004**, *42*, 1682–1692. [CrossRef]
8. Woźniak, L.; Chojnowski, J. Silyl Esters of Phosphorous-Common Intermediates in Synthesis. *Tetrahedron* **1989**, *45*, 2465–2524. [CrossRef]
9. Cypryk, M.; Gupta, Y.; Matyjaszewski, K. Anionic Ring-Opening Polymerization of 1,2,3,4-Tetramethyl-1,2,3,4-tetraphenylcyclotetrasila *J. Am. Chem. Soc.* **1991**, *113*, 1046–1047. [CrossRef]
10. Kim, H.K.; Matyjaszewski, K. Sonochemical Synthesis of Polysilanes. *Polym. Prepr.* **1988**, *29*, 168–169.
11. Matyjaszewski, K.; Hrkach, J.S. Modification of Polysilanes: Preparation of Comb-like Graft Copolymers. *J. Inorg. Organomet. Polym.* **1995**, *5*, 183–193. [CrossRef]
12. Allcock, H.R. Polyphosphazenes as an Example of the Element-Blocks Approach to New Materials. In *New Polymeric Materials Based on Element-Blocks*; Springer Publishing Co.: New York, NY, USA, 2018; pp. 167–188.
13. Allcock, H.R.; Crane, C.A.; Morrissey, C.T.; Nelson, J.M.; Reeves, S.D.; Honeyman, C.H.; Manners, I. "Living" Cationic Polymerization of Phosphoranimines as an Ambient Temperature Route to Polyphosphazenes with Controlled Molecular Weights. *Macromolecules* **1996**, *29*, 7740–7747. [CrossRef]
14. Deng, M.; Laurencin, C.T.; Allcock, H.R.; Kumbar, S.G. Polyphosphazenes as Biomaterials. In *Polymeric Biomaterials: Structure and Function*; CRC Press: Boca Raton, FL, USA, 2013; Volume 1, pp. 84–126.
15. Teasdale, I.; Bruggemann, O. Polyphosphazenes: Multifunctional, Biodegradable Vehicles for Drug and Gene Delivery. *Polymers* **2013**, *5*, 161–187. [CrossRef] [PubMed]
16. Andrianov, A.K. *Polyphosphazenes for Biomedical Applications*, 1st ed.; John Wiley & Sons Ltd.: Hoboken, NJ, USA, 2009; p. 480.

17. Summe Ullah, R.; Wang, L.; Yu, H.; Haroon, M.; Elshaarani, T.; Naveed, K.; Fahad, S.; Khan, A.; Nazir, A.; Xia, X.; et al. Synthesis of polyphosphazene and preparation of microspheres from polyphosphazene blends with PMMA for drug combination therapy. *J. Mater. Sci.* **2019**, *54*, 745–764. [CrossRef]
18. Andrianov, A.K.; Svirkin, Y.Y.; LeGolvan, M.P. Synthesis and Biologically Relevant Properties of Polyphosphazene Polyacids. *Biomacromolecules* **2004**, *5*, 1999–2006. [CrossRef] [PubMed]
19. Honeyman, C.H.; Manners, I.; Morrissey, C.T.; Allcock, H.R. Ambient Temperature Synthesis of Poly(dichlorophosphazene) with Molecular Weight Control. *J. Am. Chem. Soc.* **1995**, *117*, 7035–7036. [CrossRef]
20. Montague, R.A.; Green, J.B.; Matyjaszewski, K. The Conversion of Phosphoranimines to Polyphosphazenes in the Presence of Electrophiles. *J. Macromol. Sci. Pure Appl. Chem.* **1995**, *32*, 1497–1519. [CrossRef]
21. Allcock, H.R.; Nelson, J.M.; Reeves, S.D.; Honeyman, C.H.; Manners, I. Ambient Temperature Direct Synthesis of Poly(organophosphazenes)via the "Living" Cationic Polymerization of Organo-Substituted Phosphoranimines. *Macromolecules* **1997**, *30*, 50–56. [CrossRef]
22. Allcock, H.R.; Nelson, J.M.; Prange, R.; Crane, C.A.; de Denus, C.R. Synthesis of Telechelic Polyphosphazenes via the Ambient Temperature Living Cationic Polymerization of Amino Phosphoranimines. *Macromolecules* **1999**, *32*, 5736–5743. [CrossRef]
23. Wilfert, S.; Henke, H.; Schoefberger, W.; Bruggemann, O.; Teasdale, I. Chain-End Functionalized Polyphosphazenes via a One-Pot Phosphine-Mediated Living Polymerization. *Macromol. Rapid Commun.* **2014**, *35*, 1135–1141. [CrossRef]
24. Allcock, H.R.; de Denus, C.R.; Prange, R.; Nelson, J.M. Synthesis of Trifluoromethyl- and Methylphosphazene Polymers: Differences Between Polymerization and Initiator/TerminatorProperties. *Macromolecules* **1999**, *32*, 7999–8004. [CrossRef]
25. Allcock, H.R.; Powell, E.S.; Maher, A.E.; Prange, R.L.; de Denus, C.R. Telechelic Polyphosphazenes: Reaction of Living Poly(dichlorophosphazene) Chains with Alkoxy and Aryloxy Phosphoranimines. *Macromolecules* **2004**, *37*, 3635–3641. [CrossRef]
26. Allcock, H.R.; de Denus, C.R.; Prange, R.; Laredo, W.R. Synthesis of Norbornenyl Telechelic Polyphosphazenes and Ring-Opening Metathesis Polymerization Reactions. *Macromolecules* **2001**, *34*, 2757–2765. [CrossRef]
27. Taylor, T.J.; Soto, A.P.; Huynh, K.; Lough, A.J.; Swain, A.C.; Norman, N.C.; Russell, C.A.; Manners, I. Synthesis of Poly(Alkyl/Arylphosphazenes) via Ambient Temperature Phosphite-Mediated Chain Growth Polycondensation of N-Silylbromophosphoranimines. *Macromolecules* **2010**, *43*, 7446–7452. [CrossRef]
28. Allcock, H.R.; Reeves, S.D.; Nelson, J.M.; Crane, C.A.; Manners, I. Polyphosphazene Block Copolymers via the Controlled Cationic, Ambient Temperature Polymerization of Phosphoranimines. *Macromolecules* **1997**, *30*, 2213–2215. [CrossRef]
29. Allcock, H.R.; Prange, R. Properties of Poly(phosphazene-siloxane) Block Copolymers Synthesized via Telechelic Polyphosphazenes and Polysiloxane Phosphoranimines. *Macromolecules* **2001**, *34*, 6858–6865. [CrossRef]
30. Krogman, N.R.; Steely, L.; Hindenlang MDNair, L.S.; Laurencin, C.T.; Allcock, H.R. Synthesis and Characterization of Polyphosphazene-block-polyester and Polyphosphazene block-polycarbonate Macromolecules. *Macromolecules* **2008**, *41*, 1126–1130. [CrossRef]
31. Liu, X.; Tian, Z.; Chen, C.; Allcock, H.R. Synthesis and Characterization of Brush-Shaped Hybrid Inorganic/Organic Polymers Based on Polyphosphazenes. *Macromolecules* **2012**, *45*, 1417–1426. [CrossRef]
32. Suarez, S.S.; Soto, D.P.; Carriedo, G.A.; Soto, A.P.; Staubitz, A. Experimental and Theoretical Study of the Living Polymerization of N-Silylphosphoranimines. Synthesis of New Block Copolyphosphazenes. *Organometallics* **2012**, *31*, 2571–2581. [CrossRef]
33. Montague, R.A.; Matyjaszewski, K. Synthesis of Poly[bis(trifluoroethoxy)phosphazene] Under Mild Conditions Using a Fluoride Initiator. *J. Am. Chem. Soc.* **1990**, *112*, 6721–6723. [CrossRef]
34. Matyjaszewski, K.; Green, J.B.; Montague, R.A. Fluoride-Initiated Polymerization of Fluoroethoxy Phosphoranimine. *ACS Polym. Prepr.* **1992**, *33*, 174–175.
35. Matyjaszewski, K.; Cypryk, M.; Dauth, J.; Montague, R.; White, M. New Synthetic Routes towards Polyphosphazenes. *Makromol. Chem. Macromol. Symp.* **1992**, *54*, 13–30. [CrossRef]
36. White, M.L.; Montague, R.A.; Matyjaszewski, K.; Pakula, T. The Thermal Properties of Polyphosphazenes Synthesized by the Anionically-Initiated Polymerization of Phosphoranimines. *Polymer* **1995**, *36*, 3493–3502. [CrossRef]
37. Matyjaszewski, K.; Moore, M.K.; White, M.L. Synthesis of Polyphosphazene Block Copolymers Bearing Alkoxyethoxy and Trifluoroethoxy Groups. *Macromolecules* **1993**, *26*, 6741–6748. [CrossRef]
38. Wood, C.E.; Samuel, R.; Kucera, W.R.; Angelov, C.M.; Neilson, R.H. New Synthetic, Catalytic, and Structural Studies Related to Poly(Alkyl/Aryl Phosphazenes). *ACS Polym. Prepr.* **1993**, *34*, 263.
39. Chapman, R.D.; Welker, M.F.; Kreutzberger, C.B. Polyalkoxyphosphazenes by Room-Temperature Polymerization of an Electronegative Phosphoranimine Monomer. *J. Inorg. Organomet. Polym.* **1996**, *96*, 267–275. [CrossRef]
40. Gallazi, C.M.; Freddi, G.; Sanvito, G.; Viscardi, G. Polydialkylphosphazenes: New Synthetic Efforts and Protonation Reactions. *J. Inorg. Organomet. Polym.* **1996**, *6*, 277–300. [CrossRef]
41. Steinke, J.H.G.; Greenland, B.W.; Johns, S.; Parker, M.P.; Atkinson RC, J.; Cade, I.A.; Golding, P.; Trussell, S.J. Robust and Operationally Simple Synthesis of Poly(bis(2,2,2-trifluoroethoxy)phosphazene) with Controlled Molecular Weight, Low PDI, and High Conversion. *ACS Macro Lett.* **2014**, *3*, 548–551. [CrossRef]
42. Matyjaszewski, K.; Franz, U.; Montague, R.A.; White, M.L. Synthesis of Polyphosphazenes from Phosphoranimines and Phosphine Azides. *Polymer* **1994**, *35*, 5005–5011. [CrossRef]
43. Staudinger, H.; Meyer, J. Über neue organische phosphorverbindungen III. Phosphinmethylenderivate und phosphinimine. *Helv. Chim. Acta* **1919**, *2*, 635. [CrossRef]

44. Matyjaszewski, K.; Montague, R.; Dauth, J.; Nuyken, O. Synthesis of Poly(phenyltrifluoroethoxyphosphazene) by Direct Reaction of Trimethylsilyl Azide with Bis(2,2,2-Trifluoroethyl)Phenylphosphonite. *J. Polym. Sci.* **1992**, *30*, 813–818. [CrossRef]
45. Tebby, J.C. *Handbook of Phosphorus-31 Nuclear Magnetic Resonance Data*; Tebby, J.C., Ed.; CRC Press, Inc.: Boca Raton, FL, USA, 1991.
46. Denney, D.B.; Denney, D.Z.; Hammond, P.J.; Wang, Y.-P. Preparation and Chemistry of Penta- and Hexacoordinated Phosphorus Compounds Containing Trifluoroethoxy Groups. *J. Am. Chem. Soc.* **1981**, *103*, 1785–1789. [CrossRef]
47. Muetterties, E.L.; Mahler, W. Donor-Acceptor Function in Organofluorophosphoranes. *Inorg. Chem.* **1965**, *4*, 119–121. [CrossRef]
48. Kirby, A.J.; Warren, S.G. *The Organic Chemistry of Phosphorus*; Elsevier Pub. Co.: Amsterdam, The Netherlands, 1967.
49. Matyjaszewski, K.; Dauth, J.; Montague, R.; Reddick, C.; White, M. Polyphosphazenes by Anionic Polymerization. *ACS Polym. Prepr.* **1991**, *32*, 305–306.
50. Montague, R.A. Synthesis of Polyphosphazenes via the Catalyzed Polymerization of Phosphoranimines. In *Chemistry*; Carnegie Mellon University: Pittsburgh, PA, USA, 1993; p. 357.
51. Flindt, E.-P.; Rose, H.; Marsmann, H.C. Synthese N-Silylierter Phosphinimine. *Z. Anorg. Allg. Chem.* **1977**, *430*, 155–160. [CrossRef]
52. Koziara, A.; Zwierzak, A. Iminophosphorane-Mediated Transformation of Tertiary Alcohols into Tert-Alkylamines and Their N-Phosphorylated Derivatives. *Tetrahedron Lett.* **1987**, *28*, 6513–6516. [CrossRef]
53. Curtius, T.; Ehrhart, G. Decomposition of benzyl azide in indifferent media and in malonic ester. *Ber. Dtsch. Chem. Ges.* **1922**, *55*, 1559. [CrossRef]
54. Fuchs, E.; Breit, B.; Bergstrasser, U.; Hoffman, J.; Heydt, H.; Regitz, M. Organophosphorus Compounds. Phosphatriafulvenes and Their Reactions with Electrophiles. *Synthesis* **1991**, *12*, 1099–1107. [CrossRef]
55. Neilson, R.H.; Wisian-Neilson, P. Poly(Alkyl/arylphosphazenes) and Their Precursors. *Chem. Rev.* **1988**, *88*, 541–562. [CrossRef]
56. Montague, R.A.; Burkus, F., II; Matyjaszewski, K. Chain Terminators for Polyphosphazenes. *ACS Polym. Prepr.* **1993**, *34*, 316–317.

MDPI
St. Alban-Anlage 66
4052 Basel
Switzerland
Tel. +41 61 683 77 34
Fax +41 61 302 89 18
www.mdpi.com

Molecules Editorial Office
E-mail: molecules@mdpi.com
www.mdpi.com/journal/molecules

www.ingramcontent.com/pod-product-compliance
Lightning Source LLC
Chambersburg PA
CBHW041018270926
51850CB00024B/1382
9783036522531